Radar Systems, Peak Detection and Tracking

This book is dedicated to my best friend and my wife,
Dr Marjorie Helen Kolawole.
Your unfailing support has been a constant source of joy and strength.
You are the loveliest of women.

Radar Systems, Peak Detection and Tracking

Michael O. Kolawole, PhD

Newnes

OXFORD AMSTERDAM BOSTON LONDON NEW YORK PARIS
SAN DIEGO SAN FRANCISCO SINGAPORE SYDNEY TOKYO

Newnes
An imprint of Elsevier Science
Linacre House, Jordan Hill, Oxford OX2 8DP
200 Wheeler Road, Burlington, MA 01803

First published 2002

British Library Cataloguing in Publication Data
A catalogue record for this book is available from the British Library.

ISBN 0 7506 57731

Typeset by Integra Software Services Pvt. Ltd, Pondicherry, India
www.integra-india.com
Printed and bound in Great Britain by Biddles Ltd *www.biddles.co.uk*

Contents

Preface

This book is written to provide continuity to the reader on how radar systems work, how the signals captured by the radar receivers are processed, parameterized and presented for tracking, and how tracking algorithms are formulated. Continuity is needed because most radar systems books have been written that concentrate on certain specialized topics assuming a prior knowledge of the reader to background principles. In most cases extended references are given to the understanding of the topics in question. This can be frustrating to practitioners and students sorting through books to understand a simple topic. Hence this book takes a thorough approach to ramping up the reader in the topical foundations. Advanced topics are certainly not ignored. Throughout, concepts are developed mostly on an intuitive, physical basis, with further insight provided through a combination of applications and performance curves.

The book has been written with science and engineering in mind, so that it should be more useful to science and communications professionals and practising electrical and electronic engineers. It could also be used as a textbook suitable for undergraduate and graduate courses. As a practitioner and teacher, I am aware of the complexity involved in the presentation of many technical issues associated with the topic areas. This is the main reason why the book

- builds up gradually from a relatively low base for the reader to have a good grasp of the mathematics, and the physical interpretation of the mathematics wherever possible before the reader reaches the advanced topics, which are certainly not ignored but necessary in the formulation of tracking algorithms;
- gives sufficient real-life examples for the reader to appreciate the synergy involved and have a feel for how physical abstractions are converted to quantifiable, real events or systems;
- where real-life matters cannot be linked directly to physical derivations, gives further insight through a combination of applications and performance curves. In most cases, those seeking qualitative understanding can skip the mathematics without any loss of continuity. Professionals

in the field would greatly appreciate the background knowledge mathematics, sufficient for them to follow the advanced sections with very little difficulty;
- presents a number of new ideas which may deserve further investigation.

In general, readers of this book will gain an understanding of radar systems' fundamental principles, underlying technologies, architectures, design constraints and real-world applications. To be able to cover all relevant grounds, the book contains 12 chapters, divided into four parts. Each part represents topics of comparable relevance.

Part I contains five chapters. The chapters are structured in a way that gives the reader a continuum in the understanding of radar systems. Each chapter is somehow self-sufficient. However, where further knowledge can be gained, applicable references are given.

Chapter 1 provides the essential functional relations, concepts, and definitions that are relevant to radar systems' development and analysis and signal peak detection. This approach is taken to provide the basic groundwork for other concepts that are developed in subsequent chapters. The areas covered are sufficiently rich to provide a good understanding of the subject matter for non-specialists in radar systems and associated signal processing.

The next four chapters concentrate on radar systems. Discussions on radar systems evolve from basic concept and gradually increase to a more complex outlook. The author believes that mastering the fundamentals permits moving on to more complex concepts without great difficulty. In so doing, the reader would learn the following:

- The basic architecture of radar systems, receiver sensitivity analysis, and data acquisition and/or compression issues as well as the applications of radar in Chapter 2.
- Chapter 3 examines the physics of an antenna, which is a major item in radar systems design. It starts from the perspective of a simple radiator, the division of radiation field in front of an antenna into quantifiable regions and further discusses the principle of pulse compression. Pulse compression allows recognition of closely spaced targets as well as enabling range measurements when transmitting with signal pulses and a train of pulses.
- Chapter 4 shows how the extension of the simple radiator's radiation property to an array of radiators including slot antennas can achieve a higher gain as well as the freedom to steer the array antenna in any preferred direction.
- Chapter 5 explains how radar equations are developed recognizing the effect of the environment on the conventional, laser and secondary radar performance and the detection of targets of variable radar cross-sections and mobility.

Part II comprises two chapters: 6 and 7. When a wave traverses the regions comprising the atmosphere it results in the degradation of signal-target information due to spatial inhomogeneities that exist and vary continuously with time in the atmosphere. The spatial variations produce statistical bias errors, which are an important consideration when formulating and designing a high frequency (HF) skywave radar system. Chapter 6 explains how these errors are quantified including the polarization rotational effect on the traversing wave. Chapter 7 explains the design consideration and performance of the skywave radar.

The issue of what the true nature of data is and what to do with data acquired by radar becomes relevant after the data, which might have been corrupted prior to being processed, has been processed. Data processing involves the transformation of a set of coordinated physical measurements into decision statistics for some hypotheses. These hypotheses, in the case of radar, are whether targets with certain characteristics are present with certain position, speed, and heading attributes. To test the trueness of the hypotheses requires knowledge of probability and statistical and decision theory together with those espoused in Chapter 1 – the reader will therefore be in a better position to know the other processes involved in signal-peaks detection. Hence, Part III is structured into three chapters: 8, 9 and 10.

Chapter 8 reviews some of the important properties and definitions of probability theory and random processes that bear relevance to the succeeding topics in Part IV. By this approach, the author consciously attempts to reduce complex processes involved in synthesizing radar system signals to their fundamentals so that their basic principles by which they operate can be easily identified. The basic principles are further built on in Chapter 12 to solve more complex, technical tracking problems.

Chapter 9 investigates one type of optimization problem; that is, finding the system that performs the *best*, within its certain class, of all possible systems. The signal-reception problem is decoupled into two distinct domains, namely detection and estimation. Detection problem forms the central theme of Chapter 10 while estimation is discussed in Part IV, Chapter 11. Detection is a process of detecting the presence of a particular signal, among other candidate signals, in a noisy or cluttered environment.

Part IV contains two chapters – 11 and 12 – covering parameter estimation and radar tracking. Estimation is the second type of optimization problem and exploits the several parallels with the decision theory of Chapter 9. Three estimation procedures are considered, namely maximum likelihood, *a posteriori*, and linear estimation.

Tracking is the central theme of Chapter 12 and it brings to the fore all the concepts discussed in previous chapters. For example, target tracking now turns the tentative decision statistics, discussed in Chapters 9 and 11, into more highly refined decision statistics. The probability theory discussed in Chapter 8 is expanded on to solve the problem of uncertainty in track initiation and establishment as well as data association.

I understand during my years of engineering practice and teaching that many readers learn more by examples, which I have relied on in explaining difficult concepts. For those readers wishing to test their level of understanding several problems are written at the end of each chapter.

Acknowledgements

This book is possible because of my professional colleagues who encouraged me to write a book that demystifies the complexities associated with radar systems. To them, I am greatly indebted.

I acknowledge the effort of my colleague Mr John Bombardieri, whose blend of theoretical and practical insight is reflected in his criticism of this book in its formative period. I am also grateful to my other very valuable friend, Professor Ah Chung Tsoi of the University of Wollongong, Wollongong, Australia, for his encouragement. My greatest thanks go to my family, whose unfailing support has been my constant source of strength, especially Dr Marjorie Helen Kolawole, my wife, for allowing me to go unhindered to achieve my goals. My sweet love, my special thanks.

Notations

The symbols have been chosen as carefully as possible to prevent confusion. In a few instances, the same symbol was used. When this occurs, a clear distinction is made in their meaning and where used in the text is indicated.

Symbols	Meaning
A	Current potential in Chapter 3, or fundamental matrix in Chapter 12
A_c	Clutter illuminated surface area
A_d	Attenuation due to absorption by electromagnetic waves
A_e	Effective aperture area of the receiving antenna
A_{eb}	Effective aperture area of the beacon antenna
A_L	Insertion loss
A_m	Searched area
A_0	Signal amplitude
A_r	Rain attenuation
A_s	Area to be searched
A_t	Target area
a	Proportionality constant, or acceleration in Chapter 12
aa	Notation that relates to the radar and vehicle dynamics
a_k	Axial ratio of elliptical polarization
a_n	Fourier series coefficient
B	Receiver beamwidth, or Bayes risk in Chapter 9
B_n	Noise bandwidth
B_{na}	Available bandwidth for integration
B_w	Bandwidth of the radar signal
b	Proportionality constant
b_n	Fourier series coefficient
CW	Continuous wave
C	Speed of light
C_{ij}	Cost function
C_r	Pulse compression ratio
c_c	Level parameters of clutter model

c_k	Cumulant of the kth order
c_n	Series spectral density
c_{rr}	Weight modifier for beam shaping operation
cov[ ]	Covariance matrix of [ ]
D	A layer of the ionosphere used for radio wave propagation in Chapter 6, or aperture diameter
D_M	A body of data to be encoded
D_L	Laser lens diameter
D_r	Largest dimension of the antenna, or directive gain (also called directivity)
D_x	Detectability factor
D_y	Dynamic range
d	Allowable spacing between array elements in Chapters 4 and 7, or statistical Euclidean distance in Chapter 12
$d_{\max}$	Maximum spacing between array elements
d_n	Distance between radiators of log periodic antenna
d_u	Duty cycle
d_v	Maximum fraction of the interpulse interval available for target reception or clear region duty cycle
E	Electric charge in Chapter 3, or a layer of the ionosphere used for radio wave propagation in Chapter 6
$E[x(t)]$	Expectance (or μ mean) of the variable x, sampled at time t
E_o	Amplitude of the plane wave
E_ϕ, E_θ, E_r	Electric intensity in the ϕ, θ, r direction
e_e	Charge of an electron
e_r	Receiver sensitivity
erf(x)	Error function of (x)
F	Ratio of the resultant field at the target in the presence of surface reflection coefficient ρ in Chapter 5, or force exerted in Chapters 3 and 6
F1, F2	F layers of the ionosphere subdivided into two: F1, F2
F_1	Field pattern of a single point source radiator
F_2	Array factor for the n radiators
F_a	Noise density factor
F_I	Stage noise factor
F_n	Noise density factor
F_k	Discrete form of Fourier series sampler
F_N	Noise figure
FOM	Figure of merit
f	Frequency
$f(\theta)$	Pattern factor
$f(t)$	Function of a signal at time t
f_c	Correlation frequency in Chapter 5, or critical frequency in Chapters 6 and 7

f_d	Doppler shift
f_0	Nyquist frequency or folding frequency (in Chapter 1), cut-off frequency (in Chapter 2), or sampling frequency
foE, foF1, foF2	Frequency of maximum response at E, F1, F2 layers
f_p	Plasma frequency
$f_x(x_1, x_2, \ldots, x_n)$	Joint density function of, or probability distribution function of, a set of data $x_1, x_2, \ldots x_n$
G	Gain
G_b	Gain of the beacon antenna
G_i	Stage i gain or antenna gain of the interrogating radar
G_r	Antenna gain of receiving radar
G_t	Antenna gain of transmitting radar
g	Gravitational constant
$g\Delta$	Number of sunspot group
g_t	Gating threshold
$\mathbf{H}$	Magnetic field vector
H	Entropy in Chapter 2, magnitude of the magnetic field intensity at any point on the earth in Chapter 6, or measurement transition matrix in Chapters 11 and 12
$\hat{H}$	Scaled, or normalized height
H_{op}	Transfer function of an impulse h_{op}
H_z, H_y	Magnetic field intensity of the wave along z, y direction
h	Planck's constant, or height of a reflecting layer in the ionosphere in Chapter 6
h_a	Antenna height above datum
h_c	Height of the radar antenna above the clutter surface
$h_{\max}$	Height of maximum ionization density
h_{op}	Impulse of the optimum linear filter
$h_{m\mathrm{F2}}$	Height of the peak density of the F2 layer
h_t	Target height above datum
h_v	Virtual height
I	Alternating current
IF	Intermediate frequency
I_{F2}	Ionospheric index
IFF	Identify friendly or foe
IP_n	Intercept point of the nth order
I_0	Modified Bessel function of first kind, zero order in Chapters 1 and 5, or amplitude of the alternating current in Chapter 3
i	Total current density
i_c	Convectional current density
i_D	Displacement current density
J	Jacobian function
K	Scale or correction factor K to effect the conversion to the scale originated by Wolf for sunspot number

K_a	Acceleration steady-state variance reduction ratio
K_v	Velocity steady-state variance reduction ratio
K_x	Position steady-state variance reduction ratio
k	Boltzmann's constant
k_χ	Index of an elliptically polarized antenna
k_d	Wind direction adjustment factor
k_e	Number of degrees of freedom describing a target function
k_g	Grazing angle adjustment factor
k_p	Polarization adjustment factor
k_s	Sea state adjustment factor
k_θ	Aperture illumination constant
L	Path length of the intervening rain in Chapter 5, or likelihood function in Chapter 11
L_f	Steady-state apparent fluctuation loss
L_p	A category of norms in Chapter 1, or polarization loss between an antenna elliptically and linearly polarized in Chapter 5
L_{pi}	Propagation losses in clutter patches
L_n	Pattern constant
L_s	System loss
L_{tot}	Total losses
l	Separation distance between the electric charges
l_i	The *i*th length of the periodic antenna element
M	Moment of the dipole in Chapter 3, or complex index of refraction in Chapter 6
m out of n	m peaks selected out of n detections
m_e	Mass of an electron
m_k	*k*th moment
N	Iteration limit number
N_{amb}	Number of ambiguities that can be folded, or mapped, into a particular cell
N_B	Background interference
N_c	Number of parallel channels
N_e	Electron density
N_i	Laser radar noise power
$N_{m\mathrm{F2}}$	F2-peak density
N_n	Number of densities of neutral particles
$N_\pm$	Number of densities of positive and negative ions
N_0	Total noise at the output of the receiver or maximum electron density in Chapter 6
N_p	Number of samples coherently processed
$N_{thermal}$	Thermal noise or Johnson noise
N_2	Molecular nitrogen
$N(\mu, \sigma^2)$	Normal distribution of mean μ and variance σ^2

n	Index of refraction in Chapter 6, or iteration limit in other chapters
n_b	Number of beams
n_c	Number of cells to be searched
n_d	Number of Doppler filters
n_e	Number of cells or number of independent pulses integrated during N-pulse transmission
n_o	Refractive index of the ordinary wave in Chapter 6
n_x	Refractive index of the extraordinary wave in Chapter 6
n_p	Number of photoelectron emissions
O_2	Molecular oxygen
P	Power radiated by a dipole in Chapter 3, or covariance matrix in Chapters 11 and 12
$\mathbf{P}$	Error covariance vector
$P(x)$	Probability of variable x
PDA	Probabilistic data association
P_b	Power output of the beacon antenna
P_c	Clutter power
P_d	Probability of detection
P_e	Probability of error
P_{fa}	Probability of false alarm
P_g	Gate probability
P_o	Probability that a target can be observed
P_r	Received signal power
P_t	Transmit power of the interrogating radar
$P_{o,x}$	Polarization of the ordinary 'o', and extraordinary 'x' wave
PRI	Pulse repetition interval
Pr{}	Probability of {}
$P(x \mid y)$	Probability of x given y
$p(x)$	Probability density function of x
$p(xy)$	Joint probability density function of two variables x and y
pdf	Probability density function
Q	Obliquity factor in Chapter 6, number of channels occupied by signals greater than specified threshold in Chapter 7, or noise covariance matrix in Chapters 11 and 12
q	Oscillating charge
$+q, -q$	Positive, negative point charge
$\mathfrak{R}$	Limit of field boundary
$\mathbf{R}$	Measurement noise covariance vector
R	Generally range or noise covariance matrix in Chapters 11 and 12

$\bar{R}$	Average range
RF	Radio frequency
R_{012}	Direct radar range
R_{12}	Yearly smoothed relative sunspot number
R_c	Clutter range, being the distance from the radar to the centre range gate
R_{eq}	System equivalent impedance
R_n	Sunspots occurrence measurement
R_{rad}	Radiation resistance
R_{un}	Unambiguous range
R_{xy}, R_{xx}	Cross-correlation of the signals x and y, autocorrelation function of same signal x
$\dot{r}$	Rate of change, or first derivative, of r (range)
$\ddot{r}$, $\dddot{r}$	Second, third derivative of r
r'	Elliptical distance observed at a point not at the equator
r_e	Radius of the earth at the equator
r_i	Target position in the ith scan
r_m	Measured range
r_p	Predicted range
r_r	Rain rate
S	Sea state index in Chapter 5, received signal power in Chapter 7, or residual covariance matrix in Chapters 11 and 12
$S_{b\min}$	Minimum detectable signal of the beacon receiver
S_i	Radar input signal
$S_{\min}$	Minimum detectable signal of the radar receiver
S/N	Signal-to-noise ratio
S_o	Signal power at the output of the receiver
S_T	Target power
s	Number of observed individual sunspots
s_i	Matched filter input signal
s_o	Matched filter output response
T	Record length in Chapter 1, data interval (sampling period) in Chapters 11 and 12, or temperature elsewhere
T_Δ	Duration of waveform
T_e	Temperature of electron ions
T_f	Frame time
T_i	Integration time
T_n	Temperature of neutral particle
T_0	Ideal standard temperature
T_p	Pulse repetition period
T_s	Dwell time
T_t	Track
tr{.}	Trace of {.}
t_c	Target correlation time

t_0	Measurement interval time or time dwelled on target
t_s	Time required by the laser radar to search a field (also called laser frame time)
UV	Ultraviolet ray
$\mathbf{u}$	Plant noise vector
u	Shape parameter of clutter model
V	Electric potential between two charges
V_c	Rain clutter volume
V_{cc}	Proportion of clutter in validation volume
V_p	Propagation wave phase velocity
V_g	Propagation wave group velocity
V_t	Proportion of target peaks in validation volume
V_v	Volume of the validation region
$\hat{v}(k)$	Smoothed velocity
$\mathbf{v}$	Measurement noise vector
v	Effective angular collision frequency in Chapter 6, or velocity in Chapter 12
v'_n, $v'_{\pm}$, $v'_{\pm n}$, $v'_{\pm\mp}$	Collision frequencies of electrons with neutral particles, electrons with ions, ions with neutral particles, and ions with ions respectively
$v(\phi_b)$	Orthogonal beams in ϕ domain
v_e	Clutter amplitude or threshold voltage
$\mathbf{w}$	Weight vector
W	Weight factor
w_k	Window function
w_p	Complex weighting on the received data from pth element of the array antenna that is beamformed
$X_r'^T$, $X_r''^T$	Test signal distribution across the receiver inputs, response within the processor
x_p	Received data from pth element of the array antenna that is beamformed in Chapter 7, or forecast (predicted) position in Chapter 12
Y_n	Day number starting on 1 January
$\tilde{\mathbf{y}}$	Innovation or residual vector
$\hat{y}$	Estimate of y
$y(k)$	Measurement recorded on the k radar scans
$y_{m\mathrm{F2}}$	Semi thickness associated with height of the peak density of the F2 layer
y_k	Beam output
Z_d	Impedance of the dipole
Z_s	Impedance of the complementary slot
$z(k)$	Observations on the k radar scans
α	Reference part of the propagation coefficient in Chapter 6, or position damping factor in Chapter 12
α_g	Apparent elevation angle

α_m	Signal modulation factor
α_n	Neuvy constant
α_0	Apparent ionospheric elevation angle or threshold value in Chapter 9 for Neyman–Pearson rule
α^0	Electron density gradient
$\alpha\beta$	Two-point extrapolator filter
$\alpha\beta\gamma$	Three-point extrapolator filter
β	Phase angle or the quadrature component of the propagation coefficient, or velocity damping factor in Chapter 12
β_i	Event probability
β_τ	Geometric spacing between adjacent elements of log period antenna
β_n	Neuvy constant
$\bar{x}$	Sea reflectivity
χ_p	Angle between linear polarization and the ellipse's major axis
$\bar{x}_{\text{ref}}$	Reference reflectivity
$\chi(\tau,f_d)$	Two-dimensional function in delay, τ, and Doppler shift, f_d; called uncertainty function, correlation function, or an ambiguity function
χ_n^2	Chi-squared distribution
δ	Delta function, or solar declination in Chapter 6
δ_{kj}	Kronecker symbol
$\delta\delta$	Phase progression angle
Δf_d	Doppler shift
Δ	Proportionality constant of uniformly distributed random disturbances
$\Delta\alpha$	Refraction error angle
$\Delta\alpha_{ref}$	Measurement elevation-angle error, or refraction-angle error
$\Delta\alpha_T$	Angle between the ray path and the direct path at the target location
$\Delta\phi$	Phase caused by path difference
$\Delta\phi_0$	The phase difference of direct and reflected fields reaching the target of equal intensity
Δ_f	Filter spacing
ΔH	Vertical extent of the beam in the rain or height of the radar resolution cells (whichever is lesser)
Δ_h	Hour angle of the sun measured westward from apparent noon
Δ_{lat}	Geographic latitude
Δ_{gla}	Geomagnetic latitude
Δn	Difference in the refractive index of two magneto-ionic components

$\Delta\theta$	Width of transmitter beam
$\Delta\dot{R}_{\min}$, $\Delta\ddot{R}_{\min}$	Nominal range-rate resolution, nominal resolution in acceleration
ΔR	Time delay or range error
ΔR_d	Path difference between direct and reflected waves
ΔS	Difference between the required signal level and that of undesired distortion
Δt	Steering time delay
ΔV	Error introduced in the target Doppler velocity
Δx	Pulse width spacing
Δx_{if}	Range extent at a particular operating frequency
Δw	Width of the illuminated area
$\mathfrak{I}$	Characteristic function of a random variable
$\in$	Error
ε_r	Relative permittivity
ε_0	Permittivity of free space
ζ	Solar zenith angle
ϕ	Angle between the surface normal and incident radar signal (for laser radar in Chapter 5), or total angular excursion
Φ_0	Average noise floor
Γ	Gamma function, or functional form in Chapter 9
Γ_Δ	Input reflection coefficient of the antenna
η	Detector quantum (or optical) efficiency in Chapter 5, or apex angle of log period antenna in Chapter 7
η_0	Characteristic impedance of free space
η_v	Rain reflectivity
$\bar{\eta}_v$	Mean rain reflectivity for each cell
κ	Proportionality constant in determining rain reflectivity in Chapter 5, or weighting factor in Chapters 11 and 12
λ	Wavelength
λ^*	Radian length
λ_c	Spatial density of false (clutter) measurement
λ_D	Characteristic length or Debye length
λ_Δ	Manoeuvre correlation coefficient
λ_t	Spatial density of true target measurement
λ_v	Approximate spatial density of false and target measurements
γ	Proportionality constant for sea and land reflectivity in Chapter 5, propagation coefficient in Chapter 6, or acceleration damping factor in Chapter 12
γ_a	Decision threshold
γ_{fa}	Desired threshold or biased value for nominally accepted probability of false alarm
γ_f	Phase angle of the reflection coefficient

ξ	Solar zenith angle in Chapter 7 or significance test level in Chapter 9
μ	Arithmetic mean
μ_t, μ_c	Target, clutter distribution function
θ	Antenna elevation angle
$\dot{\theta}$, $\ddot{\theta}$	First, second derivative of θ (bearing)
θ_a	Azimuth beamwidth, or antenna elevation angle in Chapter 5
θ_{BW}	Beamwidth (laser radar)
θ_e	Antenna elevation beamwidth
θ_H	Horizontal beamwidth
θ_0	Scanning or steering angle
θ_v	Vertical beamwidth
θ_a	Depression angle
θ_{ag}	Level parameters of clutter model
θ_t	Target elevation angle
σ	Target radar cross-section in Chapter 5, or conductivity in Chapter 6
σ_c	Land, or sea, clutter cross-section
σ^0	Land, or sea, reflectivity
σ_x	Standard deviation of signal/data x
σ_x^2	Second-order moment, or variance, of signal/data x
σ_z	Root-mean-square of wave height
σ_m^2	Predicted measurement variance
$\sigma_{x_s}^2$	Position measurement variance
σ_{vr}^2	Measurement noise variance in range
$\sigma_{v\theta}^2$	Measurement noise variance in bearing
σ_{ac}^2	Variance of target acceleration
ρ	Surface reflectivity or surface reflection coefficient
τ	Pulse width, delay in range or time required for changes at the dipole to travel a distance in Chapter 3, or log-periodic antennas' geometric ratio in Chapter 7
τ_{fa}	Average time between false target peaks
Λ	Log normal distribution-model constant, or likelihood test in Chapter 9
$\cup$	Union
ψ	Total phase difference of the radiating fields from the adjacent elements in Chapter 4, grazing angle in Chapter 5, or orientation of the target velocity in Chapter 6
ψ_t	Transitional angle beyond which an adjustment factor is applied
ψ_0	Mean value of Rayleigh component of clutter
Ψ_m, Ψ_p	Measured, predicted bearing
Ω	Sample space

Ω_{bs}	Laser radar search solid angle
Ω_c	Critical region
θ	Variable bearing
$\dot{\theta}$	Variable bearing rate
θ_m	Measured bearing
θ_p	Predicted bearing
(θ_0, ρ_0)	Cartesian coordinate of point of intersection
Θ_i	Event density function
ϑ	Target speed
ν	Gate size
Λ_*	Modifying function
$\otimes$	Convolution

Part I

Radar Systems

This part contains five chapters. The chapters are structured in a way that provides continuity to the reader in the understanding of radar systems. Each chapter is somehow self-sufficient. However, where further knowledge can be gained, applicable references are given.

Chapter 1 provides the essential functional relations, concepts, and definitions that are relevant to radar system's development and analysis and signal peak detection. This approach is taken to provide the basic groundwork for other concepts that are developed in subsequent chapters. The areas covered are sufficiently rich to provide a good understanding of the subject matter for non-specialists in radar systems and associated signal processing.

The next four chapters concentrate on radar systems. Discussions on radar systems evolve from a basic concept and gradually increase to a more complex outlook. The author believes that mastering the basic fundamentals permits moving on to more complex concepts without great difficulty. In so doing, the reader would learn:

- the basic architecture of radar systems, receiver sensitivity analysis, and data acquisition and/or compression issues as well as what radars are used for in Chapter 2;
- the physics of an antenna, which is a major item in radar systems design, from the perspective of a simple radiator, the division of radiation field in front of an antenna into quantifiable regions, the principle of pulse compression that allows recognition of closely spaced targets, as well as range measurements for signal pulse and train pulses in Chapter 3;
- that by extending the simple radiator's radiation property to an array of radiators including slot antennas, a higher gain can be achieved, and the array can be steered in any preferred direction in Chapter 4;
- how the radar equations are developed recognizing the effect of the environment on the conventional, laser and secondary radar performance and detection of targets of variable radar cross-sections and mobility in Chapter 5.

I understand during my years of engineering practice and teaching that many readers learn more by examples, which I have relied on in explaining difficult concepts. For those readers wishing to test their level of understanding several problems are written at the end of each chapter.

1

Essential relational functions

The chapter begins with the study of frequency analysis of signals with the representation of continuous time periodic and aperiodic signals by means of Fourier series and Fourier transform, respectively. The Fourier transform is one of several mathematical tools that are useful in the design and analysis of linear time-invariant systems. The properties of the Fourier transform are discussed and a number of time-frequency dualities presented. An analogous treatment of discrete-time periodic and aperiodic signals follows.

Other topics covered include convolution, correlation, window functions and a generalized category of norms, L_p-norm – used for scaling of data as well as for noise and error estimation.

1.1 Fourier analysis

Fourier transform is a process whereby a given function $f(t)$ can be expressed in terms of a trigonometric series. For instance, if a periodic or aperiodic function can be expressed in the form

$$f(t) = \sum_{n=0}^{\infty} a_n \cos(nt) + b_n \sin(nt) \tag{1.1}$$

such a series is known as the Fourier series of the function $f(t)$ and the constants a_n and b_n are the Fourier coefficients. Any trigonometric functions can be scaled to possess a period of $2l$, say. Thus for a function $f(t) = \cos(\omega t)$, its period is $(2\pi/\omega)$. For the period $2l$, $\omega = \pi/l$, and the function $f(t) = \cos(\pi t/l)$ is still of period $2l$, and its Fourier series will assume the form

$$f(t) = \sum_{n=0}^{\infty} a_n \cos\left(\frac{\pi n t}{l}\right) + b_n \sin\left(\frac{\pi n t}{l}\right) \tag{1.2}$$

Equation (1.2) is the general definition of a Fourier series of the function $f(t)$ of period $2l$. In determining the Fourier series of a function, certain assumptions are made: the series exists; and the series uniformly converges within the given interval. The convergence premise provides the options of integrating the series term by term so that the values of the coefficients a_n, b_n can be determined. The interval of the integration could be any of the following $(-l, l)$ or $(0, 2l)$, or $(-\pi, \pi)$ or $(0, 2\pi)$. Where the particular interval is taken, however, makes no difference when the function $f(t)$ is periodic. It is often convenient to take the interval of integration from $-T/2$ to $+T/2$ in order to recognize possible symmetry conditions.

To develop suitable expressions for a_n, b_n, begin by integrating (1.2) with respect to t, term by term, over the interval $(-l, l)$:

$$\begin{aligned}\int_{-l}^{l} f(t)dt &= \int_{-l}^{l} \sum_{n=0}^{\infty} a_n \cos\left(\frac{\pi nt}{t}\right) + b_n \sin\left(\frac{\pi nt}{l}\right) dt \\ &= \int_{-l}^{l} a_0 dt + \sum_{n=0}^{\infty} \int_{-l}^{l} a_n \cos\left(\frac{\pi nt}{t}\right) + b_n \sin\left(\frac{\pi nt}{l}\right) dt\end{aligned} \tag{1.3}$$

It follows therefore that when $n = 0$,

$$a_0 = \frac{1}{2l} \int_{-l}^{l} f(t)dt \tag{1.4}$$

which is the mean value (MV) of $f(t)$ over a period $(-l, l)$. By definition, the MV of a function, say, $f(t)$, is given as $\text{MV} = 1/b - a \int_a^b f(x)dx$.

For the other cases of $n \geq 1$, the Fourier coefficients a_n and b_n are obtained by multiplying both sides of (1.2) by $\cos(n\pi t/l)$ and $\sin(n\pi t/l)$ respectively and then integrating the result term by term. By Aboaba (1975), the trigonometric functions $\cos(n\pi t/l)$ and $\sin(n\pi t/l)$ are used because they have important properties:

- that enable a *minimum mean square error* between the signal and the approximate value derived from the Fourier technique; and
- that are orthogonal enabling the coefficients to be determined independently of one another.

Thus, by multiplying (1.2) by $\cos(n\pi t/l)$:

$$\int_{-l}^{l} f(t) \cos\left(\frac{\pi nt}{l}\right) dt = \sum_{n=0}^{\infty} \int_{-l}^{l} a_n \cos^2\left(\frac{\pi nt}{l}\right) + b_n \cos\left(\frac{n\pi t}{l}\right) \sin\left(\frac{\pi nt}{l}\right) dt \tag{1.5a}$$

with the *sine* terms of this equation vanishing and leaving only the a_n terms; that is,

$$a_n = \frac{1}{l} \int_{-l}^{l} f(t) \cos\left(\frac{n\pi t}{l}\right) dt \quad n = 1, 2, 3, \ldots \tag{1.5b}$$

which corresponds to 2MV of $f(t)\cos(\pi nt/l)$.

Similarly, multiplying (1.2) by $\sin(n\pi t/l)$ leaves only the b_n terms because the *cosine* terms vanish. So

$$b_n = \frac{1}{l} \int_{-l}^{l} f(t) \sin\left(\frac{n\pi t}{l}\right) dt \quad n = 1, 2, 3, \ldots \tag{1.6}$$

which corresponds to 2MV of $f(t)\sin(\pi nt/l)$. Combining (1.4), (1.5b) and (1.6) together, the resulting series is called the Fourier series of $f(t)$ and the coefficients so defined are the Fourier coefficients. In order to express the coefficients uniformly as being 2MV of the respective function, the Fourier series of a periodic function $f(t)$ over the interval $(-l, l)$ is sometimes written as

$$f(t) = \frac{a_0}{2} + \sum_{n=1}^{\infty} a_n \cos\left(\frac{n\pi t}{l}\right) + b_n \sin\left(\frac{n\pi t}{l}\right) \tag{1.7}$$

where, in this instance, $a_0 = $ 2MV of $f(t)$ over a period $(-l, l)$. The sum, represented by (1.7), of the Fourier series does not necessarily equal to the function from which it is derived and the conditions under which the Fourier series converge because (1.7) depends very much on the form of the particular function chosen.

The expression in (1.7) may be represented in terms of exponential terms as

$$f(t) = \sum_{n=1}^{\infty} S_n e^{\frac{-n\pi t}{l}} \tag{1.8}$$

where

$$S_n = \frac{1}{2}(a_n - jb_n)$$

noting that $S_{-n} = S_n^*$ where the asterisk denotes a complex conjugate.

The quantum leap to this generalization is left to the reader to verify given that

$$\begin{aligned}
\cos(u) &= \cos(-u) \\
\sin(-u) &= -\sin(u) \\
\cos(u) &= \frac{1}{2}(e^{ju} + e^{-ju}) \\
\sin(u) &= \frac{1}{2j}(e^{ju} - e^{-ju}) \\
\sum_{n=1}^{\infty} a_n e^{\frac{-jn\pi t}{l}} &= \sum_{n=-1}^{-\infty} a_n e^{\frac{jn\pi t}{l}} \\
\sum_{n=1}^{\infty} jb_n e^{\frac{-jn\pi t}{l}} &= -\sum_{n=-1}^{-\infty} jb_n e^{\frac{jn\pi t}{l}}
\end{aligned} \tag{1.9}$$

Equation (1.8) is commonly quoted in the literature as the complex Fourier series. From the preceding Fourier series discussion, another important term can be introduced, namely Fourier transform which is discussed next.

1.1.1 Fourier transform

The Fourier transform of signal $s(t)$ is defined as

$$S(f) = F[s(t)] = \int_{-\infty}^{\infty} s(t)e^{-j2\pi ft}dt \tag{1.10}$$

as the period T tends to infinity. The symbol $F[\ ]$ denotes Fourier transform of [ ]. Physically, the Fourier transform $S(f)$ represents the distribution of signal strength with frequency; that is, it is a density function. Fourier transform has inversion property.

1.1.2 Inverse Fourier transform

The Inverse Fourier transform of signal $s(t)$ is defined as

$$s(t) = F^{-1}[S(f)] = \frac{1}{2\pi}\int_{-\infty}^{\infty} S(f)e^{j2\pi ft}df \tag{1.11}$$

By comparing (1.10) with (1.11) it could be seen that a transform pair exists:

$$s(t) \leftrightarrow S(f)$$

where $\leftrightarrow$ denotes a Fourier transform pair. Other Fourier transform pairs can be developed as summarized in Table 1.1.

Table 1.1 Fourier transform pairs

(i)	Basic pair	$f(\lambda) \leftrightarrow F(u)$
(ii)	Complex argument	$f^*(\lambda) \leftrightarrow F^*(u)$
(iii)	Negative argument	$f(-\lambda) \leftrightarrow F(-u)$
(iv)	Scaling by Δ	$f(\Delta\lambda) \leftrightarrow 1/\lvert\Delta\rvert F(\frac{u}{\Delta})$
(v)	Multiplication by constant κ	$\kappa f(\lambda) \leftrightarrow \kappa F(u)$
(vi)	Additive	$f_1(\lambda) + f_2(\lambda) \leftrightarrow F_1(u) + F_2(u)$
(vii)	Shift	$\begin{cases} f(\lambda)e^{j2\pi u_1\lambda} \leftrightarrow F(u-u_1) \\ f(\lambda-\lambda_1) \leftrightarrow e^{-j2\pi u_1\lambda}F(u) \end{cases}$
(viii)	Integration	$\int f(\lambda)d\lambda \leftrightarrow F(u)/ju$
(ix)	Commutative convolution	$\int_{-\infty}^{\infty} f_1(\lambda)f_2(\lambda)d\lambda \leftrightarrow F_1(u)F_2(u)$
(x)	Autocorrelation	$\int_{-\infty}^{\infty} f(\lambda_1)f^*(\lambda_1+\lambda)d\lambda_1 \leftrightarrow F(u_1)F^*(u_1)$
(xi)	Parseval theorem	$\int_{-\infty}^{\infty} f^2(\lambda)d\lambda \leftrightarrow 1/2\pi \int_{-\infty}^{\infty} [F(u)]^2 du$
(xii)	Dirac delta at pulse time $t = 0$ and $t = t_0$	$\begin{matrix} \delta(t) \\ \delta(t-t_0) \end{matrix} \leftrightarrow \begin{matrix} 1 \\ e^{-j2\pi ft_0} \end{matrix}$
(xiii)	Gaussian pulse	$e^{-\pi t^2} \leftrightarrow e^{-\pi f^2}$

1.1.3 Orthogonal relations

The following so-called orthogonal relations satisfy both the Fourier's circular and complex exponential functions:

$$\frac{1}{2\pi}\int_{-\pi}^{\pi}\cos(mx)\cos(nx)dx = \begin{cases} 0 & m \neq n \\ \frac{1}{2} & m = n > 0 \\ 1 & m = n = 0 \end{cases} \tag{1.12a}$$

$$\frac{1}{2\pi}\int_{-\pi}^{\pi}\sin(mx)\sin(nx)dx = \begin{cases} 0 & m \neq n \\ \frac{1}{2} & m = n > 0 \\ 0 & m = n = 0 \end{cases} \tag{1.12b}$$

$$\frac{1}{2\pi}\int_{-\pi}^{\pi}\cos(mx)\sin(nx)dx = 0 \qquad \text{for all } n \text{ and } m \tag{1.12c}$$

$$\frac{1}{2\pi}\int_{-\pi}^{\pi} e^{jmx}e^{-jnx}dx = \frac{1}{2\pi}\int_{-\pi}^{\pi} e^{j(m-n)x}dx = \begin{cases} 0 & m \neq n \\ 1 & m = n \end{cases} \tag{1.12d}$$

In these relations, m and n are integers and the intervals $\{-\pi, \pi\}$ may be replaced by any other interval of length 2π.

The next two examples give the reader some feeling for the general properties that might be expected.

Example 1.1 Obtain the Fourier series of $f(x)$ defined by

$$f(x) = \begin{cases} 0 & -t \leq x < 0 \\ \sin\left(\frac{\pi x}{t}\right) & 0 \leq x \leq t \end{cases} \tag{1.13}$$

The function is shown in Figure 1.1.

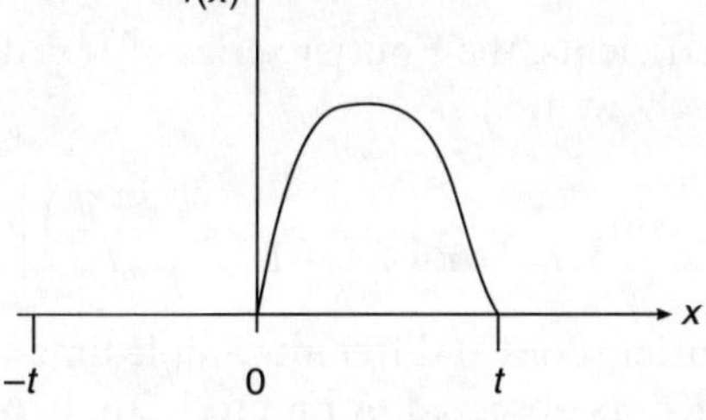

Figure 1.1 Graph of $f(x)$

Solution

In view of (1.7), and using the function defined above, the Fourier series coefficients may be expressed as follows:

$$a_0 = \frac{1}{t}\int_0^t \sin\left(\frac{\pi x}{t}\right) dx = \frac{1}{\pi} \tag{1.14}$$

$$\begin{aligned} a_n &= \frac{1}{t}\int_0^t \sin\left(\frac{\pi n x}{t}\right)\cos\left(\frac{\pi n x}{t}\right) dx \\ &= \frac{1}{2t}\int_0^t \sin\left(\frac{\pi(n+1)}{t}\right) - \sin\left(\frac{\pi(n-1)x}{t}\right) dx \\ &= \frac{1}{2\pi}\left[\left\{\frac{1}{n-1}\cos\left(\frac{\pi(n-1)x}{t}\right)\right\} - \left\{\frac{1}{n+1}\cos\left(\frac{\pi(n+1)x}{t}\right)\right\}\right]_0^t \\ &= \frac{1}{2\pi}\left[\frac{1}{n+1} - \frac{1}{n-1} + \frac{\cos(n-1)\pi}{n-1} - \frac{\cos(n+1)\pi}{n+1}\right] \end{aligned} \tag{1.15a}$$

It is difficult to extract the coefficient a_n when $n = 1$, from this solution because of the divide-by-zero term occurring. So, the case of $n = 1$ can be solved directly by putting $n = 1$ at the first integral of (1.15a); that is,

$$a_n = \frac{1}{t}\int_0^t \sin\left(\frac{\pi x}{t}\right)\cos\left(\frac{\pi x}{t}\right) dx = 0 \tag{1.15b}$$

using the orthogonal relation of (1.12c). For other cases when $n > 1$, the corresponding a_n terms are obtained, using the resulting expression of (1.15a), as

$$a_n = \begin{cases} 0 & n = \text{odd} \\ -\frac{1}{\pi}\left(\frac{2}{n^2-1}\right) & n = \text{even} \end{cases} \tag{1.15c}$$

And consequently for the b_n terms using the orthogonal relation of (1.12b):

$$b_n = \frac{1}{t}\int_0^t \sin\left(\frac{\pi n x}{t}\right)\sin\left(\frac{\pi n x}{t}\right) dx = \frac{1}{2} \tag{1.15d}$$

Collating all the coefficients, the Fourier series of $f(x)$ described by (1.13), or Figure 1.1, is concisely written as

$$F(x) = \frac{1}{\pi} + \frac{1}{2}\sin\left(\frac{\pi x}{t}\right) - \frac{2}{\pi}\left[\frac{1}{4r^2-1}\cos\left(\frac{2\pi x r}{t}\right)\right] \quad r \geq 1 \tag{1.16}$$

The plots of the Fourier series at different sample times, i.e. $t = 2, 5, 10$ s, are shown in Figure 1.2. It is observed in Figure 1.2(a, b, c) that as the function period t increases, the main lobe width widens and the side lobes, which are prominent at short sampling periods, vanish.

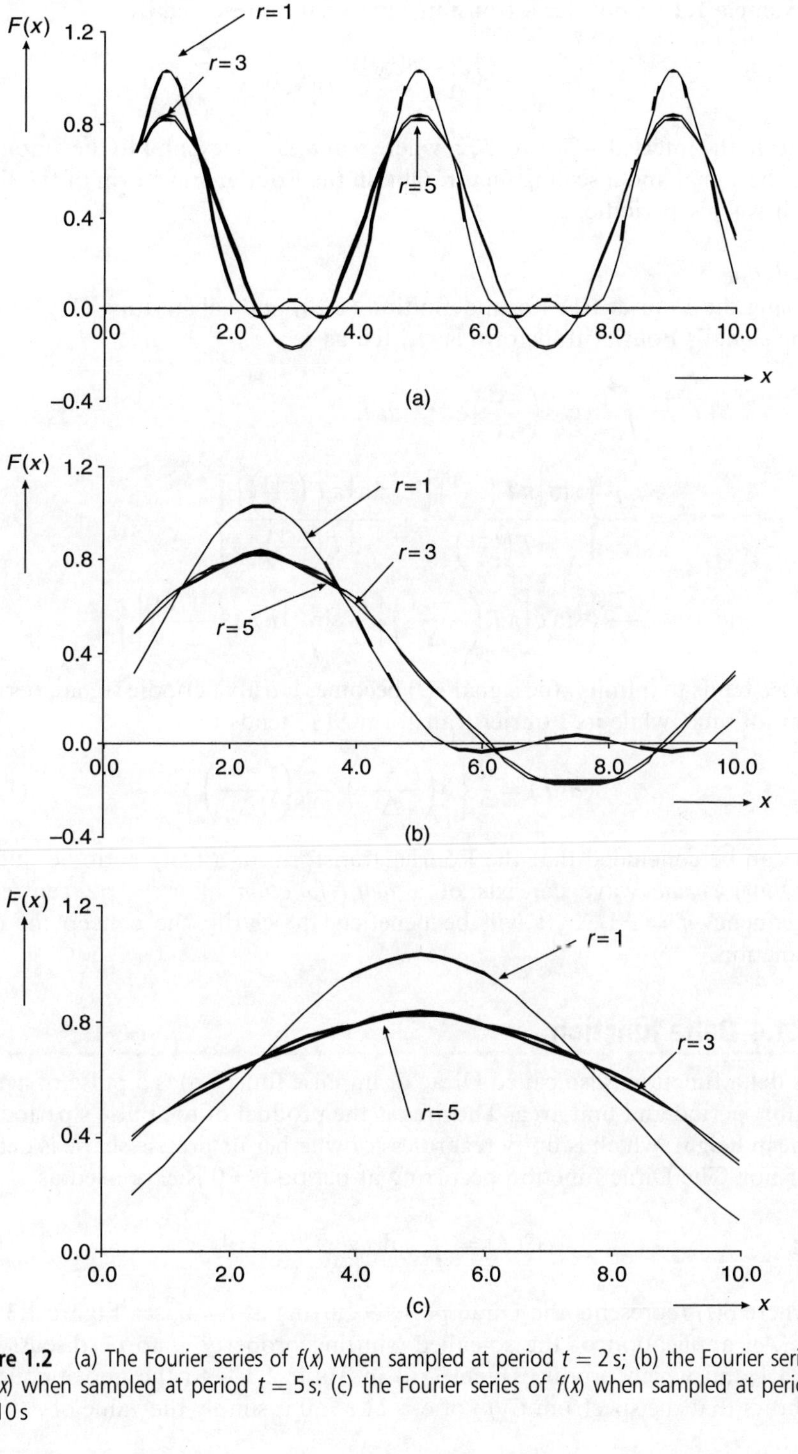

Figure 1.2 (a) The Fourier series of $f(x)$ when sampled at period $t = 2$ s; (b) the Fourier series of $f(x)$ when sampled at period $t = 5$ s; (c) the Fourier series of $f(x)$ when sampled at period $t = 10$ s

Example 1.2 Consider a transmitting signal represented by

$$s(t) = \begin{cases} \alpha \cos(\frac{2\pi t}{\Delta}) & -\frac{T}{2} \leq t \leq \frac{T}{2} \\ 0 & |t| > \frac{T}{2} \end{cases} \tag{1.17a}$$

within the interval $-T/2$ to $T/2$, where α and Δ correspond to the amplitude of the signal and a scaling factor. Obtain the Fourier transform of the signal if it is truly periodic.

Solution

Using the Fourier transform definition (1.10) and substituting (1.17a) in it, the signal's Fourier transform is written as

$$\begin{aligned} S(f) &= \int_{\frac{-T}{2}}^{\frac{T}{2}} \alpha \cos\left(\frac{2\pi t}{\Delta}\right) e^{-j2\pi ft} dt \\ &= \frac{\alpha T}{2} \left\{ \frac{\sin\left[\pi T\left(\frac{f-1}{\Delta}\right)\right]}{\pi T\left(\frac{f-1}{\Delta}\right)} + \frac{\sin\left[\pi T\left(\frac{f+1}{\Delta}\right)\right]}{\pi T\left(\frac{f+1}{\Delta}\right)} \right\} \\ &= \frac{\alpha T}{2} \left\{ \sin c\left[\pi T\left(\frac{f-1}{\Delta}\right)\right] + \sin c\left[\pi T\left(\frac{f+1}{\Delta}\right)\right] \right\} \end{aligned} \tag{1.17b}$$

As T tends to infinity, the signal $s(t)$ becomes a truly periodic signal; periodic for all time, while its Fourier transform $S(f)$ tends to

$$S(f) = \frac{\alpha}{2} \left\{ \delta\left(\frac{f-1}{\Delta}\right) + \delta\left(\frac{f+1}{\Delta}\right) \right\} \tag{1.17c}$$

It can be concluded that the Fourier transform of a truly periodic (infinite extent) *cosine* wave consists of a *delta function* of area $\alpha/2$ centred at frequency $f = \pm 1/\Delta$. It will be beneficial to clarify the concept of delta function.

1.1.4 Delta function

A delta function (also called Dirac or impulse function) is a pulse of acutely short period and unit area. The area is the product of the pulse's period and mean height, which is unity regardless of whether its precise shape is defined or not. The Dirac function occurring at period $t = 0$ is expressed as

$$G(f) = \int_{-\infty}^{\infty} \delta(t) e^{-j2\pi ft} dt = 1 \tag{1.18}$$

where $\delta(t)$ represents the Dirac pulse occurring at $t = 0$, see Figure 1.3.

An application of the so-called 'shifting property' – to be discussed in 1.3.1 and which produces item (xii) in Table 1.1 – to the above equation shows that the spectrum $G(f)$ of $\delta(t)$ at $t = 0$ is simply the value of $e^{-j2\pi ft}$ at

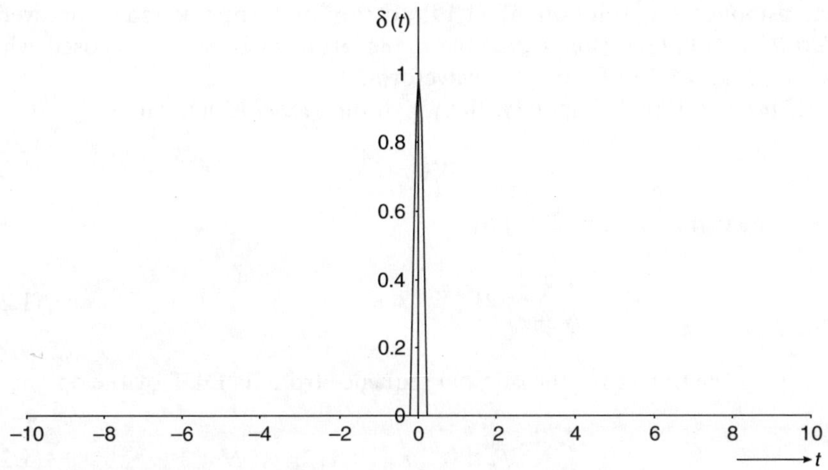

Figure 1.3 Delta function

$t = 0$; which is unity. The result implies that all frequencies are equally represented by cosine components. Suffice it to say that for a large number of cosines of equivalent amplitude but of different frequencies when added together tend to cancel each other out everywhere except at $t = 0$ where they all reinforce. In short, as higher and higher frequencies are included, the resultant becomes an extremely narrow pulse centred on $t = 0$.

The preceding discussion has focused on Fourier transforms with continuous time series signals. Fourier transforms can also be expressed in discrete form.

1.2 Discrete Fourier transform

Digital systems may accept discrete signals in the form of a train of pulses introduced by a sampler operation, or generate a sequence of numbers representing the system output. The sampler may digitize the continuous input signal f_r at equal intervals of r seconds. This type of sampler is called a periodic or uniform-rate sampler. If a total of N data points is required within the finite period t, then the sampler's Fourier coefficients can be expressed in discrete form as

$$F_k = \frac{1}{N}\sum_{r=0}^{N-1} f_r e^{\frac{-j2\pi kr}{N}} \quad k = 0, 1, 2, \ldots, N-1 \tag{1.19}$$

where k and r correspond to n and t of the continuous case. The expression in (1.19) gives the Fourier coefficients in the discrete case, appropriately called the *discrete Fourier transform* (DFT). In general, the sampling scheme may be non-uniform aperiodic or a cyclic-variable sampling rate. An extension of the solution of (1.19) to non-uniform aperiodic, or cyclic-variable, sampling rate is possible if the problem is carefully posed while taking cognizance of the input waveform.

If for notational simplicity, the weighting *kernel* is defined as

$$W_N = e^{\frac{j2\pi}{N}} \tag{1.20}$$

then one can represent (1.19) by

$$F_k = \frac{1}{N}\sum_{r=0}^{N-1} f_r W_N^{-kr} \quad k = 0, 1, 2, \ldots, N-1 \tag{1.21}$$

It is possible to recover the original sequence from its DFT by the operation

$$f_r = \frac{1}{N}\sum_{r=0}^{N-1} F_k W_N^{kr} \quad r = 0, 1, 2, \ldots, N-1 \tag{1.22}$$

This operation is called the *inverse discrete Fourier transform* (IDFT) and is valid for real terms. Since the conjugate of a product is the product of the conjugate, the complex DFT can be expressed as

$$f_r^* = \frac{1}{N}\sum_{k=0}^{N-1} F_k^* W_N^{kr} \quad r = 0, 1, 2, \ldots, N-1 \tag{1.23}$$

Alternatively,

$$f_r = \frac{1}{N}\left[\sum_{r=0}^{N-1} F_k^* W_N^{kr}\right]^* \quad r = 0, 1, 2, \ldots, N-1 \tag{1.24}$$

which shows that the inverse DFT can be computed by forward transformation.

By substituting $k = n \pm N$, or $r = n \pm N$, both the DFT and IDFT expressions become

$$F_{n\pm N} = \frac{1}{N}\sum_{r=0}^{N-1} f_r W_N^{-(n\pm N)r} \quad n = 0, 1, 2, \ldots, N-1 \tag{1.25}$$

$$f_{n\pm N} = \frac{1}{N}\sum_{r=0}^{N-1} F_k W_N^{(n\pm N)k} \quad n = 0, 1, 2, \ldots, N-1 \tag{1.26}$$

Equations (1.25) and (1.26) can be computed by a fast Fourier transform (FFT) if N is suitably factorizable. An FFT method of computation is addressed in section 1.3. The magnitude of the term $W_N^{\mp Nr}$ in (1.25), or $W_N^{\pm Nk}$ in (1.26), is always unity for all values of r (or k) showing that $F_{n\pm N}$, or $f_{n\pm N}$, is periodic; that is, repeating itself outside the 0: $N-1$ limit.

This periodicity invokes the concept of *aliasing*, which one frequently encounters in radar signal processing and estimation.

1.2.1 Aliasing

The phenomenon of an aliasing arises in a number of practical contexts, for example the wheels of a stagecoach, movie films, stroboscope and tracking. Let us discuss how this phenomenon works in the case of the wheels of a stagecoach. The wheels start accelerating from zero appear to rotate in the correct direction with increasing speed, then they appear to be rotating in the opposite direction with decreasing speed until they stop, then begin to rotate with increasing speed in the forward direction, and so on. They appear to fold over to the next speed after a particular instant or frequency. This concept can be discussed further by formalization.

It is noted in (1.19) that the DFT of the series $\{x_r\}$, where $r = 0, 1, \ldots, N-1$, is defined by

$$X_k = \frac{1}{N}\sum_{r=0}^{N-1} x_r e^{\frac{-j2\pi kr}{N}} \quad k = 0, 1, \ldots, N-1 \tag{1.27}$$

Let us attempt to calculate values for X_k for all cases when k is greater than $N-1$. Putting $k = N + L$ and upon substitution in (1.27):

$$\begin{aligned} X_{N+L} &= \frac{1}{N}\sum_{r=0}^{N-1} x_r e^{\frac{-j2\pi(N+L)r}{N}} \\ &= \frac{1}{N}\sum_{r=0}^{N-1} x_r e^{\frac{-j2\pi Lr}{N}} e^{-j2\pi r} \end{aligned} \tag{1.28}$$

which, since the magnitude of $e^{-j2\pi r}$ is always equal to unity whatever the value of r, the resulting waveform repeats itself periodically. So,

$$X_{N+L} = X_L \tag{1.29}$$

Furthermore, it is easy to see from (1.27) that if the terms in series $\{x_r\}$ are real, then

$$X_{-L} = X_L^* \tag{1.30}$$

which is in agreement with the Fourier transform of x_k demonstrated by (1.8). Hence

$$|X_{-L}| = |X_L| \tag{1.31}$$

indicating that the response of X_k will be symmetrical about the zero frequency position. For sampling time interval 'd' seconds, the unique part of this response occupies the frequency range $|\omega| \leq 2\pi/d$ (rad/s). Beyond this, several spurious Fourier coefficients occur would appear as repetitions of the original which apply at frequencies below $2\pi/d$. Suffice to say therefore that the X_k coefficients calculated by the DFT are only correct for Fourier coefficients up to

$$\omega_k \leq \frac{2\pi k}{Nd} \quad k = 0, 1, \ldots, \frac{N}{2} \tag{1.32}$$

If there are frequencies above $2\pi/d$ present in the original spectrum, the high-frequency components will introduce a distortion called *aliasing*. In essence, the high-frequency components contribute to the series $\{x_r\}$ and regrettably falsely distort the Fourier coefficients calculated by the DFT for frequencies below $2\pi/d$.

If ω_0 is the fundamental and maximum frequency component present in the series $\{x_r\}$, then aliasing can be avoided by guaranteeing that the sampling interval d is small enough such that

$$f_0 < \frac{\omega_k|_{k=\frac{N}{2}}}{2\pi} < \frac{1}{2d} \quad \text{(Hz)} \tag{1.33}$$

This frequency is called the *Nyquist* frequency (or sometimes called the *folding* frequency), which is the maximum frequency that can be uniquely identified from data sampled at time spacing d.

Example 1.3 An FFT processor is employed to spectrally analyse a randomly generated real signal. The following requirements are given:

Desired resolution between frequencies $\leq$2.5 Hz
Maximum frequency in signal $\leq$1.75 kHz

If the points permitted by the processor are an integer power of two, determine (a) the minimum record length, (b) maximum allowable time between signals and (c) the minimum number of sampling points in a record.

Solution

(a) The minimum record length is equivalent to the desired resolution, so

$$d_{\text{min}} \geq \frac{1}{f_k} \geq \frac{1}{2.5} = 0.4\,\text{s}$$

(b) From (1.33), the maximum time between sampling must be confined to

$$d_{\text{max}} \leq \frac{1}{2f_d} \leq \frac{1}{2 \times 1.75} \leq 0.28571\,\text{ms}$$

(c) From (1.32), the minimum number of sampling points in a record N can be estimated when $k = 1$:

$$N \geq \frac{1}{d_{\text{max}} f_k} \geq 700$$

In conclusion, the phenomenon of aliasing is most important when analysing real-time data. The sampling frequency f_0 must be high enough to cover the full bandwidth of the continuous time series. Otherwise the spectrum from equally spaced samples will differ from the true spectrum primarily due to aliasing. In certain instances, the only way of avoiding aliasing is to intentionally filter out the higher-frequency components of the time series before the analysis begins.

1.3 Other useful functions

This section briefly discusses the concept of convolution, correlation and translation of signal from one domain description to another. For examples, a space formed by $[\omega, F(\omega)]$ is called the *frequency domain* while the space formed by $[t, f(t)]$ is called the *temporal* or *time* domain if the independent variable t represents time. The time domain can also be called the *spatial domain* if t represents a spatial variable. The subsequent properties will enhance the understanding of some of the Fourier transform pairs listed in Table 1.1.

1.3.1 Shifting in time and frequency domain

The shifting theorem states that if the function $f(t)$ has a Fourier transform given by $F(\omega)$, then:

> The Fourier transform of the function $f(t \pm t_0)$ is given by $F(\omega)e^{\pm j\omega t_0}$ and The function $g(t) = f(t)e^{\pm j\omega_0 t}$ has a Fourier transform given by $F(\omega - \omega_0)$

Proof

(i) Following (1.10), the Fourier transform of $f(t)$ is

$$F\{f(t)\} = \int_{-\infty}^{\infty} f(t)e^{-j\omega t}\,dt$$

which follows that

$$F\{f(t \pm t_0)\} = \int_{-\infty}^{\infty} f(t \pm t_0)e^{-j\omega t}\,dt \tag{1.34}$$

By letting $t = \tau \pm t_0$ and substituting it in (1.34), and changing the variable of integration,

$$\begin{aligned} F\{f(\tau)\} &= \int_{-\infty}^{\infty} f(\tau)e^{-j\omega(\tau \pm t_0)}\,d\tau \\ &= e^{\mp j\omega t_0}\int_{-\infty}^{\infty} f(\tau)e^{-j\omega\tau}\,d\tau \end{aligned} \tag{1.35}$$

The integral component of (1.35) is by definition $F\{\tau\}$. Hence

$$F\{f(\tau)\} = e^{\mp j\omega t_0}F\{\tau\} \tag{1.36}$$

This expression shows that if a signal whose function is delayed in time by t_0, the magnitude spectral density of the signal remains unchanged but an additional term $\mp\omega t_0$ is added to the phase spectral in each of its frequency components.
(ii) Following (1.10),

$$\begin{aligned} F\{g(t)\} &= \int_{-\infty}^{\infty} f(t)e^{-j\omega_0 t}e^{\pm j\omega_0 t}dt \\ &= \int_{-\infty}^{\infty} f(t)e^{-j(\omega\mp\omega_0)t}dt \\ &= F(\omega \pm \omega_0) \end{aligned} \tag{1.37}$$

This implies that a signal multiplied by a time function $e^{\pm j\omega_0 t}$ causes its spectral density to be translated in frequency by $\pm\omega_0$.

1.3.2 Convolution

The convolution theorem states that if three time functions $h(t)$, $f_1(t)$ and $g(t)$ have Fourier transforms $H(\omega)$, $F_1(\omega)$ and $G(\omega)$ respectively, and if

$$G(\omega) = H(\omega) \cdot F_1(\omega)$$

then the *multiplication* of these two frequency functions $H(\omega)$ and $F_1(\omega)$ is equivalent to the *convolution* of their corresponding time functions. That is,

$$g(t) = h(t) \otimes f_1(t)$$

where '$\otimes$' denotes convolution.

Example 1.4 The question devised for this example is an abridged version of Stanley (1975). Consider two three-port aperiodic functions $x(n)$ and $g(n)$, represented by

$$\begin{aligned} x(n) &= 2[\delta(n) + \delta(n-1) + \delta(n-2)] \\ g(n) &= \delta(n) + 2\delta(n-1) + 3\delta(n-2) \end{aligned}$$

Plot the convolution $y(n)$ of the functions.

Solution
The convolution of the functions can be written as

$$y(n) = x(n) \otimes g(n) = \sum_{k=0}^{n} x(k)g(n-k)$$

With the expression, Figure 1.4 is drawn by letting $n = 1$.

Our understanding of signal-analysis procedure is enhanced by the application of the convolution theorem. An illustration of such benefit is that of windowing in signal data processing, particularly in spectral estimation, when a reduction in sidelobes of such data is desired or gating in tracking

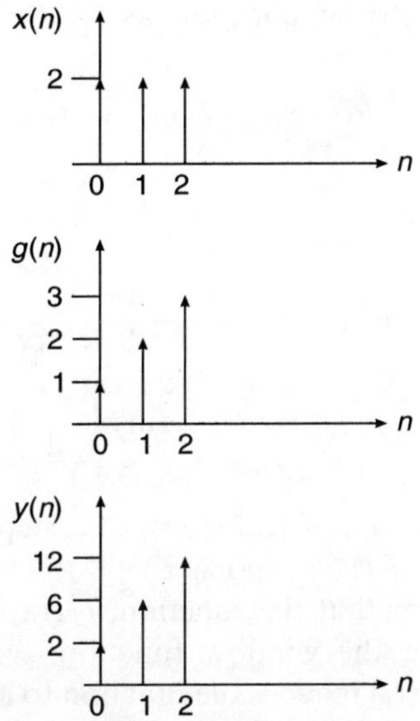

Figure 1.4 The convolution plot of functions $x(n)$ and $g(n)$

specifically in range filtering when using Doppler filters and still preserving range information. For clarity, these new terms – gating and windowing – are defined as follows.

A gate is simply a switch that opens and closes at preset times. Gating is a term also used in radar tracking as a screening technique used in cutting down the number of unlikely tracks postulated for a target. For instance, if a range gate is set to pass echoes from all targets between x_1 and x_2 kilometres away, the gate will open, say for t_a microseconds after the transmitted pulse $t_a(=kx_1,$ assuming $k\mu$ sec/km) and will close $t_b[=k(x_2-x_1)]$ microseconds later. More is said of tracking in Chapter 12.

Windowing involves multiplying a desired impulse response by a finite duration window. It reduces abrupt changes at the beginning and end of acquired data. For example, if one assumes that a finite duration T function, represented by $f_T(t)=f(t)w_T(t)$ is Fourier transformable, then in view of (1.10):

$$\begin{aligned}F_T\{f_T(t)\} &= F\{f(t)w_T(t)\}\\ &= \int_0^T\left[\frac{1}{2\pi}\int_{-\infty}^{\infty}F(\alpha)e^{j\alpha t}d\alpha\right]e^{-j\omega t}dt\\ &= \frac{1}{2\pi}\int_{-\infty}^{\infty}F(\alpha)\left[\int_0^T e^{j(\alpha-\omega)t}dt\right]d\alpha\\ &= \frac{T}{2\pi}\int_{-\infty}^{\infty}F(\alpha)e^{\frac{j(\alpha-\omega)t}{2}}\frac{\sin(\alpha-\omega)\frac{T}{2}}{(\alpha-\omega)\frac{T}{2}}d\alpha\end{aligned} \tag{1.38}$$

This expression can be written concisely as

$$F(\omega) = \frac{1}{2\pi} F(\omega) \cdot w_T(\omega) \tag{1.39}$$

where

$$\begin{aligned} w_T(\omega) &= Te^{\frac{-j\omega t}{2}} \frac{\sin \frac{\omega T}{2}}{\frac{\omega T}{2}} \\ &= Te^{\frac{-j\omega t}{2}} \sin c\left(\frac{\omega T}{2}\right) \end{aligned} \tag{1.40}$$

which translates to the Fourier transform of a step function window representable by $F\{w_T(t)\} = F\{u(t) - u(t - T)\}$.

From (1.39) it is clear that the function, $f_T(t)$, is the convolution of the true function, $f(t)$, and the window function, $w_T(t)$. Suffice to say that windowing of function $f(t)$ reduces the function to a signal of finite duration, which has the potential effect of spreading out the estimate $F(\omega)$ with the Fourier transform of the window function. It can be seen in (1.40) that the window function has an infinite range of frequencies. As such, one will always obtain signal spectrum illustrations that are not band limited when windowing a continuous signal of a finite length.

Windowing can be performed in either the time or frequency domain. Either function has its limitations. Windowing in the frequency domain introduces leakages (or distortions) in the time domain in the same way that windowing in the time domain causes spreading or leakage of the spectrum into adjacent frequencies and sidelobes in the frequency domain. This happens because multiplication in the frequency domain is similar to convolution in the time domain.

The benefit of windowing is that it reduces leakage in spectrum analysis: considerable reduction in the function's sidelobes and as well as reduction in the filter's sidelobes. In the light of this benefit, to improve the frequency response of a truncated time series, therefore, one can use a number of window functions, which will readily modify the impulse response in a prescribed way.

1.3.3 Window functions

A brief discussion of some of the commonly used window functions in practice is given in this subsection and their ability to pick out peaks (resolvability), using a similar input $s(t)$ whose Fourier transform consists of three delta functions centred at f_0, f_1 and f_2 (see Figure 1.5).

In Figures 1.6, 1.7 and 1.8, the frequency spacing $(f_2 - f_1)$ was chosen to be $1/T$. By this selection, the length of each window and each window's

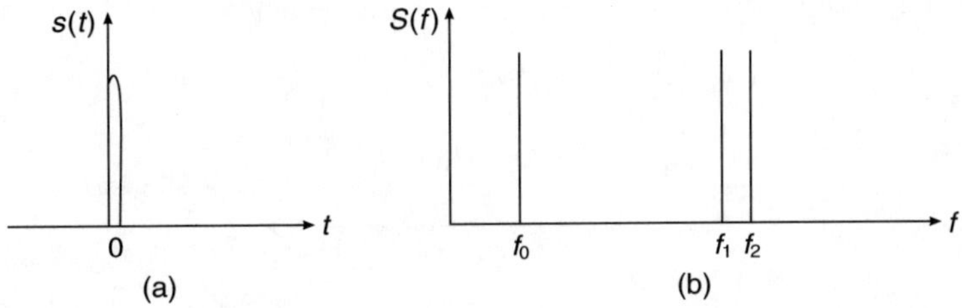

Figure 1.5 Input signal $s(t)$ and its amplitude spectrum: (a) input signal (delta function); (b) Fourier transform of input signal centred at f_0, f_1 and f_2

ability to pick out peaks can be investigated. More is said about target resolvability in Chapter 3.

1.3.3.1 Rectangular window

Equation (1.40) typifies the spectral window of the data window function defined by (1.41a) and Figure 1.6(a). Equation (1.41b) and Figure 1.6(b) can define another data window shape, also rectangular.

$$w_k(t) = \begin{cases} 1 & |t| \leq \frac{T}{2} \\ 0 & |t| > \frac{T}{2} \end{cases} \tag{1.41a}$$

$$w_k(t) = \begin{cases} 1 & |t| \leq T \\ 0 & |t| > T \end{cases} \tag{1.41b}$$

Figure 1.6(a) shows that with a rectangular data window of length T, it is impossible to distinguish the two peaks at f_1 and f_2. But with a rectangular data window of length $2T$, as in Figure 1.6(b), the peaks are easily distinguishable. It can thus be deduced that, for the rectangular data window, to separate two peaks at frequencies f_1 and f_2 it is necessary to use a record length T of order

$$T \geq \frac{1}{f_2 - f_1} \tag{1.42}$$

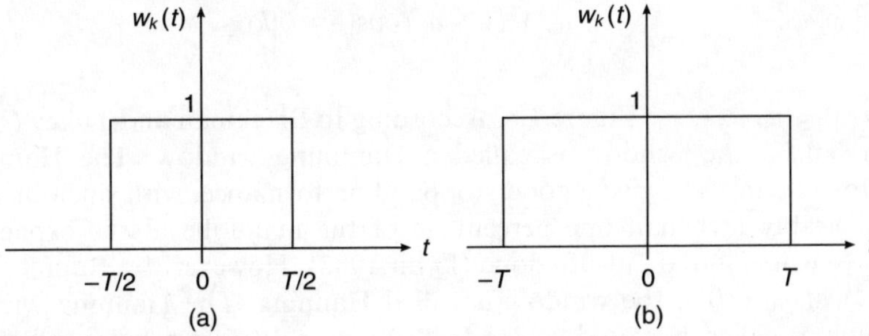

Figure 1.6 Rectangular windows

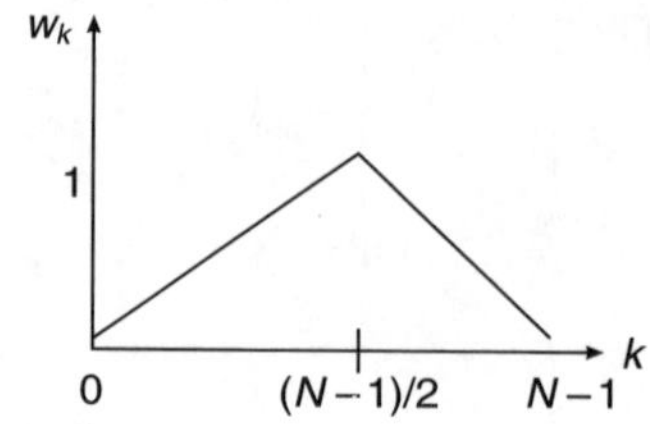

Figure 1.7 A triangular window

For non-rectangular windows – for example, triangular and Hamming, Hanning and Blackman shapes in sections 1.3.3.2 and 1.3.3.3 – to separate two peaks at frequencies f_1 and f_2 will require a record length T of the order

$$T > \frac{2}{f_2 - f_1} \tag{1.43}$$

The reader can verify these assertions by (a) finding the Fourier transform of each window function and (b) plotting each of the window's amplitude spectra (Fourier transforms) at these frequency centres f_0, f_1 and f_2 and observe the plots at frequencies f_1 and f_2.

1.3.3.2 Triangular or Bartlett window

Following Parsen (1962), a triangular window (also called the Bartlett window), depicted in Figure 1.7, is defined by the function

$$w_k = \begin{cases} \frac{2k}{N-1} & 0 \le k \le \frac{N-1}{2} \\ 2 - \frac{2k}{N-1} & \frac{N-1}{2} \le k \le N-1 \end{cases} \tag{1.44}$$

1.3.3.3 Hamming, Hanning and Blackman window

Following Jones (1962), the generalized Hamming window function is given by

$$w_k = \begin{cases} a_0 + (1 - a_0)\cos(\frac{\pi k}{N}) & |k| \le N \\ 0 & |k| > N \end{cases} \tag{1.45}$$

where $0 < a_0 < 1$, see Figure 1.8. According to Blackman and Tukey (1958), if $a_0 = 0.54$, the window is called a Hamming window. The Hamming window attempts to give a good stopband performance, with sidelobe levels considerably less than one percentage of the mainlobe at the expense of slightly worse initial cut-off slope (Lynn 1982). However, by Rabiner *et al.* (1974), if $a_0 = 0.5$, the window is called Hanning. The Hanning window, sometimes called a 'raised cosine bell' function, strikes a balance between passband and stopband performance.

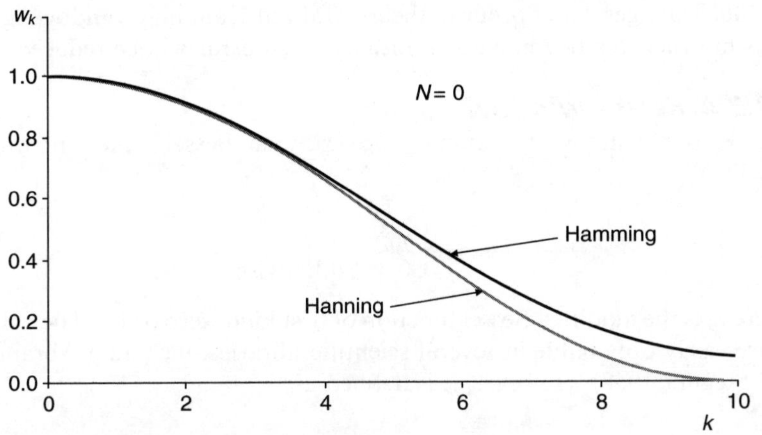

Figure 1.8 Hamming and Hanning windows

The Blackman window can be thought of as being an extension of the generalized Hamming window, defined as follows

$$w_k = \begin{cases} \sum_{m=0}^{N/2} (-1)^m \beta_m \cos(\frac{2\pi mk}{N}) & |k| \leq N \\ 0 & |k| > N \end{cases} \tag{1.46}$$

for $N = 4$, the constants become $\beta_0 = 0.42$, $\beta_1 = 0.50$, and $\beta_2 = 0.08$.

Harris further expands the Blackman window function, hence the term Blackman–Harris window. Harris used a gradient search method to find the third and fourth terms of (1.46) that either minimized the maximum sidelobe level for fixed mainlobe width, or that swapped mainlobe width with maximum sidelobe level. Typical values are shown in Table 1.2.

In summary, the generalized Hamming window functions have decaying sidelobes and are easy to generate. Often these window functions are utilized in beamforming (Hamming), sidelobes cancellation (Blackman) and range forming (Hanning) operations. Briefly, the terms beamforming and range forming are defined as follows. Beamforming is the ability of the receiving device (e.g. radar) to resolve received data in azimuth. The concept of beamforming is discussed in Chapter 7, section 7.3. It should be noted that

Table 1.2 Parameter values for the Blackman–Harris window function

Number of terms, N	Peak sidelobes level (dB)	Values of β parameters			
		β_0	β_1	β_2	β_3
6	−70.83	0.4232	0.4975	0.0792	—
6	−62.05	0.4496	0.4936	0.0568	—
8	−92.00	0.3588	0.4883	0.1413	0.0117
8	−74.39	0.4022	0.4970	0.0989	0.0019

sidelobe leakages could occur in the generalized Hamming windowing functions but their relative impact on measurement error will be reduced.

1.3.3.4 Kaiser window

The Kaiser window function is basically a Bessel function window. Specifically,

$$w_k = \begin{cases} \frac{I_0(\chi)}{I_0(\varsigma)} & |t| \le \frac{T}{2} \\ 0 & \text{otherwise} \end{cases} \tag{1.47}$$

where I_0 is the modified Bessel function of first kind, zero order. The values of I_0 are easily obtainable in several scientific libraries, including Abramowitz and Stegun (1968). However, it is defined as

$$I_0(x) = \frac{1}{2\pi}\int_0^{2\pi} e^{x\cos\theta}\,d\theta \tag{1.48a}$$

$$\chi = \Lambda * \sqrt{\frac{T^2}{2} - t^2} \tag{1.48b}$$

$$\varsigma = \eta\frac{T}{2} \tag{1.48c}$$

$\Lambda* =$ modifying parameter, typically in the range

$$\frac{8}{T} < \Lambda* < \frac{18}{T} \tag{1.48d}$$

which corresponds to a range of sidelobe peak heights of 3.1 per cent down to 0.04 per cent. Lynn (1982) demonstrated that the Kaiser window function offers excellent sidelobe suppression, at the expense of a slightly inferior initial cut-off slope. Reduction in Kaiser window's sidelobes depends on the choice of the modifying parameter.

1.3.3.5 Summary of window functions

The windows described above display a symmetrical tapering away from the centre, except for the rectangular window. Windowing technique can be applied for sidelobe reduction. An increase in the 3 dB filter bandwidth and associated decrease in the signal-to-noise ratio gain accompany the downside of the reduction. The window function quintessentially became very popular with the discovery of FFT. Hamming and Hanning windows can easily be formed after an unweighted FFT (Rabiner and Gold 1975) because a cosine in the time domain corresponds to pulses in the frequency domain. Childers and Durling (1981) and Oppenheim and Schafer (1975) describe other design discussions of windowing and effects on sampling, which lie outside the scope of this book. See also Helms and Rabiner (1972) for detailed discussion on Dolph–Chebyshev window functions.

1.3.4 Correlation functions

Correlation is a mechanism for signal comparison. It is a process of determining the mutual relationships that exist between several functions or signals. Correlation functions are measurements of the statistical dependence of one random signal upon another, or upon itself. A measure of the average self that exists within a signal is called the *autocorrelation* function while that which exists between signals is called *cross-correlation* function. Signal features such as periodicity and correlation times can be obtained through the autocorrelation operation. Cross-correlation has great utility in the study of linear systems particularly in radar applications. For example, range information is contained in the time delay between the transmission and reception of a pulse.

The cross-correlation coefficient R_{xy} of two functions $x(t)$ and $y(t)$ may be defined as

$$R_{xy}(\tau) = \frac{1}{T}\int_T x(t)y(t+\tau)dt \tag{1.49}$$

as $T \to \infty$, where τ is called the delay operator. Alternatively

$$R_{xy} = \frac{\int_{-\infty}^{\infty} x(t)y(t)dt}{\left(\int_{-\infty}^{\infty} x^2(t)dt \int_{-\infty}^{\infty} y^2(t)dt\right)^{\frac{1}{2}}} = \frac{\text{cov}[xy]}{\sqrt{\text{var}(x)\text{var}(y)}} \tag{1.50}$$

where 'cov' and 'var' correspond to covariance and variance of the functions $x(t)$ and $y(t)$. The expression in (1.50) is also called the normalized correlation coefficient or normalized cross-correlation coefficient. Often in signal processing, the unnormalized correlation coefficient is used. The cross-correlation coefficient can be interpreted as a measure of the average values of $x(t)$ with $y(t)$ displaced τ seconds. If R_{xy} is zero, the two functions $x(t)$ and $y(t)$ are said to be uncorrelated. If R_{xy} is ± 1, the functions $x(t)$ and $y(t)$ have perfect positive or negative relationship. The immediate value gives partial relationships.

The autocorrelation function R_{xx} of signal $x(t)$ is a measure of the signal with its delayed or shifted version. It is a special case of the unnormalized cross-correlation function. It applies only to one time series. The autocorrelation function of $x(t)$ may be written as

$$R_{xx}(\tau) = \frac{1}{T}\int_T x(t)x(t+\tau)dt \tag{1.51}$$

as $T \to \infty$. The frequency-domain characteristics of autocorrelation can be obtained through the application of the Fourier operator. Assume that the time series $x(t)$ has a Fourier coefficient c_n and can be expressed as

$$x(t) = \sum_n c_n e^{\frac{j2\pi nt}{T}} \tag{1.52}$$

then in view of (1.51) and (1.52), the autocorrelation function of series $x(t)$ can be written as

$$\begin{aligned} R_{xx}(\tau) &= \frac{1}{T}\int_T \left(\sum_m c_m e^{\frac{j2\pi mt}{T}}\right)\left(\sum_n c_n e^{\frac{j2\pi n(t-\tau)}{T}}\right) dt \\ &= \frac{1}{T}\sum_n c_n e^{\frac{-j2\pi n\tau}{T}} \sum_m c_m \int_T e^{\frac{j2\pi(m+n)t}{T}} dt \\ &= \frac{1}{T}\sum_n c_n e^{\frac{-j2\pi n\tau}{T}} \sum_m c_m \sin c\left[\frac{\pi}{T}(m+n)\right] \end{aligned} \tag{1.53}$$

where

$$\sin c\left[\frac{\pi}{T}(m+n)\right] = \frac{\sin\left[\frac{\pi}{T}(m+n)\right]}{\frac{\pi}{T}(m+n)} = \begin{cases} 0 & m \neq -n \\ T & \text{otherwise} \end{cases} \tag{1.54}$$

Using the principle of superposition[1] and the Fourier coefficients relationship of (1.8), the autocorrelation function

$$R_{xx}(\tau) = \sum_n c_n c_{-n} e^{\frac{-j\pi nt}{T}} = \sum_n |c_n|^2 e^{\frac{-j\pi nt}{T}} \tag{1.55}$$

Noting that by the Parseval theorem, the sum of the energy in one period is

$$\sum_{n=-\infty}^{\infty} |c_n|^2 = \frac{1}{T}\int_{-\frac{T}{2}}^{\frac{T}{2}} |x(t)|^2 dt \tag{1.56}$$

The series $|c_n|^2$ is the *power spectral density* of $x(t)$.

Autocorrelation function is widely used in signal analysis. It is especially useful for the detection or recognition of signals that are masked by additive noise because white noise has infinite extent in the frequency domain, and therefore its autocorrelation function has negligible extent in the time domain. This observation is important in the recognition of white noise, particularly in radar receivers, in the sense that any waveforms at the input of the receivers that are subject to white noise can alternatively be considered as being subject to independent but identical noise prob-

[1] The output waveform from a simple linear time-invariant system is the convolution of the input waveform and the impulse-response of the system. Suppose a linear system with an input $v(t)$, having an integral or sum of impulsive elements at time τ and of strength $v(\tau)$, can be expressed in the form

$$v(t) = \int_{-\infty}^{\infty} v(\tau)\delta(t-\tau)d\tau$$

If each of the system's impulsive elements $\delta(\tau)$ can be replaced by the response it provokes, say $u(t)$, then the output waveform of the system becomes

$$h(t) = \int_{-\infty}^{\infty} v(\tau)u(t-\tau)d\tau = v \times u$$

which is the convolution of v and u, already discussed in section 1.3.2.

ability distributions at each distinguishable point in the time domain. Suffice to say that, although noise whiteness leads to the independence property, it does not guarantee that individual temporal noise distribution will be identical. In practical cases it is reasonable to make this assumption occasionally.

1.4 Fast Fourier transform

The fast Fourier transform (FFT) is an efficient algorithm for the numerical computation of the discrete Fourier transform (DFT) with a minimum computation time. An algorithm is a systematic technique of performing a series of computations in sequence. The FFT algorithm developed below is due to Cooley–Tukey (1965) and Weaver (1983).

Suppose there is an N-point sequence denoted by $f(k)$, and N is an integer divisible by 2. Our interest is finding the DFT of $f(k)$. Since the N-point is divisible by two, two new albeit disjointed sequences – $f_1(k)$ and $f_2(k)$ with periodicity p – can be formed, and defined as

$$\begin{aligned} f_1(k) &= f_1(2k) \\ f_2(k) &= f_2(2k+1) \end{aligned} \qquad k = 0, 1, 2, \ldots, p \quad \text{where } p = \frac{N}{2} \tag{1.57}$$

Following (1.21), an N-point sequence DFT can be expressed as

$$F(r) = \frac{1}{N}\sum_{k=0}^{N-1} f(k)W_N^{-kr} \qquad r = 0, 1, 2, \ldots, N \tag{1.58}$$

This expression can be described in terms of two formed sequences:

$$F(r) = \frac{1}{N}\sum_{k=0}^{p-1} f_1(2k)W_p^{-2kr} + \frac{1}{N}\sum_{k=0}^{p-1} f_2(2k+1)W_p^{-(2k+1)r} \tag{1.59}$$

A closer examination of (1.59) reveals that

$$\begin{aligned} F(r) &= \frac{1}{N}\sum_{k=0}^{p-1} f_1(k)w_p^{-kr} + \frac{w_N^{-r}}{N}\sum_{k=0}^{p-1} f_2(k)w_p^{-kr} \\ &= \frac{1}{2}[F_1(r) + w_N^{-r}F_2(r)] \end{aligned} \tag{1.60}$$

Using the definition in (1.20) and noting that the translation of kernels in N to p, for example, the following equations can be written:

$$w_N^{-2kr} = e^{\frac{-j2\pi(2k)r}{N}} = w_p^{-kr} \tag{1.61a}$$

$$w_N^{-(2k+1)r} = e^{\frac{-j2\pi(2k+1)r}{N}} = e^{\frac{-j2\pi(2k)r}{N}}e^{\frac{-j2\pi r}{N}} = w_p^{-kr}w_N^{-r} \tag{1.61b}$$

It is evident in (1.60) that the FFT technique lies in the relationship between DFT of split sequences with the DFT of a full sequence. Also, the computation of the sequence DFT requires operations involving complex multiplication, additions and subtractions.

Following the repeated sequence demonstrated above an algorithm can be developed for 8-point sequence. Thus

$$\begin{aligned} F(r) &= \frac{1}{2}\begin{bmatrix} F_1(r) + w_N^{-r}F_2(r) + w_N^{-2r}F_3(r) + w_N^{-3r}F_4(r) \\ +w_N^{-4r}F_5(r) + w_N^{-5r}F_6(r) + w_N^{-6r}F_7(r) + w_N^{-7r}F_8(r) \end{bmatrix} \\ &= \frac{1}{2}\sum_{n=1}^{N-2} F_n(r) w_N^{-(n-1)r} \end{aligned} \tag{1.62}$$

The preceding discussion has so far treated the case of an N-point sequence divisible by 2, and by deduction 4, 8, etc. Instead, suppose there is an N-point sequence divisible by 3. Three new sequences with periodicity p are formed as

$$\begin{aligned} f_1(k) &= f_1(3k) \\ f_2(k) &= f_2(3k+1) \quad k = 0, 1, 2, \ldots, p \quad \text{where } p = \frac{N}{3} \\ f_3(k) &= f_3(3k+2) \end{aligned} \tag{1.63}$$

Splitting (1.58) into three sequences gives

$$\begin{aligned} F(r) = \frac{1}{N}\sum_{k=0}^{p-1} f_1(3k) W_p^{-3kr} + \frac{1}{N}\sum_{k=0}^{p-1} f_2(3k+1) W_p^{-(3k+1)r} \\ + \frac{1}{N}\sum_{k=0}^{p-1} f_3(3k+2) W_p^{-(3k+2)r} \end{aligned} \tag{1.64}$$

Following the weighting kernels expansion similar to (1.61), expression (1.64) can be reconstituted as

$$\begin{aligned} F(r) &= \frac{1}{N}\sum_{k=0}^{p-1} f_1(k) w_p^{-kr} + \frac{w_N^{-r}}{N}\sum_{k=0}^{p-1} f_2(k) w_p^{-kr} + \frac{w_N^{-2r}}{N}\sum_{k=0}^{p-1} f_3(k) w_p^{-kr} \\ &= \frac{1}{3}\left[F_1(r) + w_N^{-r}F_2(r) + w_N^{-2r}F_3(r)\right] \quad r = 0, 1, \ldots, p \end{aligned} \tag{1.65}$$

The preceding FFT algorithms can be programmed for use on the computers. Examples of FFT programs can be found in Childers and Durling (1981) and Fraser (1979). The Fraser's program is reproduced in Appendix 1A with permission. Although the program is not optimum, it, however, provides the reader with an avenue to follow step by step as to how the program works as well as optimizing the program. The number of operations necessary to form a spectrum of N sequences or channels in an FFT is $N/2 \log_e N$ complex multiplication, additions and subtractions (Bergland 1969). More application of FFT to radar measurement is covered in Chapter 2, section 2.1.3.

Example 1.5 Determine (a) closed form expression for the DFT of $x(n) = 1$, for $0 \leq n \leq N-1$, (b) the energy contained in the time signal, and (c) verify Parseval's theorem for this function.

Solution
(a) the DFT of $x(n)$ is

$$X(m) = \sum_{n=0}^{N-1} x(n) w_N^{mn} = \sum_{n=0}^{N-1} (1) w_N^{mn} \tag{1.66}$$

which constitutes a finite geometric series expressable as

$$X(m) = \frac{1 - w_N^{mN}}{1 - w_N^{m}} = \frac{1 - e^{-j2\pi m}}{1 - e^{-j\left(\frac{2\pi m}{N}\right)}} \qquad 0 \leq m \leq N-1 \tag{1.67}$$

This expression is zero for all integer values within this limit except at $m = 0$. Upon an application of L'Hospital's rule to (1.67) as $m = 0$; that is,

$$\lim_{x \to 0} \left(\frac{e^x}{e^{\frac{x}{N}}} \right) = N \tag{1.68}$$

The solution to (1.67) at $m = 0$ is

$$X(0) = N \tag{1.69}$$

(b) The energy contained in the time series can be expressed as

$$\sum_{n=0}^{N-1} x^2(n) = \sum_{n=0}^{N-1} (1) = N \tag{1.70}$$

(c) By Parseval's theorem, from (1.56),

$$\sum_{m=0}^{N-1} |c_n(m)|^2 = \frac{|X(0)|^2}{N} = \frac{N^2}{N} = N \tag{1.71}$$

It is observed that (1.70) and (1.71) are the same indirectly proving the Parseval's theorem as a measure of power spectral density.

1.5 Norm of a function

One category of norms that is regularly used is the set called L_p-norms. The L_p-norm, denoted by $\|x(t)\|_p$, of a continuous function $x(t)$ defined over an interval [0, 1], can be written as

$$L_p = \|x(t)\|_p = \left[\int_0^t |x(t)|^p dt \right]^{\frac{1}{p}} \tag{1.72}$$

Three values of p are of special interest:

$$\text{(a) } p = 1: \quad L_1 = \|x(t)\|_1 = \int_0^t |x(t)| dt \tag{1.73}$$

$$\text{(b) } p = 2: \quad L_1 = \|x(t)\|_2 = \left[\int_0^t |x(t)|^2 dt\right]^{\frac{1}{2}} \tag{1.74}$$

which is the expression for the energy of the function $x(t)$.

$$\text{(c) } p = \infty: \quad L_\infty = \|x(t)\|_\infty = \max_{0 \le t \le 1} |x(t)| \tag{1.75}$$

This expression is called the Chebyshev's norm.

Example 1.6 If $f(x) = 1/\sqrt[3]{x}$ exists in the Lebesgue sense (that is, integrable) within (0, 1), find the norm of the function $f(x)$.

Solution
From (1.74),

$$\|f(x)\| = \sqrt{\int_0^1 |f(x)|^2 dx} = \sqrt{3} \tag{1.76}$$

A good discussion on the overall design problem and the design of optimum filters that approximate a given frequency response in the L_∞ sense can be found in Rabiner *et al.* (1974). In real life, norms are employed to measure approximately the discrepancy between a function $f(x)$ and the function $F(x)$ being approximated. For example, if the norm is L_2, the least square method would be a convenient approximation and in Chebyshev's sense if the norm is L_∞. By introducing a real positive weighting function of $w(x)$, the difference between functions $f(x)$ and $F(x)$ can be generalized, in the L_p sense, as

$$\|f(x) - F(x)\|_p = \left[\int_0^t |f(x) - F(x)|^p w(x) dx\right]^{\frac{1}{p}} \tag{1.77}$$

Some obvious applications of these expressions include calculating filter coefficients, scaling internal data in memories, noise estimation, and optimum error estimation between design and desired response in an ordered one-dimensional case.

1.6 Summary

This chapter has covered some of the basic principles necessary for understanding radar signal processing and the subsequent chapters.

Time series signals have been expressed in terms of Fourier series. The continuous and discrete signals have been Fourier analysed. It is often useful to establish the essential relational functions of Fourier transform pairs, examples are convolution and correlation, which were also discussed. A direct result of these properties is that the Fourier transform reduces convolution operations to simple multiplication. Furthermore, an efficient algorithm for the numerical computation of the discrete Fourier transform (DFT) called the fast Fourier transform (FFT) was introduced. The essence of the FFT technique lies in the relationship between the DFT of split sequences and the DFT of a full sequence.

The concept of windowing as a tool in spectral estimation was discussed as well as some commonly used window functions. Windows attempt to reduce spectral sidelobes due to abrupt truncation of randomly processed data, which causes spectral distortion. Finally, one category of norms, L_p-norm, which is frequently employed in radar signal processing and tracking, was also discussed.

Appendix 1A A fast Fourier transform computer program

This program has been reproduced by permission of Associate Professor D. Fraser (1979), School of Electrical Engineering, Australian Defence Force Academy, Canberra, Australia.

This appendix lists five Fortran subroutines designed to perform some of the operations most frequently used in spectral analysis. In writing, the subroutines are kept simple at the expense of efficiency in order that the reader can understand them easily. As long as they are used for problems within the limits prescribed, there is no excessive wastage of time and storage. For those who wish to start experimenting with spectral analysis techniques the subroutines should make things very convenient. Once the reader gets into serious data analysis he/she would want to write his/her own, more efficient, and more specialized computer programs. Even then, the availability of these subroutines should facilitate the reader's programming effort. No detailed explanation of these subroutines will be given here as each has its own comments. A short list is given below:

1. FFT: for both forward and inverse transform of complex vectors.
2. FFTR: for the forward transform of a real vector or its recovery from its DFT. (X_i for $i = 0, 1, \ldots, 1/2N$ only.)
3. PERIOD: computes the periodogram of a real vector at half integer frequencies, $i/2$, $i = 0, 1, \ldots, N - 1$.
4. AUTCOR: computes the autocorrelation estimate of N given values up to time delay M.
5. COTRAN: computes the Fourier transform of a real, even vector, also known as a cosine transform. It returns the power spectrum if given the autocorrelation function.

```
      Subroutine FFT(A,M,IS)
C FFT of complex array A, of 2^M elements, IS = +1 or -1 sign
  of CEXP
C (Note that initial data in array A is replaced by its
  Fourier transform)
```

```
C First part is bit-reversed permutation using recursive
  algorithm, which increments a
C reversed index when needed for each bit position final
  part, from label 7, is base-2 FFT
C computation, which requires minimum different W, gener-
  ated recursively
      COMPLEX A(I),TEMP,W,D
      INTEGER IRA(I6),NR(16),5PAN,STEP
      DATA PI/3.141592653589793/
      N=2**M
      DO 1 J-1,M
      IRA(J)=0
1     NR(J)=2**(J-1)
C Reversed index sets (for each bit position) initialized
      IF=1
2     IR=1RA(M)+1
      IF(IR.LE.IF)GO TO 3
C Prevents nullifying double swap
      TEMP=A(IF)
      A(IF)=A(IR)
      A(IR)=TEMP
C Reversed index pair swapped
3     IF=IF+1
C Increment forward index IF
      IF(IF.GT.N)GO TO 7
      J=N
4     IF(IRA(J).LT.NR(J))GO TO 5
C Alternate increment of IRA(J), must go back one bit
      J=J-1
      GO TO 4
5     IRA(J)=IRA(J)+NR(J)
C Simple, alternate increment of reversed index
      IF(J.EQ.M)GO TO 2
      IRA(J+1)=IRA(J)
C Work forward through reversed index bit set
      J=J+1
      GO TO 6
C Array is now in bit-reversed order, M computing passes follow
7     DO 9 J1=1,M
      SPAN=2**(J1-1)
      STEP=2*SPAN
C Span between elements in pair, step to next pair with same W
      W=(1.,0.)
      D=CEXP(CMPLX(0.,PI/SPAN))
      IF(IS.LT.0)D=CONJG(D)
```

```
C Starting phase adjuster W, modifier D
      DO 9 J2=1,SPAN
      DO 8 J=J2,N,STEP
      K=J+SPAN
      TEMP=A(K)*W
      A(K)=A(J)-TEMP
8     A(J)=A(J)+TEMP
C Inner loop arithmetic - two point transforms
9     W=W*D
C Recursive modification of phase adjuster W
      RETURN
      END

      Subroutine FFTR(A,M,IS)
C Real-to-Complex (or vice versa half-length FFT of array
  A. (Note that initial data in array
C A is replaced by its Fourier transform). Real data
  assumed packed alternately as real and
C imaginary values, most easily achieved by equivalencing
  real and complex array names
C 2**M real elements (+2 dummies), or 2**(m-1)+1 complex
  elements IS=+1 or -1 sign
C of CEXP and direction (+1=real-to-complex, -1 reverse).
  Uses scramble/unscramble
C algorithm and call to half-length complex FFT
      COMPLEX A(1),TA,TB,W,D
      DATA PI/3.141592653589793/
      MH=M-1
      N=2**MH
      INCNT=N/2+1
      W=(1.,0.)
      D=CEXP(CMPLX(0.,PI/N))
C Starting phase adjuster W, modifier D for scramble/
  unscramble
      IF(IS.LT.0)GO TO 2
C Real-to-complex FFT follows, half-length complex FFT first
      CALL FFT(A,MH,IS)
      A(N+1)=A(1)
      DO 1 J=1,INCNT
      K=N+2-J
      TA=(A(J)+CONJG(A(K)))*0.5
      TB=CONJG(A(J))+A(K)
      TB=CMPLX(AIMAG(TB),REAL(TB))*W*0.5
      A(J)=TA+TB
      A(K)=CONJG(TA-TB)
```

```
1     W=W*D
C Elements unscrambled, W recursively modified
      RETURN
C Complex-to-real FFT follows
2     D=CONJG(D)
      DO 3 J=1,INCNT
      K=N+2-J
      TA=A(J)+CONJG(A(K))
      TB=(A(J)-CONJG(A(K)))*W
      TB=CMPLX(AIMAG(TB),REAL(TB))
      A(J)=TA-CONJG(TB)
      A(K)=CONJG(TA)+TB
3     W=W*D
C Elements scrambled, W recursively modified
      CALL FFT(A,MH,IS)
C Half-length complex FFT finishes complex-to-real FFT
      RETURN
      END

      SUBROUTINE PERIOD(N,DATA,PDGRAM)
C This subroutine accepts N input values and returns their
  periodogram.
C N must not exceed 512. The method is bad for large N.
      DIMENSION DATA(N),PDGRAM(N),FIXCOS(513),FIXSIN(513)
      DATA NSAVE/0/
      NN=N+1
      N2=N*2
      NN2=NN*2
      REC=1./FLOAT(N)
C The loop below stores values of sine and cosine between 0
  and π.
C IF NSAVE=N, then the subroutine has been called earlier
  with the same N and so must
C already contain correct FIXCOS and FIXSIN.
      IF(NSAVE.EQ.N)GO TO 10
      REC2=REC*4.*ATAN(1.)
C This is π/N.
      DO 5 1=1,NN
      ARG=FLOAT(I-1)*REC2
      FIXCOS(I)=COS(ARG)
      FIXSIN(I)=SIN(ARG)
5     CONTINUE
10    CONTINUE
      REC=REC*REC
      DO 20 I=1,N
```

```
C TEMP1 and TEMP2 will be the real and imaginary parts of the
  DFT of data
    TEMP1=DATA(1)
    TEMP2=TEMP1
    II=I-1
C K is the value of I*J after subtraction of multiples of 2N.
    K=1
    DO 15 J=2,N
    K=K+II
    IF(K.GT.N2)K=K-N2
    IF(K.GT.NN)GO TO 12
C Argument of sine and cosine not over π.
    A=FIXCOS(K)
    B=FIXSIN(K)
    GO TO 13
C  Argument  of  sine  and  cosine  more  than  π.  Use
  SIN(ARG)=-SIN(2*PI-ARG),
C COS(ARG)=COS(2*PI-ARG)
12  KK=NN2-K
    A=FIXCOS(KK)
    B=-FIXSIN(KK)
13  D=DATA(J)
    TEMP1=TEMP1+D*A
    TEMP2=TEMP2+D*B
15  CONTINUE
C Square real and imaginary parts and add to give power.
    PDGRAM(I)=(TEMP1*TEMP1+TEMP2*TEMP2)*REC
20  CONTINUE
    NSAVE=N
    RETURN
    END

    SUBROUTINE AUTCOR(N,M,DATA,COR)
CN=the number of input data, M=the number of autocorrela-
  tions needed.
C M should not be more than 256. The method is bad for large M.
    DIMENSION DATA(N),COR(M)
    REC=1./FLOAT(N)
    DO 10 I=1,M
    TEMP=0.
    DO 5 J=1,N
    JJ=J-I+1
    TEMP=TEMP+DATA(J)*DATA(JJ)
5   CONTINUE
    COR(I)=TEMP*REC
```

```
10   CONTINUE
     RETURN
     END

     SUBROUTINE COTRAN(M,COR,SPECTR)
C This subroutine accepts m autocorrelation values and
  returns the real parts of their
C Fourier Transform, i.e., unwindowed spectrum. Windowing
  may be applied either by
C multiplication before calling this subroutine, or by
  averaging neighbouring terms after
C return. M must not exceed 128. The method is bad for large M.
     DIMENSION COR(M),SPECTR(M),FIXCOS(129)
     DATA MSAVE/0/
     REC=1./FLOAT(M)
     MM=M+1
     M2=M*2
     MM2=MM*2
C The loop below stores the values of cosine between 0 and π. If
  MSAVE=0, the
C subroutine has not been called before. If MSAVE = M, then
  FIXCOS already contain
C correct values.
     IF(NSAVE.EQ.M)GO TO 10
     REC2=REC*4.*ATAN(1.)
     DO 5 I=1,MM
     ARG=FLOAT(I-1)*REC2
     FIXCOS(I)=COS(ARG)
5    CONTINUE
10   HALF=COR(1)*0.5
     REC=REC*2.
     DO 20 I=1,M
     TEMP=HALF
     II=I-1
C K is the value of I*J reduced by multiples of 2M
     K=1
     DO 15 J=2,M
     K=K+II
     IF(K.GT.M2)K=K-M2
     KK=K
C K greater than M+1 means argument of cosine is more than π.
C Use COS(ARG)=COS(2*PI-ARG).
     IF(KK.GT.MM)KK=MM2-KK
     TEMP=TEMP+COR(J)*FIXCOS(KK)
```

```
15    CONTINUE
      SPECTR(I)=TEMP*REC
20    CONTINUE
      MSAVE=M
      RETURN
END
```

Problems

1. In the interval $-\pi \le x \le \pi$, the function $f(x) = |x|$ is defined. Obtain the Fourier series for the function. Deduce from your result an expression for $\pi^2/8$ in a series form.
2. A trapezoidal wave has a period T, height $\pm h$, and a rise time from zero to h of m seconds. Select a time axis that will give a Fourier expansion with sine terms only and analyse the wave.
3. Consider the two causal finite-length sequences shown in Figure 1.9.
   (a) Form the sequence $x_n = a_n \times b_n$.
   (b) Determine the finite Fourier transforms A_k and B_k of the sequences a_n and b_n for $k = 0, 1, \ldots, 4$.
   (c) Using the Fourier transform, one can find the convolution of two sequences a_n, b_n by forming the product $A(\omega)B(\omega)$ of their corresponding Fourier transforms and then taking the inverse Fourier transforms of this product. Does this convolutional procedure work if you use finite Fourier transforms instead of Fourier transforms? Explain clearly your reasoning.
4. If two signals, of frequency components 0.9 kHz and 1.0 kHz, were required to be separated. Determine the sampling frequency interval required distinguishing the two signals. Determine also the length of record required to distinguish the signals' peaks in the Fourier transform.
5. Find the frequency spectrum of a half-wave rectified sine wave of peak value V_m, represented by Figure 1.10.

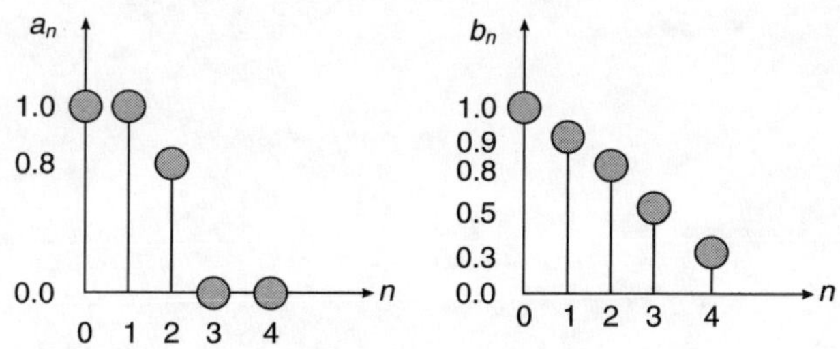

Figure 1.9 Two causal finite-length sequences

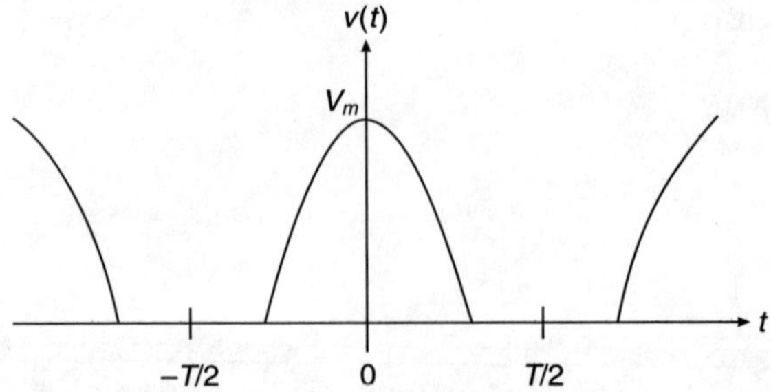

Figure 1.10 The frequency spectrum of a half-wave rectified sine wave

Table 1.3 Ionospheric data

Item, t	1	2	3	4	5	6	7	8	9	10	11	12
Index, x_t	−6	−18	−28	−12	−5	9	−20	−8	−9	18	21	12

6. An ionospheric sounder generated the data tabulated in Table 1.3. Compute the autocorrelation and the FFT of the data using the program in Appendix 1A and computationally. Compare the results. Any differences? And why?

2

Understanding radar fundamentals

In designing any radar, for a beginner (and even a professional who needs a refresher), requires an understanding of the main issues: how radar evolves, how to analyse component parts and interpret the composite outcome in a way that becomes an operational tool. For this reason, the author has used typical radar architecture to explain the radar fundamentals.

2.1 An overview of radar system architecture

Radar is an acronym derived from radio detection and ranging. Today's radar is best defined as active electromagnetic surveillance. Basically, the function of a radar is to transmit a burst of electromagnetic energy necessary to allow detection of targets intercepting the energy by its receiver.

The purpose of this section is to examine radar system architecture and explain the functions of various circuit blocks. A schematic diagram of a typical radar system is shown in Figure 2.1. It may be instructive, therefore, to walk through Figure 2.1 block by block and summarize their functionality before concentrating on the iterative procedure for determining the overall radar expressions that may enable us to estimate the radar merit and power budget.

2.1.1 Transmitter

The function of a transmitter is to amplify an RF carrier modulated with the desired signal, adding a minimum distortion to the encoded information. Essentially three prime components form the transmitter chain: a high-powered amplifier (HPA) with high-stability electron gun, waveform generator (local oscillator, LO) and timing, and an antenna (see Figure 2.2).

Unlike the antenna in Figure 2.2, which radiates electromagnetic waves from the transmitter, the simplex transceiver antenna arrangement in Figure 2.1 serves two purposes: as a radiator and as a receptor. The properties of an

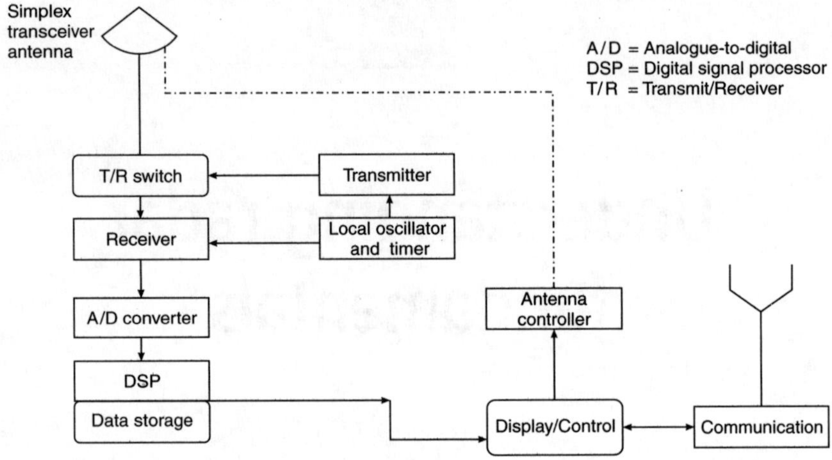

Figure 2.1 A block diagram of a radar system

antenna system when used as a transmitter are similar in nearly all aspects to the corresponding properties of the same antenna when used as a receiver to abstract energy from a passing radio wave. Therefore the relative response of the antenna to waves arriving from different directions is exactly the same as the relative radiation in different directions from the same antenna when excited as a transmitting antenna. These reciprocal relations between receiving and radiating properties of antenna systems make it possible to reach a conclusion on the merits of a receiving antenna from transmission tests, and vice versa. How then does one predict the type(s) of radiation patterns originating from an antenna? Chapter 3 sheds some light on this question.

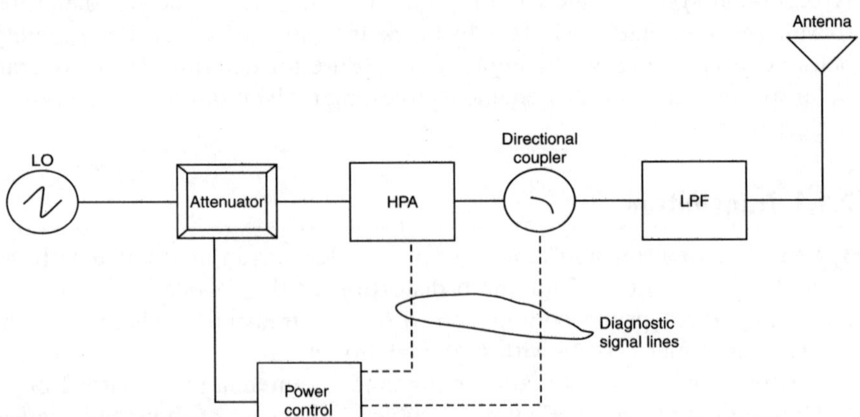

Figure 2.2 A schematic diagram of a transmitter

2.1.1.1 Local oscillator

Local oscillators (LO) are waveform generators. Like any communication and surveillance systems, radar systems require sophisticated, highly stable, synthesized LOs with low phase noise, fast frequency lock time, and low power consumption. Both transmitter and receiver require LOs, as in Figure 2.1, but the LO technology is probably dictated more by the actual application than anything else in the receiver. If the receiver's frequency is expected to be programmable, a frequency synthesizer may be required.

2.1.1.2 Attenuator

Attenuators are used to increase isolation between the oscillators and the changing load. An attenuator can be as simple as the T-section pad shown in Figure 2.3.

To design an attenuator, it is important to know the iterative impedance, Z_0, of the network. Knowing Z_0, the insertion loss, A_L, of the iterative operation can be expressed as

$$A_L = 1 + \frac{R_1}{R_2} + \frac{Z_0}{R_2} \tag{2.1}$$

In practice, the desired insertion loss is known as part of system requirements, and the pad's components can easily be estimated for a given iterative impedance.

2.1.1.3 High-powered amplifiers (HPA)

The high-powered amplifiers (HPA) could be travelling wave tubes (TWT), magnetrons, or klystrons. These amplifiers permit frequency agility and in-pulse frequency scanning which are essential features of modern radar systems. Selection of any of the tubes depends on application. A pictorial view of a klystron is shown in Figure 2.4.

A klystron is a microwave generator, typically about 1.83 m long and works as follows:

(a) The electron gun (1) produces a flow of electrons.
(b) The bunching cavity (2) regulates the speed of the electrons so that they arrive in a bunch at the output cavity.

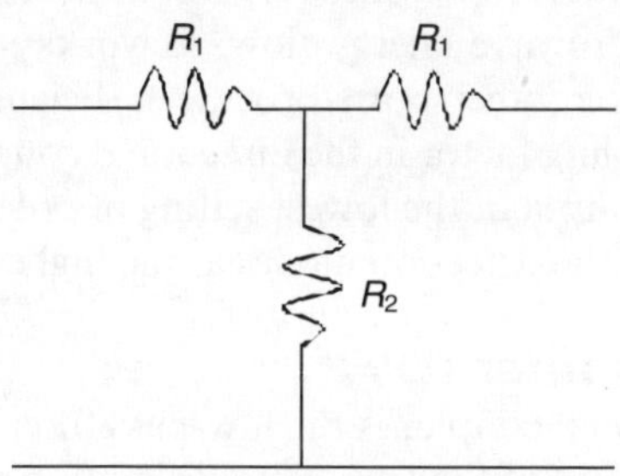

Figure 2.3 A symmetrical T-attenuator pad

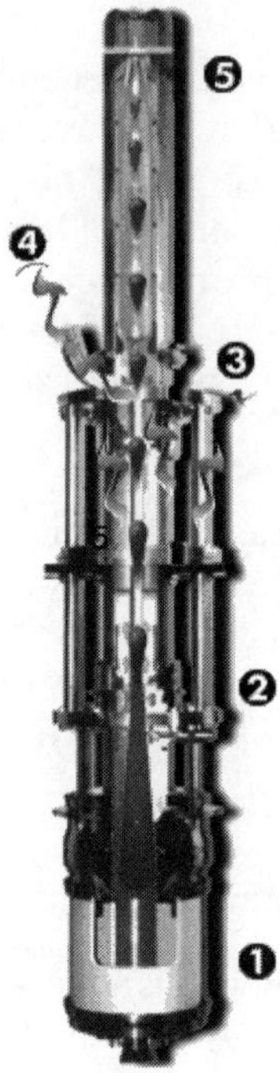

Figure 2.4 A klystron (courtesy: NASA)

(c) The bunch of electrons excites microwaves in the output cavity (3) of the klystron.
(d) The microwaves flow into the waveguide (4), which transports them to the accelerator. An accelerator is a device used to produce a high-energy high-speed beam of charged particles, such as electrons, protons or heavy ions.
(e) The electrons are absorbed in the beam stop (5).

2.1.1.4 Directional coupler

The directional coupler (or circulator) interfaces between the HPA and the RF amplifier of the transmitter. It provides very low impedance and negligible losses in the direction of microwave energy flow. It works in a way that when the assigned ports are active, other ports provide sufficient isolation from microwave energy. The coupling factor in the directional coupler must be sufficiently high to sample HPA output at the lowest setting in order to prevent harmonics from coupling back to the detection diodes at the highest power setting.

2.1.1.5 Low-pass filter (LPF)

Following the directional coupler is the low-pass filter (LPF), whose purpose is to attenuate harmonics of the transmitted signal. An LPF can be as simple as shown in Figure 2.5.

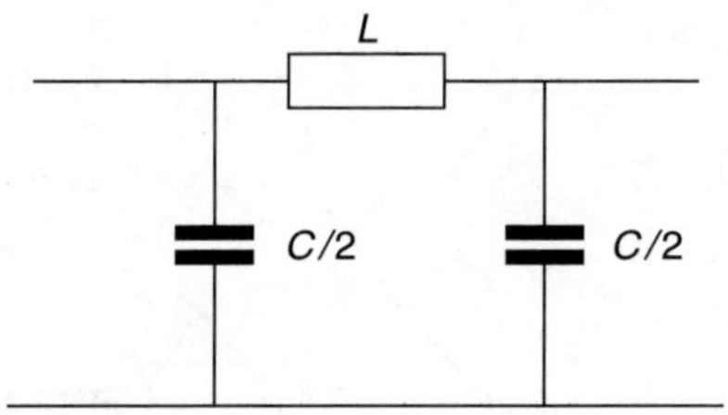

Figure 2.5 A low-pass filter

An LPF is a network designed to have zero attenuation up to a given frequency (called *cut-off frequency*, f_0) and a large attenuation above this. Theoretically, it is composed of pure reactance in order to have zero dissipation. The cut-off frequency can be written as

$$f_0 = \frac{1}{\pi\sqrt{LC}} \tag{2.2}$$

The characteristic impedance Z_0 can be expressed as

$$Z_0 = \sqrt{\frac{L}{C\left(1 - \frac{\omega^2}{\omega_0^2}\right)}} \tag{2.3}$$

where $\omega = 2\pi f$, $\omega_0 = 2\pi f_0$ and f is the propagation frequency. Since the desire is to have zero attenuation (i.e. $\alpha = 0$), above ω_0 the propagation coefficient, γ, has a reference component, and so signals are attenuated between input and output. So, the phase angle, β, between input and output, when terminated by Z_0 can be expressed as

$$\beta = \tan^{-1}\left(\frac{\omega\sqrt{LC\left(1 - \frac{\omega^2 LC}{4}\right)}}{1 - \frac{\omega^2 LC}{2}}\right) \tag{2.4}$$

Note that $\gamma = \alpha + j\beta$. The phase angle, β, will vary from 0° (when $\omega = 0$) to 180° (when $\omega = \omega_0$; that is, $\tan^{-1}(0/-1)$). Between $\omega = 0$ and $\omega = \omega_0$, β is positive, and the output *lags* behind the input. Above ω_0, β remains constant at 180° independent of frequency, see Figure 2.6. In practice, most of the LPFs are reflective. As a precautionary measure, LPFs are overdesigned to provide more rejection than would normally be necessary (Morton 1966).

The power control loop is used to slow down the transmitter turn-on and turn-off times to minimize generation of spectral components of adjacent channels. Caution must be exercised not to introduce low-frequency instability into the control loop. The diagnostic signals are intended to sense HPA final current and temperature, as well as the forward and reverse power levels of the directional coupler.

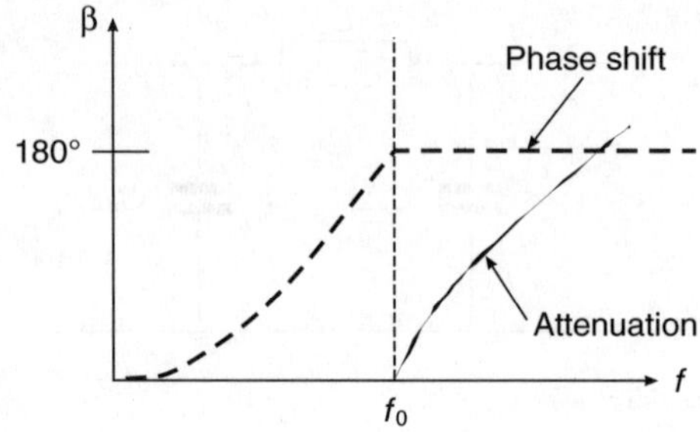

Figure 2.6 Phase response of an LPF

2.1.2 Receiver

The low energy signal, collected by the antenna, is brought through the circulator and the transmit/receive (T/R) switch tube, or isolator, and the radio frequency (RF) amplifier. A typical dual-conversion receiver is shown in Figure 2.7. It is made of a series of components, namely RF filter, amplifier, mixers and intermediate frequency (IF) amplifiers.

The received signal is mixed, in some type of non-linear device (i.e. *mixer*), with a signal from a local oscillator (LO), to produce an intermediate frequency (IF), i.e. beat frequency, from which the modulating signal is recovered (i.e. in the *detector*). The method of detection used typifies the receiver, namely direct and coherent detection receivers. Direct-detection receivers employ a square-law device, which produces an electrical signal proportional to the intensity of the incident optical signal (e.g. a photodiode), whose signal's power is measured directly.

In the case of coherent-detection, the received signal is beat against a local oscillator field of nearly the same frequency, and the output signal is proportional to the received field strength. In the ideal case, the proportionality of the beat term to the local oscillator field strength provides essentially noiseless predetection gain, so that thermal and dark-current noises inherent to the direct-detector are dwarfed by the quantum noise inherent in the signal itself. A truly coherent wave would be perfectly coherent at all points in space. In practice, however, the region of high coherence may extend over only a finite distance.

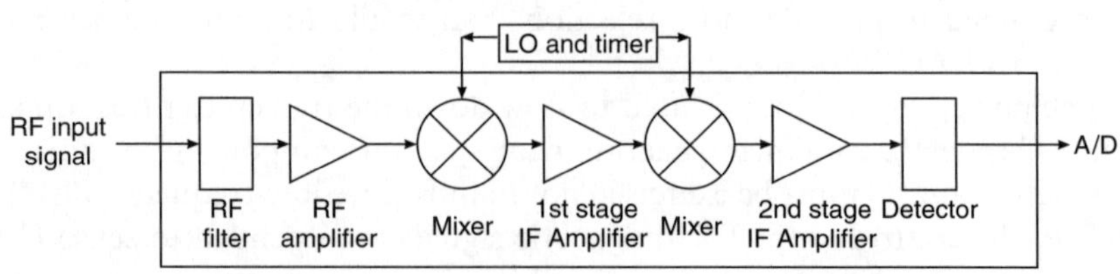

Figure 2.7 A dual-conversion radar receiver

Unlike the direct-detection, the coherent-detector is subject to thermal and dark-current noises as well as the background light incident on the detector. The coherent-detection ideally requires (i) strict conditions on the spectral purity of the source signal and (ii) that the received signal and the local oscillator have spatial phase fronts, which are nearly perfectly aligned, over the active area of the detector.

Since optical phase information is lost in the direct-detection process, it cannot be used to measure the Doppler frequency shift of the radar echo. Under ideal conditions when signal strength is limited, the coherent-detection technique provides superior sensitivity to direct-detection. However, direct-detection has advantages over coherent-detection when either source temporal coherence or the spatial phase characteristics of the received signal cannot be strictly controlled, or when complexity or cost is an important design issue.

The receiver's input RF filter performs three basic functions:

- to limit the bandwidth of the spectrum reaching the RF amplifier and mixer to minimize intermodulation distortion. Intermodulation distortion is caused by non-linearity of the system components, which upon passing through two or more signals acts as a mixer and introduces sum-and-difference products of the applied frequencies. The intermodulation distortion problem is less important in broadband RF power applications as is harmonic distortion;
- to attenuate receiver spurious image noise and half-IF responses; and
- to suppress LO energy originating in the receiver.

The drive levels of the LO permit higher intercept point performance of the mixers. The intercept point is a measure of system linearity that allows us to calculate distortion from the incoming, or outgoing, signal amplitudes. An intercept point method is used to minimize intermodulation distortion. For example, for a fixed LO power, the nth order of intercept point, IP_n, can be predicted, provided the distribution products are known for a particular input or output level, using Vizmuller (1995)

$$\mathrm{IP}_n = A_o + \frac{\Delta S}{n-1} \quad \text{(dBm)} \tag{2.5}$$

where

A_o = the input or output intercept point (dBm)
ΔS = difference between required signal level and undesired distortion (dBm)
n = order of distortion.

For detailed analysis on how the intercept point is evaluated, the reader is advised to read Vizmuller (1995).

Example 2.1 Suppose that in a radar system a certain order of spurious signals was measured. In this case, a certain (4,2) high-order spurious response was measured to be 70 dB down when the input level is −16 dBm. Calculate the distortion product for an input level of −22 dBm.

Solution
The fourth order intercept point, $\text{IP}_4 = -16 + 70/3 = 7.33\,\text{dBm}$
The input level $A_o = -22\,\text{dBm}$
From (2.5), the distortion product ΔS is

$$\begin{aligned}\Delta S &= (\text{IP}_4 - A_o)(n-1)\\ &= (7.33 + 22)(4-1)\\ &= 87.99\,\text{dB}\end{aligned}$$

Mixers are very important building blocks in any RF system. Down-conversion mixers link together the low-noise RF amplifier, local oscillator (for the first stage mixer) and IF stage (for the second stage mixer) of which the performances are interrelated. Their highly non-linear behaviour makes analysis and optimization difficult. This non-linearity behaviour can cause noise and spurious signals to move across frequencies. The sensitivity e_r (in volts) of the receiver can be predicted whether the receiver is limited by *thermal* or *non-thermal* noise using:

(1) For thermal limited receiver noise:

$$e_r^2 = kF_T B_n T(\text{SNR})R_{\text{eq}} \tag{2.6}$$

(2) For non-thermal limited receiver noise:

$$e_r^2 = k\left[T_{\text{eq}} + T_a\right]B_n(\text{SNR})R_{\text{eq}} \tag{2.7}$$

where

F_T = total noise figure. Note that this noise figure should be the total device noise, which should include the channel noise factor, the noise derived from image frequency stage noise figure(s) and the noise figure from the local oscillators
T_{eq} = equivalent system temperature = $(F_T - 1)T_s$
T_s = system temperature (K). This temperature is often taken as the standard ambient temperature in accordance with IEEE Standard 145-1983 (IEEE Standard 145-1983), where T is 17 °C, equating to 290 K
T_a = antenna temperature (K)
R_{eq} = system equivalent impedance (Ω)
B_n = noise bandwidth (Hz)
k = Boltzmann's constant, 1.38×10^{-23} (W/Hz – K)
SNR = signal-to-noise ratio (linear unit).

How the noise figure is obtained is described fully in Chapter 5, section 5.1.6. RF amplifier noise figure, gain and intercept-point are set by the receiver performance requirements.

Example 2.2 A system's overall equivalent noise factor and bandwidth are given as 14.87 and 12 kHz respectively. The received signal at the detector output is 6 dB. Calculate the sensitivity of the system across 50 Ω impedance if:

(a) it operates at room temperature; and
(b) the antenna temperature is constantly above the room temperature with an average value of 18.2 °C.

Solution

The solution to (a) is found by using (2.6), given that
SNR = 6 dB, converting it to linear unit to have SNR $= 10^{0.6} = 3.981$

$R_g = 50\,\Omega$

$F_T = 14.87$

$B_n = 12\,000$

$k = 1.38 \times 10^{-23}$

$T = 273 + 17 = 290$

Substituting numerical values in (2.6),

$$e_v = 0.377\,\mu\text{V}$$

By using (2.7), the solution to part (b) of the question is solved. Replace T with $[T_{eq} + T_a] = (14.87 - 1)290 + (18.2 + 273) = 4313.5$, to obtain

$$e_v = 0.3771\,\mu\text{V}$$

2.1.3 Data processing

A digital signal processor (DSP) for data processing buffers the output of the analogue-to-digital (A/D) converter. The DSP attempts to extract information from radar echoes, with a view to classifying targets and characterizing geophysical phenomena. Signal processing is handled by a DSP operating under algorithms tailored to the requirements of the radar. Many modern radars perform a signal spectrum analysis function in a DSP using fast Fourier transform (FFT) – already discussed in Chapter 1, section 1.4.

More important properties of FFT are discussed briefly at this instance to allow the reader a feel of the properties in their application to radar system. The input of an FFT is a sequence of 2^m time samples, where m is an integer. The output, on the other hand, is 2^m complex numbers having in-phase and quadrature components representing the frequency spectrum. The output is analogous to a bank of uniformly spaced filters covering the frequency region from zero up to the transmitter *pulse repetition frequency* (PRF), as shown in Figure 2.8. As such the filter spacing, Δ_f, can be expressed as

$$\Delta_f = \frac{\text{PRF}}{2^m} \tag{2.8}$$

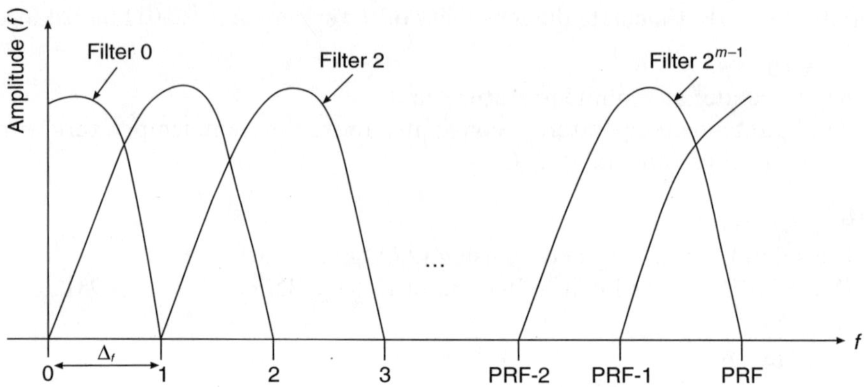

Figure 2.8 FFT with 2^m filters

It should be noted that if the input consists of complex samples, then the frequency region from zero to PRF is unambiguous. Conversely, if the input samples are real, the unambiguous region is simply half the PRF; that is, PRF/2. The number of Doppler filters depends solely on the number of time samples. If it is possible to eliminate blind speeds, or resolve Doppler ambiguities, the frequency (and also bandwidth) spacing of the filters would automatically adjust. Regardless of whether the Doppler frequency is unambiguous, target returns would move from one filter to another as PRF changes.

Recently, signal processing has assumed higher-order statistical analysis with a view to extracting more information from the radar echoes (Cover and Thomas 1991). Some aspects of signal processing and applicable algorithms are the subjects of Part III.

2.1.4 Data compression and storage

A myriad of data is often acquired during any radar scans or sweeps. An example of this is that acquired by skywave radars, which are particularly noted for their wide-area scanning or sweeping. The unprocessed data acquired can often occupy a large facility. Pre- and post-processed data could also be large and might require large transfer and processing time. In a real-time operational situation, in particular during tracking, time is a critical element if the true-target profile under investigation is to be quickly ascertained in real time. To ensure fast transportation and delivery of data to its intended destination, a *compression process* is used.

Data compression is the process of converting an input data stream (the source stream, or the original raw data) into a smaller data stream (the output, or the compressed stream) that has a smaller size. A stream is either a file or a buffer in memory. If one can denote the input stream by D_M and the compressed stream by $\partial(D_M)$, it must be possible for the compressed data $\partial(D_M)$ to be decoded (reconstructed) back to the original body of data D_M

or some acceptable approximation. Compressed data are stored on a digital storage device (e.g. compact disc, tapes) and when retrieved from the device they are decompressed.

Data compression is a topic grounded in the field of information theory: the study of the representation, storage, transmission and transformation of data. The coding and decoding process, being part of information theory, is quite involved and entails different approaches. It is known by many names such as *entropy coding, lossless coding, data compaction coding*, or *data compression*. The inquiring reader is advised to consult Kolawole (2002), McEliece (1977), Cover and Thomas (1991) and Storer (1976).

Based on the requirements of reconstruction, data compression schemes can be divided into two broad classes: *lossless* and *lossy* compression. A lossless compression technique takes compressed data $\partial(D_M)$ and reconstructs it to the original data D_M. The lossy compression technique is the process of transforming a body of data D_M to a smaller body $\partial_i(D_M)$, where $i = 1, 2, \ldots, m$, from which an approximation of the original can be constructed. Lossy compression provides, in general, much higher compression than lossless compression. Often reconstruction requirements dictate the type of compression schemes to use. A generalized description of a class of algorithms is discussed in the following subsections.

2.1.4.1 Effective algorithms for data compression

To effectively discern real target signatures from the noise and clutter, some decision is made by setting a limit (or threshold) where anything above the limit is associated with target and anything below is those associated with noise and/or clutter. The resulting processed data may be called 'static' if the probabilities were *a priori*; that is, they are given in advance. If the radar data were collected in a 'hostile environment', which often is the case with skywave radars, it would be reasonable to dynamically threshold the data, that is, using a compression algorithm that estimates these probabilities dynamically. The Huffman (1951) and Shannon (1959)–Fano (1963) compression algorithms offer an example of how data compression can be dynamically achieved. The difference between these algorithms is that Shannon–Fano constructs its codes top to bottom (from the leftmost to the rightmost bits), while Huffman constructs a code tree from the bottom up (builds the codes from right to left). There have been intensive research activities into data compression since the papers of Huffman, Shannon and Fano. The next subsection discusses the basic Huffman coding algorithm, though there have been several enhancements to the original.

2.1.4.2 Huffman coding algorithm

Suppose that one can represent every peak associated with a target in the data map by the symbols a_k and corresponding probabilities $p(a_k)$, where $k = 0, 1, 2, \ldots, m - 1$. These symbols and their probabilities are shown in Table 2.1 as a list $\mathcal{L}$.

Table 2.1 List $\mathcal{L}$

Symbol	Probability
a_0	$p(a_0)$
a_1	$p(a_1)$
a_2	$p(a_2)$
...	...
a_{m-2}	$p(a_{m-2})$
a_{m-1}	$p(a_{m-1})$

Tidying up is done by representing the input to the encoder α by $A = \{a_0, a_1, a_2, \ldots, a_{m-1}\}$ and the codeword lengths $l_k = n(a_k)$.

For clarity, an encoder is a means of assigning one of the codewords to an input, or source, symbol. It compresses the raw data in the input stream and creates an output with compressed (low-redundancy) data. The decompressor or decoder converts in the opposite direction to the compressor. The term companding stands for 'compressing/expanding'. The original input stream denotes unencoded, raw or original data. The output content, which is a compressed stream, is the encoded or compressed data.

Using the *Kraft inequality* theorem, the prefix property ensures that there exists a necessary and sufficient condition for the Huffman code to be uniquely decodable (decipherable). This condition is mathematically expressed as

$$\sum_{k=0}^{m-1} 2^{-l_k} \leq 1 \qquad (2.9)$$

for a noiseless source code A, encoder α, and codeword lengths l_k. If the codeword lengths can be ordered as $l_0 \leq l_1 \leq l_2 \leq \ldots \leq l_{m-1}$, then a collection of codewords will represent a binary tree of depth l_{m-1}. For example, by putting $m = 4$, a binary $\{1, 0\}$ code tree is drawn as in Figure 2.9 by labelling one branch '0' and the other '1'. By convention, a '1' is normally put on the upper branch in a horizontally drawn tree and a '0' on the lower branch. The binary tree starts with a *root*, which has two *branches* extending from it. Each branch ends in a *node*; in this case as the first level nodes or depth one nodes. Nodes can extend further into branches leading to more nodes, or simply terminate. When nodes end, they are called terminal *nodes* or *leaves*.

At a further level, a node connected by a branch is said to be a *child* or *sibling* of the preceding node (called the *parent* node). There is a one-to-one correspondence between paths from the root node to the terminal node and the codewords, sometimes called *path maps*. It can be seen in Figure 2.9 that the code can be represented by a *subtree* – denoted by white circle – consisting of branches from the *root* (source) of the tree to the terminal *nodes* (or *leaves*) – denoted by a blackened circle of the subtree. The codewords correspond to the sequences of branch labels from the root of the tree to

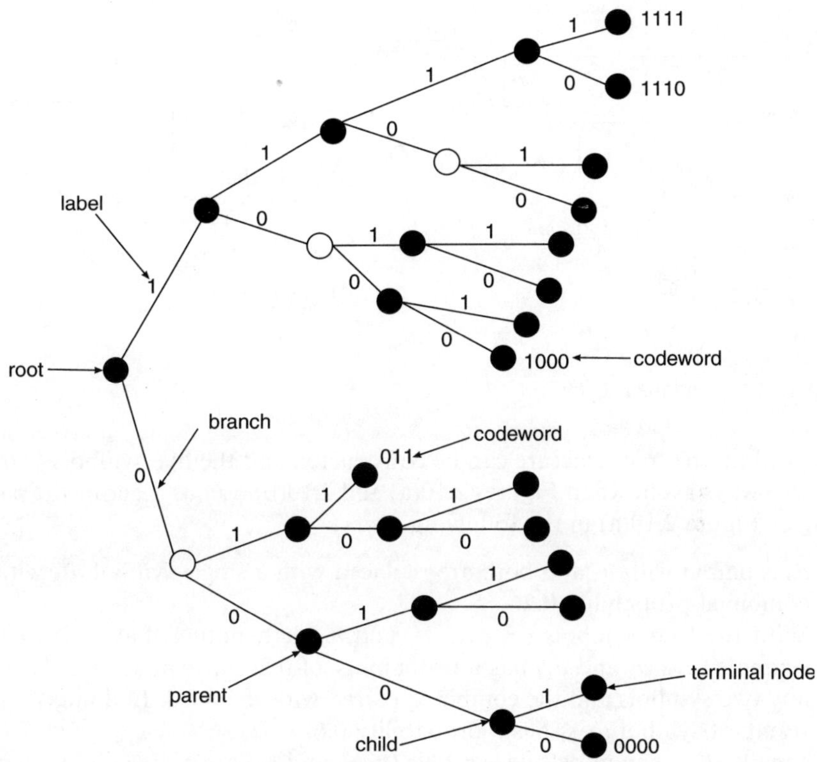

Figure 2.9 Binary tree code of variable lengths

the leaf. In summary, binary codewords of length l_{m-1} or shorter may be described as paths through the tree, or as terminal nodes of such a path.

With the background information and following Gallager (1978), the static Huffman coding algorithm is described as follows:

(a) Represent the list of the probabilities of the source that is considered to be associated with the leaves of a binary tree by $\mathcal{L}$.
(b) Take the two smallest probabilities in $\mathcal{L}$ and make the corresponding nodes siblings. Generate an intermediate node as their parents and label the branch from the parent to one of the child nodes '1' and label the branch from parent to the other child '0'.
(c) Replace the two probabilities and associated nodes in $\mathcal{L}$ by the single new intermediate node with the sum of the two probabilities.

If $\mathcal{L}$ now contains only one element, end iteration. Otherwise go to step (b).

This algorithm is best illustrated by an example.

Example 2.3 Consider a five-symbol alphabet a_0, a_1, a_2, a_3, a_4 with corresponding probabilities 0.4, 0.2, 0.2, 0.1, 0.1. Using the static Huffman

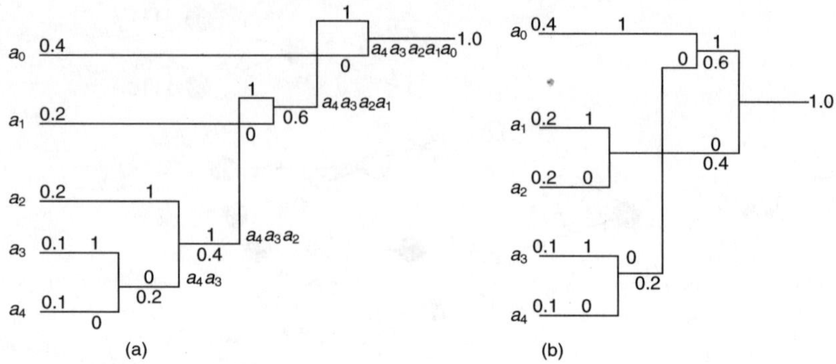

Figure 2.10 Huffman codes

algorithm, the tree structure can be constructed and the five symbols paired in two ways as shown in Figures 2.10(a) and 2.10(b). Let us describe the pairing of Figure 2.10(a) in the following order:

- a_4 is paired with a_3 and both are replaced with a single symbol a_{43} with a combined probability 0.2.
- With the four symbols (a_{43}, a_2, a_1 and a_0) left, noting that each of the symbols (a_{43}, a_2 and a_1) has a probability of 0.2, one can arbitrarily take any two symbols and the combined paired with the third. In doing so, the resultant symbol a_{4321} has a probability 0.6.
- Finally, the remaining two symbols (a_{4321} and a_0) are paired and replaced with a_{43210} with probability 1.0.

Having completed the tree, with root node on the right and the five leaves on the left, it is time to assign codes. With the labelling of every pair of edges, the resulting codewords are the codes read off from right to left for each of the symbols: 0, 10, 111, 1101 and 1100. Specifically,

$$\begin{aligned} a_0 &= 0 \\ a_1 &= 10 \\ a_2 &= 111 \\ a_3 &= 1101 \\ a_4 &= 1100 \end{aligned}$$

The number of bits $n(a_k)$ in each codeword a_0, a_1, a_2, a_3, a_4 is 1, 2, 3, 4, 4 respectively.

Similarly, for the tree structure represented by Figure 2.10(b) and assigned pairing, each symbol is encoded as

$$\begin{aligned} a_0 &= 11 \\ a_1 &= 01 \\ a_2 &= 00 \\ a_3 &= 101 \\ a_4 &= 100 \end{aligned}$$

The number of bits, $n(a_k)$, in each codeword a_0, a_1, a_2, a_3, a_4 is 2, 2, 2, 3, 3 respectively, which is different from that of the tree structure of Figure 2.10(a). The difference shows that the arbitrary decisions made when constructing the Huffman tree affect the individual codes, but not the average size of the codewords. The reader might ask which of these codes is better? To answer this question, the better code is the one with the smallest variance. Two new terms have just been introduced: 'average size' and 'variance'. How do we quantify these terms in the light of Huffman coding?

Average size $\langle l_k \rangle$ is defined by

$$\langle l_k \rangle = \sum_{k=0}^{m-1} p(a_k) n(a_k) \tag{2.10}$$

Variance is defined by

$$\sigma^2 = \sum_{k=0}^{m-1} p(a_k)[n(a_k) - \langle l_k \rangle]^2 \tag{2.11}$$

From (2.10), the average size of the codes obtained from Figures 2.10(a, b) is the same; that is, 2.2 bits/symbol in this instance. However, using (2.11), two different variances are obtained: 1.36 and 0.16 for Figure 2.10(a) and Figure 2.10(b) respectively. Hence, the code of Figure 2.10(b) is preferred. Often, the *entropy* of the code is required. Entropy, H, is the quantity of data transmitted per second, or the average self-information per transmitted symbol. The 'entropy'[1] H of symbol 'a' is defined by:

$$H = -\sum_{k=0}^{m-1} p(a_k) \log_2 p(a_k) \quad \text{bits} \tag{2.12}$$

Choosing $p(a_k) = 1/m$ for all $1 \leq a_k \leq m$ gives the maximum possible value of H for a given value of m. Equation (2.12) shows that the entropy of the data depends on the individual symbols' probabilities $p(a_k)$ and is smallest when all m probabilities are equal. This fact is used to define the redundancy $\mathcal{R}$ in the data.

[1] In analogue communication systems in which the transmitted signal is a continuous voltage waveform $v(t)$, the entropy H for each independent sample of $v(t)$ may be defined by

$$H = -\int_{-\infty}^{\infty} p(v) \log_2 p(v) dv \quad \text{bits/sample}$$

where $p(v)$ is the probability density function of $v(t)$. The form of $p(v)$ that maximizes H for a given signal power is the Gaussian distribution. When $p(v)$ is Gaussian with square mean value N, then entropy is

$$H = \ln \sqrt{2\pi e N} \quad \text{bits/sample.}$$

Redundancy is defined as the difference between the entropy and the smallest entropy:

$$R = -\sum_{k=0}^{m-1} p(a_k)\log_2 p(a_k) - \log_2 m \tag{2.13}$$

With this expression, test for fully compressed data (no redundancy) by

$$\sum_{k=0}^{m-1} p(a_k)\log_2 p(a_k) = \log_2 m \tag{2.14}$$

In practice, little is known in advance about the input stream and its associated probabilities. As such, look into ways of devising an approach that is more adaptive in spirit, which essentially builds on the 'static' approach. For example, suppose that one wishes to modify the estimates of the list's probabilities as more data arrive and to adapt the code correspondingly. A strategy similar to the previous 'static' construction could be adopted. For instance, suppose that at time $(i-1)$ the probability estimates $p_{i-1}(a_k)$ for all of the source symbols a_k are available along with the corresponding Huffman code; i.e.

$$p_{i-1}(a_k) = \frac{n_{i-1}(a_k)}{i-1} \quad i > 1 \tag{2.15}$$

where $k = 0, 1, 2, \ldots, m-1$. If the *i*th input symbol $a_i = a$ is encoded and decoded using this Huffman code and all of the probabilities updated with the new relative probabilities, then the only count for the symbol 'a' would change to

$$\begin{aligned} p_i(a) &= \frac{n_i(a)}{i} = \frac{1+(i-1)p_{i-1}(a)}{i} \\ p_i(a_k) &= \frac{n_i(a_k)}{i} = \frac{1+(i-1)p_{i-1}(a_k)}{i} \end{aligned} \tag{2.16}$$

provided $a_k \neq a$. These new and improved probabilities are made available to the encoder and decoder, which would then be used to design a new Huffman code for use on the next input symbol.

In practice, radar data are quantified by *weights*, w_k, where $k = 0, 1, 2, \ldots, m-1$. Since these weights are non-negatives, the weights can be used in place of the probabilities to design a Huffman code and to find the corresponding ordered tree.

Recent advances in technology have enabled system manufacturers to include encoding/decoding chips in their hardware, invisible to the users that perform data compression/decompression.

2.1.5 Display and communications system

In modern radar systems, the radar data is highly processed before display. The display unit provides a full-range presentation of received signals. The display device is a console, which is conceptually similar to the computer-driven monitor. Target detections are often represented by target symbols on the display unit or console. In some cases, a command from the display unit is used to trigger control signals to steer the radar antenna in the desired direction.

A communication system ensures that the internal and external communications systems meet the intended requirements including voice, text, accurate timing and location finding via the global positioning system (GPS).

2.1.6 Radar application

The application purpose of a particular radar determines its limit of operation. Theoretically, radar may be developed having capabilities that exploit a great shift in wavelength, or when precision tracking and high resolution in range, angle (azimuth), target identification and Doppler are required. New development in laser radar technology has achieved this. For instance, laser radar has combined the capabilities of conventional radar and optical systems to achieve high resolution and accurate target tracking, imaging, aim-pointing assessment, and autonomous operation. By combining laser radar systems with passive sensors, further improvement can be gained in target estimation and precision independent of time of the day or night. More is said about laser radar in Chapter 5. Radar usage varies dramatically including:

1. strategic and tactical surveillance;
2. remote atmospheric and sea-state sensing;
3. tracking and guidance; and
4. precision disaster control or monitoring.

Radar systems that operate on line-of-sight principles are called conventional radar (examples are microwave, laser and beacon), while those that see beyond the horizon are called skywave radar (to be discussed in Chapter 7). The *over the horizon radar* (OTHR) is an example of a skywave radar. An OTHR utilizes *high frequencies* (HF) unlike the conventional microwave radar, which operates between 0.2 and 40 GHz. A major difference between the HF skywave and conventional line-of-sight radar is the need to adapt the waveform and frequency of the former to the environment. The detailed design of a system for a particular application can differ significantly. It also involves compromise between cost, implementation, and operating parameters to achieve realistic performance. There may also be differences in the characteristics of the respective propagation media and in the signals processed which are reflected in the implementation used for the two systems. Despite this the fundamental principles are common.

2.1.7 Summary

This chapter has explained the fundamental architecture of a typical radar system. It also covered the issue of receiver sensitivity, data compression and radar utilization. The next chapter looks at the physics of an antenna, which is a major item in radar systems design, as well as range measurements for signal pulse and train pulses.

Problems

1. If you have any compression/decompression programs on your computer (e.g. 'StuffIt'), then perform the following exercise. Use any of your documents, say joke.doc, and drop the document on 'StuffIt'. A new document will be created in your directory called 'joke.sea'. Compare the size of uncompressed file 'joke.doc' with the compressed file 'joke.sea'. To reconstruct (or retrieve) the compressed data to the original, drop 'joke.sea' on 'StuffIt expander' to create another file called 'joke1.doc'. Compare the size of the reconstructed file 'joke1.doc' with the original 'joke.doc'. Both sizes should be approximately equal.
2. Why is it that an already-compressed data cannot be compressed further?
3. Suppose an eight-symbol list is as given in Table 2.2. Design a Huffman code for the symbols. Estimate the average length, variance, and the entropy of codes.
4. Design a Huffman code for a source with seven symbols a_k, where $k = 0, 1, 2, \ldots, 6$ with the symbols' probabilities having a functional relation given by $p(a_k) = 0.3/1.3^k$.
5. If the probabilities in problem (3) are weighted as $p(a_i) = w_i / \sum_{k=0}^{m-1} w_k$, design a corresponding Huffman code.
6. Estimate the noise bandwidth range required for a receiver's sensitivity to be maintained at 0.35 mV, the antenna is operational at temperatures between $-10\,°C$ and $45\,°C$, the effective impedance is $75\,\Omega$, and the total noise factor is 12.64 dB.
7. Will the noise bandwidth estimated in question (6) be suitable for the same receiver if the antenna were kept at room temperature?

Table 2.2 List of the eight symbols

Symbol	Probability
a_0	0.01
a_1	0.02
a_2	0.05
a_3	0.09
a_4	0.18
a_5	0.20
a_6	0.20
a_7	0.25

3

Antenna physics and radar measurements

The previous chapter has briefly explained the functionality of a practical radar system. One of the major components of radar is an antenna. The basic physics of antenna radiation and how the field in front of the antenna is divided into regions are explained in this chapter. In addition, the concept of pulse compression is investigated for a single pulse and a train of pulses. The compression filter's response is used to explain measurement ambiguities in range and Doppler as well as resolving closely spaced targets.

3.1 Antenna radiation

One of the simplest forms of radiator is the dipole antenna. A dipole (or doublet) consists of a metallic wire whose length is an appreciable portion of a wavelength. If the wire is fed at its centre by an electric source (or a transmitter or generator), equal charges of opposite signs (^-q and ^+q) are induced. A schematic representation of a dipole is given in Figure 3.1. If the values of the charges are varied harmonically in time, the dipole will radiate energy. By the nature of the generator, the current varies and moving electric charges produce radiated fields. The faster the charges accelerate the better the dipole radiates. How does one predict the radiation pattern of an antenna? The next paragraphs attempt to shed some light on this question.

By Coulomb's rule, the interaction between two-point charges ^-q and ^+q is interchangeable. These charges are assumed of equal amplitude and may be in close proximity compared to the distance in the surrounding field, say at point P. According to the *superposition rule*, two or more electric fields acting at any given point would add vectorially. Thus, the electric potential at point P can be expressed as the sum of the potentials due to the individual charges:

$$V = \frac{q}{4\pi\varepsilon}\left(\frac{1}{r_1} - \frac{1}{r_2}\right) \tag{3.1}$$

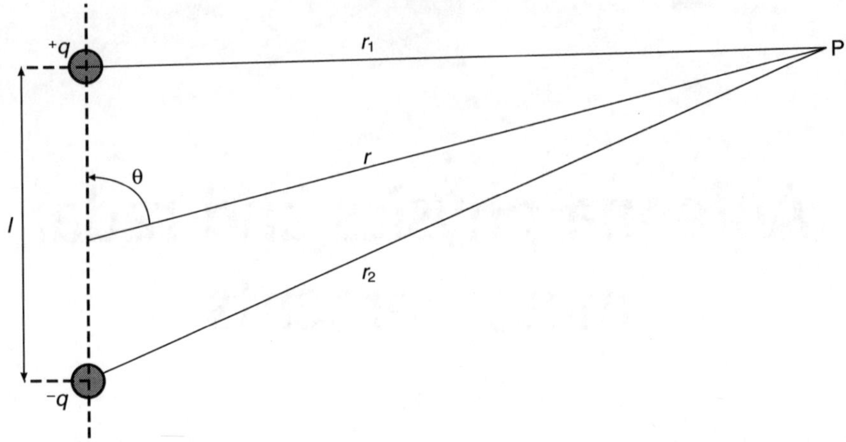

Figure 3.1 A centre-fed dipole

Since the distance r measured from the centre of the wire to the observation point P is far greater than the separation distance l between the electric charges (i.e. $r \gg l$), the approximate distances r_1 and r_2 are

$$\begin{aligned} r_1 &\approx r - \frac{l\cos\theta}{2} \\ r_2 &\approx r + \frac{l\cos\theta}{2} \end{aligned} \tag{3.2}$$

Substituting (3.2) in (3.1), and observing as $l \to 0$, that is, the point-dipole limit, the electric potential becomes exact:

$$V = \frac{ql\cos\theta}{4\pi\varepsilon r^2} \tag{3.3}$$

Alternatively,

$$V = \frac{M\cos\theta}{4\pi\varepsilon r^2} \tag{3.4}$$

where ε is a constant, called the permittivity, which depends on the medium surrounding the charge. In this instance ε is maintained constant. M is the moment of the dipole. Equation (3.4) is also valid at large distances from any finite size dipole. It can be seen in (3.4) that the electric potential V varies inversely as the square of the distance from the dipole, in contrast with the reciprocal distance law of the point charge expressed in (3.1).

Given that electrostatic fields are conserved, the electric intensity at any point is equal to the space rate of change of potential:

$$E = -\frac{\partial V}{\partial s} = -\nabla V \tag{3.5a}$$

where ∇V is the gradient of V and defines both the magnitude and the direction of the maximum rate of change of V. The minus sign arises because the work done in moving a unit charge is positive when it is done by some external force against the field.

By defining the dipole moment vector **M** directed from ^{-}q to ^{+}q as having a magnitude ql, the potential V in (3.4) can then be expressed as

$$V = \frac{\mathbf{M} \cdot r_1}{4\pi\varepsilon r^2} \tag{3.5b}$$

The components of the electric intensity E, in spherical coordinates, can be estimated by performing the gradient operation of V in (3.5b):

$$E_r = -\frac{\partial V}{\partial r} = \frac{2M\cos\theta}{4\pi\varepsilon r^3} \tag{3.6a}$$

$$E_\theta = -\frac{\partial V}{r\partial\theta} = \frac{M\sin\theta}{4\pi\varepsilon r^3} \tag{3.6b}$$

$$E_\phi = -\frac{\partial V}{r\sin\theta\partial\phi} = 0 \tag{3.6c}$$

Since E is the vector sum of all the components, the electric intensity becomes

$$\begin{aligned} E &= E_r\hat{r} + E_\theta\hat{\theta} + E_\phi\hat{\phi} \\ &= \frac{M}{4\pi\varepsilon r^3}\left(2\cos\theta\hat{r} + \sin\theta\hat{\theta}\right) \end{aligned} \tag{3.7}$$

where $\hat{r}$, $\hat{\theta}$, $\hat{\phi}$ are unit vectors in the r, θ, ϕ directions respectively. Often (3.7) is called the *static* components in the literature. This equation demonstrates that the electric intensity of a dipole falls off as the cube of the distance, in contrast to the inverse square law of the potential expressed in (3.4). A sketch of the electric intensity pattern of the point dipole is shown in Figure 3.2.

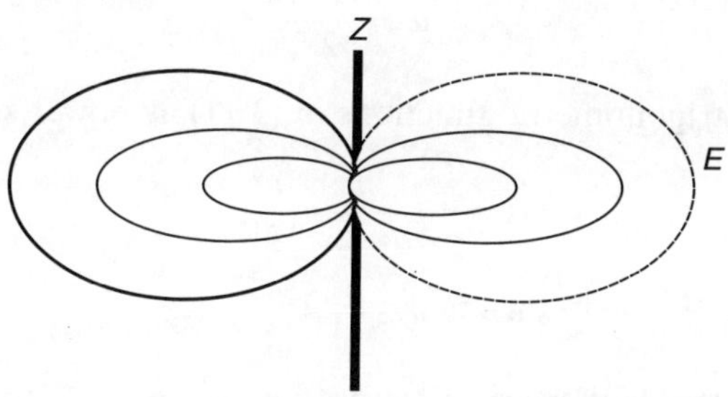

Figure 3.2 Radiation pattern of a point dipole

The preceding discussion has assumed a static dipole. As stated earlier, if the values of the electric charges are varied harmonically in time and space, the expectation is that the dipole will radiate energy. Therefore the oscillating charge, q, and its moment, M, would become

$$\begin{aligned} q &= q_0 e^{j\omega t} \\ M &= M_0 e^{j\omega t} \end{aligned} \tag{3.8}$$

where $M_0 = q_0 l$.

If a thin wire of negligible resistance is assumed and a capacitance connects the pair of charges, then an alternating current I that flows upwards from ${}^{-}q$ to ${}^{+}q$ may be expressed as

$$I = \frac{dq}{dt} = j\omega q_0 e^{j\omega t} = I_0 e^{j\omega t} \tag{3.9}$$

This expression does not take into account the time τ required for charges at the dipole to travel to observation point P leading to potential retardation at P, as do V and E, where $\tau = r/c$ and c is the velocity of light. At this point, let us examine the influence of delay on the potential V and electric field intensity E at point P.

By substituting (3.2) and (3.8) in (3.1), the electric scalar potential is written as

$$V = \frac{q_0}{4\pi\varepsilon}\left(\frac{e^{j\omega(t-\tau_1)}}{r - \frac{l\cos\theta}{2}} - \frac{e^{j\omega(t-\tau_2)}}{r + \frac{l\cos\theta}{2}}\right) \tag{3.10}$$

Which, by expansion

$$V = \frac{2q_0 e^{j\omega(t-\tau)}}{4\pi\varepsilon}\left(\frac{u\lambda_* \cos u + jr\sin u}{r^2\left[1 - \left(\frac{u\lambda_*}{r}\right)^2\right]}\right) \tag{3.11}$$

where

$$\begin{aligned} \lambda_* &= \frac{\lambda}{2\pi} \\ u &= \frac{l\cos\theta}{2\lambda_*} \end{aligned} \tag{3.12}$$

Expressing the trigonometric functions in (3.11) as power series; that is,

$$\begin{aligned} \cos u &= 1 - \frac{u^2}{2!} + \frac{u^4}{4!} - \cdots \\ \sin u &= u - \frac{u^3}{3!} + \frac{u^5}{5!} - \cdots \end{aligned}$$

By neglecting the higher-order terms, i.e. $u^2, u^3, \ldots$, the electric scalar potential is

$$V = \frac{2q_0 e^{j\omega(t-\tau)}}{4\pi r\varepsilon}\left(\frac{j + \frac{\lambda_*}{r}}{1 - \left(\frac{u\lambda_*}{r}\right)^2}\right) \tag{3.13}$$

Substituting (3.12) in (3.13), and noting that $r \gg l$,

$$V = \frac{q_0 l \cos\theta}{4\pi r\varepsilon\lambda_*}\left(1 + \frac{\lambda_*^2}{r^2}\right)^{\frac{1}{2}} e^{j\omega\left(t - \frac{r}{c} + \frac{\beta_*}{\omega}\right)} \tag{3.14}$$

where

$$\beta_* = \tan^{-1}\left(\frac{r}{\lambda_*}\right) \tag{3.15}$$

By comparing (3.4) with (3.14), one observes how the electric potential amplitude changes from r^{-2} dependence for a static dipole to r^{-1} dependence for an oscillating dipole. By letting $\omega = 0$, and $\lambda_* \to \infty$, both equations (3.4) and (3.14) agree as expected. Suffice to say that a similar behaviour can be observed for changes in the magnetic H component for a constant current to a varying current. This is left to the reader to verify.

For a non-zero frequency, the exponential term in (3.14) indicates that the potential V will propagate as a wave at a phase velocity c. This is not quite true, due to the complex $(j + \lambda_*/r)$ term. Since $r \gg \lambda_*$, $\beta_* \approx \pi/2$, which is approximately independent of r. In this instance, V can be said to have a phase velocity c. However, close to the dipole the magnitude r is not much larger than λ_*, the value of β_* becomes variable and consequently gives a phase velocity much larger than c. The quantity λ_* is called the *radian length*; the distance over which the phase of the wave changes by one radian; which is approximately $\lambda/6$.

By Maxwell theory, the electric field intensity can be obtained using

$$\mathbf{E} = -\left(\frac{\partial A}{\partial t} + \nabla V\right) \tag{3.16a}$$

where

$$\nabla V = -\left(\frac{\partial V}{\partial r}\hat{r} + \frac{\partial V}{r\partial\theta}\hat{\theta} + \frac{\partial V}{r\sin\theta\partial\phi}\hat{\phi}\right) \tag{3.16b}$$

From (3.13), we can write

$$-\nabla V = \frac{M_0 e^{j\omega(t-\tau)}}{4\pi r\varepsilon\lambda_*^2}\left(\left\{-1 + \frac{j2\lambda_*}{r}\left(1 + \frac{\lambda_*}{r}\right)\right\}\cos\theta\hat{r} + \left\{\frac{j\lambda_*}{r}\left(1 + \frac{2\lambda_*}{r}\right)\right\}\sin\theta\hat{\theta}\right) \tag{3.17a}$$

The potential due to current distribution I at the same particular moment as for the electric potential can be deduced as

$$A = \frac{\mu l I}{4\pi r}\left(\cos\theta\hat{r} - \sin\theta\hat{\theta}\right) \tag{3.17b}$$

where μ is the propagating medium permeability. Using (3.17b) and equation (3.9), the current differential

$$-\frac{\partial A}{\partial t} = \frac{M_0 e^{j\omega(t-\tau)}}{4\pi r\varepsilon\lambda_*^2}\left(\cos\theta\hat{r} - \sin\theta\hat{\theta}\right) \tag{3.17c}$$

Hence, substituting (3.17) in (3.16), the electric field intensity is written as

$$E = \frac{M_0 e^{j\omega(t-\tau)}}{4\pi r\varepsilon\lambda_*^2}\left[\frac{j2\lambda_*}{r}\left(1 + \frac{\lambda_*}{r}\right)\cos\theta\hat{r} + \left\{\frac{j\lambda_*}{r}\left(1 + \frac{2\lambda_*}{r}\right) - 1\right\}\sin\theta\hat{\theta}\right] \tag{3.18}$$

As $\lambda_* \to \infty$ and $\omega = 0$, (3.18) reverts to (3.7), the static terms.

The electric field intensity E is seen to propagate through space with a velocity c for $r \gg \lambda_*$, as does V. The situation where $r \gg \lambda_*$ is referred to as the *far field* for the doublet. More is said of the division of a radiating field in front of an antenna into regions in section 3.1.2. In the instance where $r \gg \lambda_*$, what is left in (3.18) is the radiation term. Specifically

$$E = -\frac{M_0 e^{j\omega(t-\tau)}}{4\pi r\varepsilon\lambda_*^2}\sin\theta\hat{\theta} \tag{3.19}$$

However, close to the dipole, r in (3.18) would not be much larger than λ_*, E will involve five components: two varying as r^{-2}; two varying r^{-2} but leading by $\pi/2$ (radians); and finally the r^{-1} term leading the other r^{-2} term by π (radians). Equation (3.19) demonstrates the r^{-1} dependence for an oscillating doublet ensuring conservation of energy.

Example 3.1 For free space, by substituting (3.8), (3.12) and $c = (\varepsilon_0\mu_0)^{-1/2}$ in (3.19), the magnitude of the field radiation is simplified as

$$|E| = \left|\frac{60\pi I_0 l}{\lambda}\sin\theta\right| \quad \text{(V/m)} \tag{3.20}$$

noting that $\varepsilon = \varepsilon_0 = 8.854\,\text{pF/m}$, $\mu_0 = 400\pi\,\text{pH/m}$ and $c = 3 \times 10^8\,\text{m/s}$, the speed of light. The normalized polar plot of the radiation field induced by a unitary current and $l/\lambda = 0.1$ is shown in Figure 3.3; that is, $E/60\pi$ versus θ.

It should be noted that if the radiating element were placed vertically on a plane its image would be taken into account. An example of where the ground effect is replaced by the radiator image is a vertical monopole,

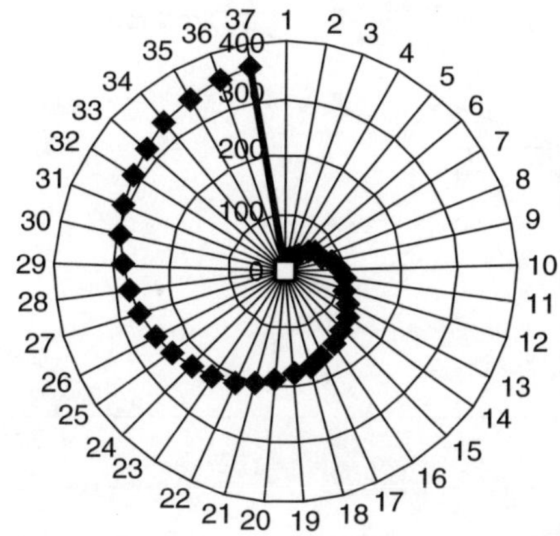

Figure 3.3 Polar plot of radiation pattern of a dipole in free space

which is briefly explained in the next section. Also note that a radiating element is not restricted to dipoles or monopoles, but may be other radiators such as slots, open-ended waveguides (or small horns) and microstrips. If radiators are similarly located at regularly spaced points, an antenna array is formed. In such a formation the array's resultant electric field will be given approximately by the sum of the fields contributed by all radiating elements. The effectiveness of such an array would depend on the operating frequency, power handling capability, polarization technique and method of feeding. More is said on the types of antenna array, the formulation of their electric intensity and applications in Chapter 4.

3.1.1 Vertical monopole

A vertical monopole is the simplest form of vertical antennae; it is grounded at the lower end. This form of antenna is commonly used as receiving elements for skywave radars (for example, over-the-horizon-radar: more is said of this type of radar in Chapter 7). When an antenna is near the ground, energy radiated toward the ground is reflected as shown in Figure 3.4.

The total field in any direction then represents the vector sum of a direct wave plus a reflected wave. For purpose of calculation, it is convenient to consider that the reflected wave is generated not by reflection but rather by a suitable image antenna located below the surface of the ground.

For clarity, the symbols θ_t, θ and ψ, in Figure 3.4, are defined as the target elevation angle, antenna elevation angle and grazing (or reflected) angle respectively, while l is the height of the antenna above the ground. In the case of a perfect ground (of infinite conductivity) the reflection coefficient is unity; that is, $\rho = 1$. The currents, I, in corresponding parts of the actual and image antennas are of the same magnitude and flow in the same direction

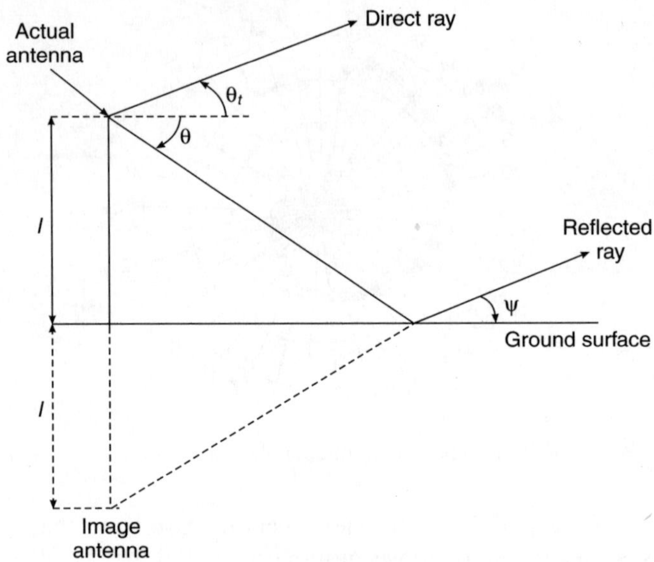

Figure 3.4 Geometry of a vertical monopole

in the vertical arm while the image current flow is opposite to that of the actual antenna in direction in the horizontal component.

For developmental purposes, consider an element dz at distant z from ground with radiated field observable at distance r from the source. In view of (3.20), the resultant electric field of the vertical monopole can be written as

$$E = \frac{60\pi}{\lambda} I_0 \int_{-l}^{l} \sin\left(\frac{\pi z}{l}\right) \sin\theta \cos\left(\frac{\Delta\phi}{2}\right) dz \tag{3.21}$$

where the phase difference $\Delta\phi$, due to path difference, is given by

$$\Delta\phi = 2\pi\left(\frac{z}{l}\right) \cos\theta \tag{3.22}$$

θ and I_0 are the antenna elevation angle and the magnitude of the current flowing in the antenna respectively.

Solving (3.21) yields

$$E = \frac{120\pi l I_0}{\lambda} \left[\frac{1 + \cos(\pi\cos\theta)}{\sin\theta}\right] \tag{3.23a}$$

The term in [.] is called the pattern factor, $f(\theta)$, for this type of arrangement. Specifically,

$$f(\theta) = \frac{1 + \cos(\pi\cos\theta)}{\sin\theta} \tag{3.23b}$$

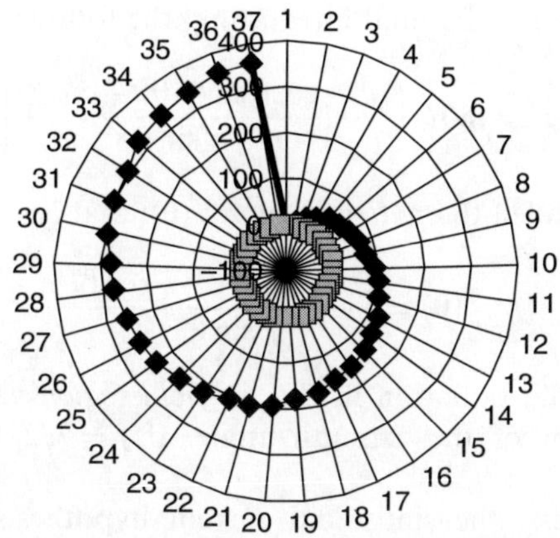

Figure 3.5 Polar plot of a vertical monopole radiation pattern

The normalized polar plot of the radiation field, that is, $E/60\pi$ versus θ induced by a unitary current, $l/\lambda = 0.1$, for a vertical monopole, is shown in Figure 3.5.

The difference between the radiation field induced by both dipole and vertical monopole is shown in Figure 3.6 as a combined plot. Besides field strength, the effect of ground contribution is visible between the two graphs when transversing from the positive phase to the next.

In general, if the dipole is symmetrical and of length $2l$ and letting $I(z)$ be the amplitude of the sinusoidal current as a function of the z-axis; that is, in the form

$$I(z) = I_0 \sin[\beta(l - |z|)] \tag{3.24}$$

Then, the far-field expression E in the spherical coordinates is given by

$$E = j\frac{60\pi}{\lambda}\frac{e^{-j\beta r}}{r}\int_{-l}^{l} I(z)e^{j\beta z\cos\theta}dz \tag{3.25}$$

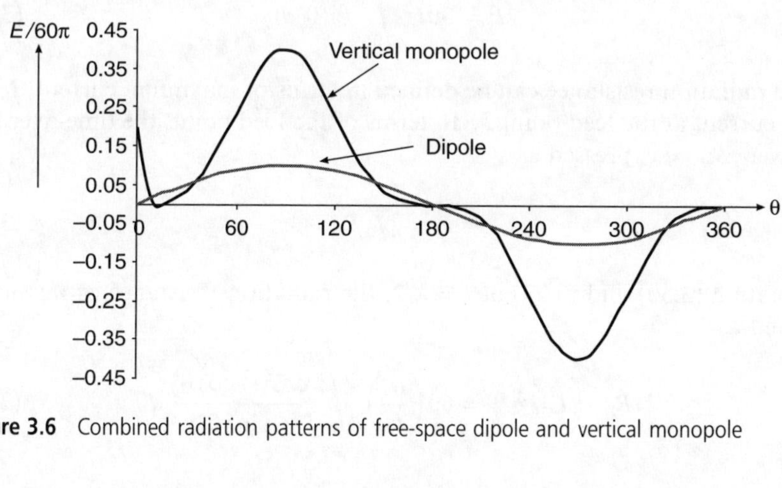

Figure 3.6 Combined radiation patterns of free-space dipole and vertical monopole

Substituting (3.24) in (3.25) and integrating, the following is found:

$$E = j60I_0 \frac{e^{-j\beta r}}{r}\left[\frac{\cos(\beta l\cos\theta) - \cos\beta l}{\sin\theta}\right] \tag{3.26}$$

The term in [.] is called the *pattern factor*, $f(\theta)$; that is,

$$f(\theta) = \frac{\cos(\beta l\cos\theta) - \cos\beta l}{\sin\theta} \tag{3.27}$$

It describes how the radiation in the *far-field region* varies with direction and is independent of the azimuth angle. If $l = \lambda/2$, $\beta l = \pi$; then (3.27) reverts to (3.23).

For completeness, the sinusoidal current hypothesis is less acceptable when the dipole is thicker and at a distance from resonance, which is certainly the case for asymmetrical dipoles. Despite these reservations, the sinusoidal hypothesis is used because it is an approximate that is simple to visualize, very practical in the far field and allows students to conceptualize the subject matter.

3.1.1.1 Radiation resistance and power

Power radiated by a dipole of length $2l$ is defined by

$$P = \frac{1}{2}\int_0^{2\pi}\int_0^{\pi} Re[\mathbf{EH}^*]r^2\sin\theta d\theta d\phi = \frac{\eta}{2}\left(\frac{I_m}{2\pi}\right)^2\int_0^{2\pi}\int_0^{\pi} f^2(\theta)d\theta d\phi \tag{3.28}$$

where I_m is the maximum current, $f(\theta)$ is the pattern factor from (3.27) and η is the characteristic impedance of the dipole. In free space, $\eta = \eta_0 = \sqrt{\mu_0/\varepsilon_0} = 120\pi\,(\Omega)$. After performing the integration, the power expression is found to be

$$P = 30I_m^2\int_0^{\pi} f^2(\theta)d\theta \tag{3.29}$$

The radiation resistance can be defined in terms of maximum current, I_m, or the current at the feed point I_0. In terms of the feed point, the time-averaged power can be expressed as

$$P = \frac{1}{2}I_0^2 R_{rad} \tag{3.30}$$

Equating (3.30) and (3.29) at $l = \lambda/2$, the radiation resistance expression is found as

$$R_{rad}\left(l = \frac{\lambda}{2}\right) = 60\left(\frac{I_m}{I_0}\right)^2\int_0^{\pi}\frac{\cos^2\left(\frac{\pi}{2}\cos\theta\right)}{\sin^2\theta}d\theta \tag{3.31}$$

If $I_m = I_0$ and let $x = \cos\theta$, and change the limits of integration accordingly, (3.31) can be recast as

$$R_{rad}\left(l=\frac{\lambda}{2}\right) = 60\int_{-1}^{1}\frac{\cos^2\left(\frac{\pi}{2}x\right)}{1-x^2}dx = 15\left\{\int_{-1}^{1}\frac{1+\cos(\pi x)}{1-x}dx + \int_{-1}^{1}\frac{1+\cos(\pi x)}{1+x}dx\right\} \tag{3.32}$$

Furthermore, put $y = \pi(1+x)$ in (3.32) and change the limits of integration accordingly,

$$R_{rad}\left(l=\frac{\lambda}{2}\right) = 30\int_{0}^{2\pi}\frac{1-\cos(y)}{y}dy \tag{3.33}$$

This expression can be related to a well-known function $Cin(x)$ defined by Abramowitz and Stegun (1968):

$$Cin(x) = \int_{0}^{x}\frac{1-\cos(y)}{y}dy \tag{3.34}$$

Comparing (3.34) with (3.33):

$$R_{rad}\left(l=\frac{\lambda}{2}\right) = 30Cin(2\pi) \tag{3.35}$$

Since $Cin(2\pi) = 2.438$,

$$R_{rad}\left(l=\frac{\lambda}{2}\right) = 30Cin(2\pi) = 73.14\,\Omega \tag{3.36}$$

Therefore dipoles of length that are multiples of $\lambda/2$ can readily be obtained.

It is appropriate at this stage to describe field regions and give the reader some idea of their physical dimensions.

3.1.2 Field regions

The field in front of an antenna may be divided into three regions: the reactive near-field region, the radiating near-field region (also called the *Fresnel* region), and the radiating far-field region (also called the *Fraunhofer* region). These regions are devised to identify the field structure in each. Although there are no discernible changes in the field configurations as the regions' boundaries are crossed, various criteria have been established which identify the regions. Using Figure 3.7 as a guide, these regions are defined as follows.

The *reactive near-field* region is the sector of the field immediately surrounding the antenna. By IEEE Standard 145-1983 (IEEE Standard

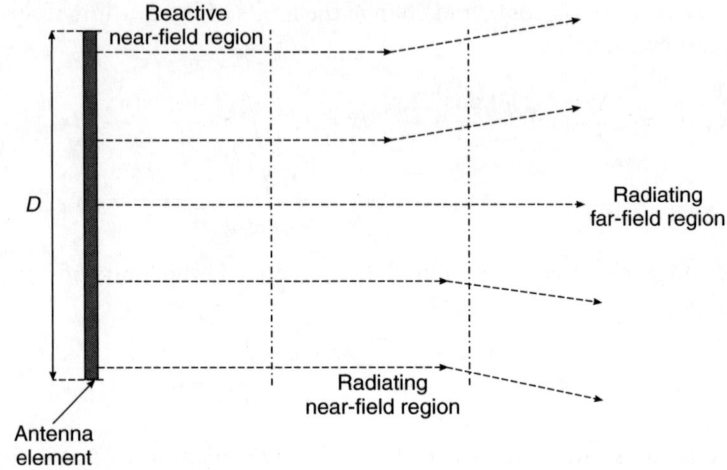

Figure 3.7 Boundaries of field regions

145-1983), for most antennas, the criterion used to define the outer boundary $\Re$ of this field is:

$$\Re < 0.62\left(\frac{D^3}{\lambda}\right)^{\frac{1}{2}} \tag{3.37}$$

where D is the largest dimension (aperture) of the antenna and λ is the wavelength: all units in metres.

The *radiating near-field* region is a sector where the angular distribution of the radiated energy is dependent on the distance from the antenna where the radial field component is significant. The radial distance where radiating near-field region exists is

$$0.62\left(\frac{D^3}{\lambda}\right)^{\frac{1}{2}} \leq \Re < \frac{D^3}{\lambda} \tag{3.38a}$$

The location of the antenna near field as a function of direction (Lewis and Newell 1985):

$$\Re = \frac{(D\cos\theta_d)^2}{8\Delta} + \frac{D}{2}\sin\theta_d \tag{3.38b}$$

where

θ_d = direction angle from the antenna plane
Δ = flatness of the field, typically $\lambda/16$.

Technically, beyond the radiating near-field region is the *radiating far* field whose outer field is at infinity. Silver (1949) suggested the immediate limits of the radiating far-field region by

$$\Re \geq \frac{2D^3}{\lambda} \tag{3.39}$$

The field components in the far field are primarily transverse to the radial distance.

3.2 Target measurements

Range is the distance between the radar and the target. Assume a radar wanting to measure a target range, R. The radar transmits a pulse and measures the elapsed time t, for the target echo to be received. The elapsed time is measured by placing *range gates* along the receive time, as in Figure 3.8. Figure 3.8 is similar to that given in Skolnik (1980).

The consecutive range gates open for the duration of a single pulse τ, beginning with each transmitted 1st pulse. The presence of a signal in these gates corresponds to the elapsed time. From basic physics, an electromagnetic wave travels at the speed of light, c. Thus, the target range can be expressed as

$$R = \frac{1}{2}ct \tag{3.40a}$$

The maximum unambiguous range, R_{un}, is related to the interpulse period, T_s:

$$R_{un} = \frac{1}{2}cT_s \tag{3.40b}$$

Rather than timing the transmit–receive pulses, another type of pulse radar (called *pulse-Doppler* radar, which uses *pulse repetitive frequency*, PRF) applies the Doppler principle to estimate the target range. The Doppler principle relates to measuring the frequency shift, or difference between the transmitted frequency and the target-return frequency, Δf: this principle is explained as follows.

When the source of fluctuation or the observer of the fluctuation is in motion, a shift in frequency will occur. This effect is called the *Doppler effect*: it forms the basis of *continuous wave* (CW) radar. For example, consider the distance between a radar and target as R. The total number of wavelengths λ contained in a round trip between the radar and target

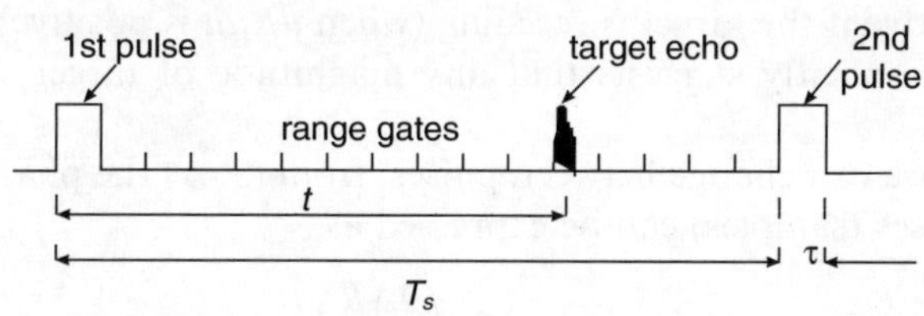

Figure 3.8 Estimating target range within pulses

would be $2R/\lambda$. Given that one λ equates to 2π (radians), the total angular excursion ϕ for the round trip will be

$$\phi = 2\pi\left(\frac{2R}{\lambda}\right) \tag{3.41}$$

R and ϕ will be continually changing if the target is in motion. From classical physics,

$$\frac{d\phi}{dt} = \omega = 2\pi\Delta f \tag{3.42}$$

Differentiating ϕ in (3.41) with respect to time, t, we have

$$\frac{d\phi}{dt} = \frac{4\pi}{\lambda}\frac{dR}{dt} = 2\frac{2\pi}{\lambda}\dot{R} \tag{3.43}$$

Equating (3.43) to (3.42):

$$\Delta f = \frac{2\dot{R}}{\lambda} \tag{3.44}$$

The range rate $\dot{R}$ is radial (i.e. relative to radar). To indicate the direction of measurement of the target, the range rate can be expressed by

$$\dot{R} = \pm\frac{\Delta f\lambda}{2} \tag{3.45}$$

The $\pm$ in the above equation specifies the target's direction relative to the radar. For instance, a negative sign implies an approaching target (decreasing range or negative range rate) while a positive sign indicates outbound direction (target moving away from the radar). The Doppler principle helps to separate echoes of non-moving targets from that of moving target echoes. It should be noted that cross-range velocity has no Doppler effect.

Doppler (frequency) shift is easily measured by mixing the original frequency f with f_d backscattered from a target with approaching radial velocity v_r. As such

$$f_d = -\frac{2}{\lambda}\left(\frac{dR}{dt}\right) = 2\frac{v_r}{\lambda} \tag{3.46}$$

This expression suggests that the Doppler shift f_d will be positive; that is, at a higher frequency if the target is approaching (when dR/dt is negative), or f_d will be negative if the target is receding (when dR/dt is positive). Expression (3.46) also implicitly suggests that any magnitude of target speed can be measured.

Since phase can change between pulses, from (3.41) the phase change $\Delta\phi$ between pulses (samples) can be expressed as

$$\Delta\phi = 2\pi\left(\frac{2\Delta R}{\lambda}\right) \tag{3.47}$$

where ΔR denotes the range change between pulses. From this expression, three certain conditions can be inferred. If:

- $\Delta\phi < 2\pi$, the Doppler frequency f_d can be unambiguously measured;
- $\Delta\phi = 2\pi$, the Doppler frequency f_d equals the PRF, i.e. $f_d = \text{PRF}$. Note that PRF is related to the time interval between pulse (PRI) in the form:

$$\text{PRF} = \frac{1}{\text{PRI}} \tag{3.48a}$$

  A shift of 2π is indistinguishable from a shift of any multiple of 2π, including zero. A moving target that moves at such a speed will appear stationary (non-moving). Its echoes will be cancelled along with echoes from fixed (stationary) targets. The speeds that cause Doppler shift to be an integral multiple of 2π are called *blind speeds*. This phenomenon is due to the presence of a large ground return at zero that frequently prevents the detection of the target of interest. More is said about *blind zones* in section 3.2.5; and finally
- $\Delta\phi > 2\pi$, the target will always be detectable but the observed Doppler frequency f_d will not correctly represent the target speed and will be incorrect by an integral multiple of PRF. Often, multiple PRFs are used to eliminate blind speed and to resolve ambiguous target speed measurements. In this instance, the observed Doppler frequency is more correctly represented by

$$f_d = \left(\frac{2v_r}{\lambda}\right) \text{modulo(PRF)} \tag{3.48b}$$

Example 3.2 For a 1 GHz base frequency, and a target radial speed of 20 knots, determine the Doppler shift.

Solution

$$v_r = 20\,\text{knots} \quad (\text{note that } 1\,\text{knot} = 1.852\,\text{km/hr})$$
$$f = 10^9\,\text{Hz} \quad c = 3 \times 10^8\,\text{m/s}$$

Using (3.46), the Doppler shift, f_d, is calculated as 68.59 Hz.

The above range measurement discussion has avoided the issue of noise and other losses and their effect on range estimation. From radar theory, the maximum range R beyond where a target cannot be seen can be calculated for low, medium and high PRFs, using the radar equation, this is discussed in Chapter 5.

Radars with a pulse repetition frequency (PRF) sufficiently low so that range is unambiguously measured are called *low*-PRF (LPRF). For example, the transmitted pulse travels to and from the range of maximum interest during the interpulse period before the transmission of the next pulse. LPRF radars do ambiguously measure Doppler shift. The unambiguous range, R_{un}, is calculated by using

$$R_{un} = \frac{c}{2\text{PRF}} \tag{3.48c}$$

Radars with a PRF sufficiently high so that all velocities (Doppler shifts) of interest are unambiguously measured are called *high*-PRF (HPRF). The maximum Doppler shift that can be unambiguously measured is given by (3.46), or rightly by (3.48b).

Radars that are ambiguous in both range and Doppler are called *medium*-PRF (MPRF). These radars appear to combine the worst of both LPRF and HPRF radars.

Pulse transmitters are peak-power-limited. When sampling is done at very short duration (in nanoseconds), it would be difficult to obtain high resolution in range. For this reason high range resolution is obtained from the received signal by a process called *pulse compression*, which reduces the response width and increases the signal-to-noise ratio (S/N) of uncompressed response to individual reflection points of the target. When two targets are closely spaced, it is often difficult to resolve them. The pulse compression technique enables closely spaced targets to be resolved from received signal.

3.2.1 Pulse compression

Radar waveforms are generally modulated in phase, or frequency, to increase the bandwidth of the transmitted pulse. The pulse may be repeated at short intervals to increase the signal duration. This pulse repetition method is conveniently used in practice to increase the signal duration without a proportionate decrease in the transmission bandwidth. In this way, shifting the carrier frequency from one pulse to another can increase the bandwidth. The enhanced signal bandwidth may be used by matched, or mismatched, filtering on receive to increase the range resolution of the radar system. This general *pulse compression* technique is frequently used in modern radar systems to simultaneously maintain a requirement range resolution while increasing average power on a single-pulse basis.

Pulse compression encompasses various signal-modulation and processing techniques utilized in radar systems, particularly in pulse Doppler radar, allowing the transmission of relatively long-duration waveforms while retaining the advantages inherent in high range resolution waveforms. Modulation is a signal processing technique.

The modulation process involves switching or keying the amplitude, frequency, or phase of the carrier in accordance with the information binary digits. There are three basic modulation schemes: amplitude shift keying (ASK), frequency shift keying (FSK), and phase shift keying (PSK). These schemes are, respectively, the binary equivalent of analogue transmission's amplitude modulation (AM), frequency modulation (FM) and pulse modulation (PM) when used to transmit data signals.

Consider an input signal $s(t)$ of duration T_s being represented by

$$s(t) = A_0 \cos(2\pi f_c t + \phi_n) \quad (n-1)T_s \leq t \leq nT_s \tag{3.49}$$

where A_0 and ϕ_n correspond to amplitude and nth phase of the signal.

For the ASK scheme, the signal's amplitude A_0 is varied while the phase, ϕ_n, and carrier frequency, f_c, remain constant. In the FSK, only the frequency f_c is varied with A_0 and ϕ_n remaining constant. In the case of PSK, A and f_c are kept constant with the ϕ_n varied. PSK, compared with the other schemes, has excellent protection against noise because the information is contained within its phase. Noise mainly affects the amplitude of the carrier. For radar modulation schemes, transmitted waveform is maintained at constant amplitude; thereby leaving modulation to either frequency or phase.

The range resolution achievable with a given radar system is given by (3.40). In a pulse compression system, the transmitted waveform is modulated in phase or frequency so that the bandwidth is allowed to be far greater than the reciprocal of transmitted pulse duration; that is, $B \gg 1/T_s$. This effectively allows an equation of the effective pulse length of the system after compression to that of the pulse width upon substitution in (3.40a); specifically

$$R = \frac{1}{2} c\tau \tag{3.50}$$

where $t = \tau$. This equation demonstrates that a pulse compression radar can use a transmit pulse of duration T_s and still achieve range resolution equivalent to that of a simple pulse system with a pulse of duration τ, where $T_s \gg \tau$.

If a pulse compression scheme is incorporated in a radar system of low peak power and long-duration pulse, it could be concluded that, by selecting an appropriate modulation scheme, the compressed (effective) pulse length of the resulting waveform would have a range resolution and detection performance of an equivalent short pulsed, high peak power system.

The ratio of T_s to τ is called the *pulse compression ratio* C_r; that is,

$$C_r = \frac{T_s}{\tau} \tag{3.51a}$$

Alternatively as a time-bandwidth product:

$$C_r = BT_s \tag{3.51b}$$

Often, pulse compression systems are characterized by their time-bandwidth products. This characterization will become obvious to the reader in the next section.

Of all pulse compression techniques, linear *frequency modulation* (FM) is the oldest and best developed. It is used to improve detection performance while maintaining range resolution. It is particularly useful for detection of moving targets, since it can provide broad Doppler coverage even with long-duration transmit waveform. The variety of pulse compression waveforms is too large for a comprehensive treatment in this book. However, a linear FM technique is used to investigate the output response of the pulse compression filter, which is treated next.

3.2.2 Pulse compression processing technique

Figure 3.9 illustrates a conceptual implementation technique of a radar system with a pulse compression processor. Consider a chirp signal from an RF generator of width τ. A chirp signal owes its importance to the fact that constant-amplitude waveforms place the least requirements on radar transmitters. This chirp signal is passed through a dispersive-delay block of pulse duration T_s, which is assumed to be far greater than the chirp's width from the generator. The signal from this dispersive-delay block is amplified and transmitted through the directional switch to the antenna, if one assumes that the signal received is properly processed with a minimum loss, and is further amplified and passed through the pulse compression filter. For brevity, the effective bandwidth of this filter is matched to the transmitted waveform, hence, having an effective bandwidth of $1/\tau$ (Hz). More is said about the concept of 'matched filtering' in Chapter 10, section 10.3.

The resulting waveform of the matched filter is a compressed pulse of effective duration of τ (sec), with a time extent of the order of $2T_s$. With this, the system would be capable of resolving targets separated in range by at least $c\tau/2$ (m). The filter output has some sidelobes, called *range sidelobes*, at $|t| < \tau$ (sec). These sidelobes must be controlled because in a given range bin, they may appear in adjacent range bins as signals. The compression concept,

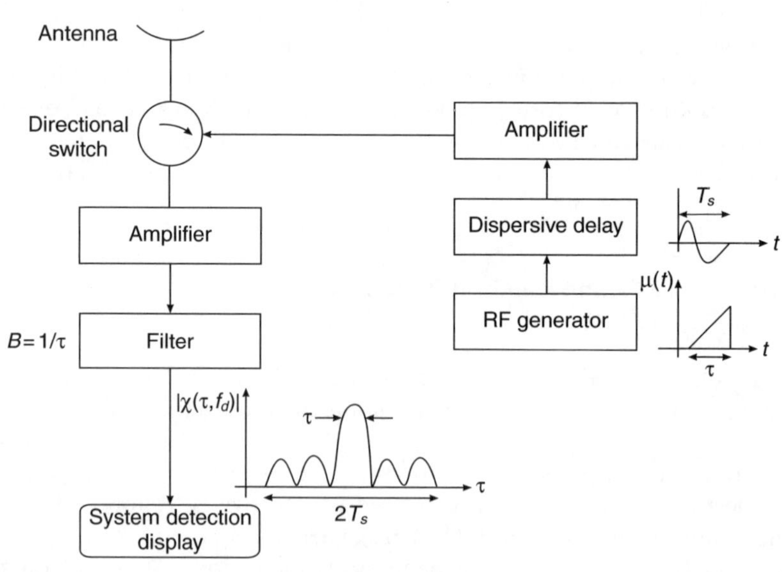

Figure 3.9 A block diagram of a pulse compression processing scheme in a radar system

described above, has been developed mostly on an intuitive basis. To complement this intuitive approach, a physical basis containing basic mathematical development is provided next. Those seeking only a qualitative understanding may skip the next two subsections without loss of continuity.

If one assumes that the chirp signal has a constant-amplitude envelope in the time domain, but in the frequency domain the rectangular envelope is only approximated, the signal generated would be by carrier frequency modulation of a constant-amplitude pulse. Thus, it can conveniently be expressed that the normalized form of the chirp signal is in the form

$$\mu(t) = \frac{1}{\sqrt{T_s}} \text{rect}\left(\frac{\tau}{T_s}\right) e^{j\pi b t^2} \tag{3.52}$$

The duality between time and frequency produces an analogous result in frequency; a form like (3.52) can be written as

$$\mu(t) = \frac{1}{\sqrt{B}} \int_{-B/2}^{B/2} e^{j2\pi f t} e^{-j\pi b f^2} df \tag{3.53}$$

where b is a factor that determines the slope of the chirp signal, measured in radian/sec^2. For example, if an instantaneous frequency is swept over the band B during the signal duration T_s, the absolute value of k is given as

$$|b| = \frac{B}{T_s} \tag{3.54}$$

The matched filter response can be obtained by correlating a signal with its Doppler-shifted and time-translated version: a function that describes the interplay between measurement ambiguity and target resolution. If the range bin width were considered the same as nominal range resolution, the matched filter response would determine the shape and size of a resolution cell. If $\chi(\tau, f_d)$ represents a two-dimensional correlation function in delay, τ and Doppler shift, f_d and be defined by

$$\chi(\tau, f_d) = \int_{-\infty}^{\infty} \mu(t)\mu^*(t-\tau)e^{j2\pi f_d t} dt \tag{3.55}$$

where $\mu^*(t)$ is the complex conjugate of $\mu(t)$. By this definition the response of the processed chirp can be investigated. The function of actual interest is the real envelope of the response, which is simply $|\chi(\tau, f_d)|$. In the literature, $\chi(\tau, f_d)$ has been called by different names, such as an *uncertainty function*, a *correlation function*, or an *ambiguity function*. In this book, it is simply called the pulse compression filter function having a response $|\chi(\tau, f_d)|$. (See Appendix 3A for the derivation of the ambiguity function of a chirp signal using the non-normalized approach.)

By changing the signs of τ and f_d in (3.55), one observes that

$$\chi(-\tau, -f_d) = \chi^*(\tau, f_d)e^{j2\pi f_d \tau} \tag{3.56}$$

As earlier indicated, the quantity of interest is the envelope of the function. Since

$$|\chi(\tau, f_d)| = |\chi^*(\tau, f_d)|, \text{ by symmetry}$$

$$|\chi(-\tau, -f_d)| = |\chi(\tau, f_d)| \tag{3.57}$$

So, upon substitution of (3.52) in (3.55), the two-dimensional pulse compression filter function is written as

$$\chi(\tau, f_d) = \frac{1}{\sqrt{T_s}} \int_{-\infty}^{\infty} \text{rect}\left[\frac{t-\tau}{T_s}\right] e^{j\pi b\left[t^2-(t-\tau)^2\right]} e^{j2\pi f_d t} dt \tag{3.58}$$

For positive τ, this expression becomes

$$\chi(\tau, f_d) = \frac{1}{T_s} e^{j\pi b\tau^2} \int_{\tau-\frac{T_s}{2}}^{\frac{T_s}{2}} e^{j2\pi t(b\tau+f_d)} dt \tag{3.59}$$

By factoring out the exponent term $e^{j\pi(b\tau+f_d)\tau}$, the integral can be readily solved as

$$\chi(\tau, f_d) = \left(1 - \frac{\tau}{T_s}\right) \frac{\sin \pi T_s\left(1 - \frac{\tau}{T_s}\right)(b\tau + f_d)}{\pi T_s\left(1 - \frac{\tau}{T_s}\right)(b\tau + f_d)} e^{j\pi f_d \tau} \quad 0 \le \tau \le T_s \tag{3.60a}$$

Or

$$\chi(\tau, f_d) = e^{j\pi f_d \tau}\left(1 - \frac{\tau}{T_s}\right) \sin c\left[\pi T_s\left(1 - \frac{\tau}{T_s}\right)(b\tau + f_d)\right] \quad 0 \le \tau \le T_s \tag{3.60b}$$

Similarly, an expression can be written for negative τ. In this case, the limits of integration will change to $(\tau + T_s/2, -T_s/2)$. Although the expression in (3.58) can be solved with the new limits, instead the relation in (3.57) is used with some minor modification, specifically

$$\chi(\tau, f_d) = \begin{cases} e^{j\pi f_d \tau}\left(1 - \frac{|\tau|}{T_s}\right) \sin c\left[\pi T_s\left(1 - \frac{|\tau|}{T_s}\right)(b\tau + f_d)\right] & |\tau| \le T_s \\ 0 & |\tau| > T_s \end{cases} \tag{3.61}$$

Alternatively, in view of (3.54),

$$\chi(\tau, f_d) = \begin{cases} e^{j\pi f_d \tau}\left(1 - \frac{|\tau|}{T_s}\right) \sin c\left[\pi B T_s\left(1 - \frac{|\tau|}{T_s}\right)\left(\frac{\tau}{T_s} + \frac{f_d}{B}\right)\right] & |\tau| \le T_s \\ 0 & |\tau| > T_s \end{cases} \tag{3.62}$$

The pulse compression filter's response is simply the amplitude (magnitude) of (3.62):

$$|\chi(\tau, f_d)| = \begin{cases} \left|\left(1 - \frac{|\tau|}{T_s}\right) \sin c\left[\pi B T_s\left(1 - \frac{|\tau|}{T_s}\right)\left(\frac{\tau}{T_s} + \frac{f_d}{B}\right)\right]\right| & |\tau| \le T_s \\ 0 & |\tau| > T_s \end{cases} \tag{3.63}$$

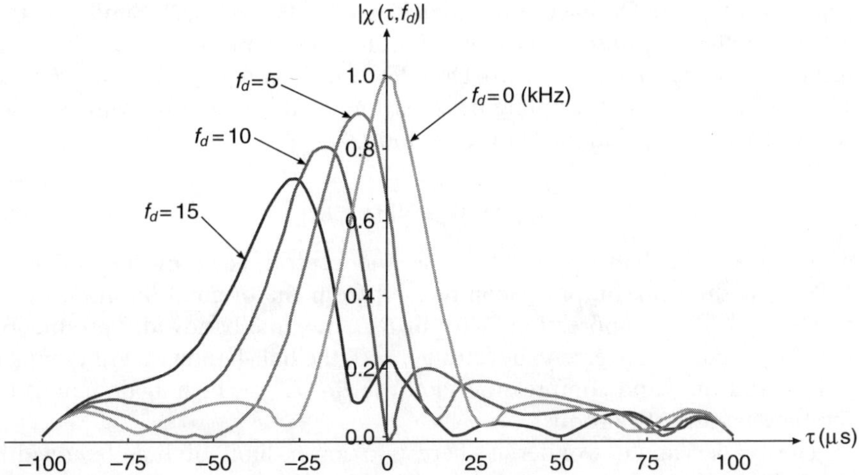

Figure 3.10 Pulse compression filter response

This expression shows the time-bandwidth (BT_s) product of the chirp signal. Note also that this time-bandwidth product is the pulse compression ratio. A plot of (3.63) gives the waveform of the matched filter, shown in figure 3.10, for a 50 kHz bandwidth and time extent of 0.1 ms. As seen in the figure, sidelobes or subsidiary ridges are of diminishing amplitudes (magnitudes) and surround the mainlobe, main ridge, of the filter's response, which is consistent with $|\sin(x)/x|$, or $|\sin c(\mathrm{x})|$, profile. The compressed pulse is of effective duration of τ with time extent of the order of $2T_s$, which is consistent with the intuitive description given earlier in this section of the filter output.

A quick look at (3.63) reveals that the term $(1 - |\tau|/T_s)$ only attempts to slowly decrease the amplitude as one moves away from the Doppler axis, f_d. In fact, its effects near the origin, both proceeding and within the sin c function, are negligible. This explains the effect of the finite signal duration and the incomplete overlap between $\mathrm{rect}(\tau/T_s)$ and $\mathrm{rect}(t - \tau/(T_s)$.

In essence, the relative delay τ shortens the effective signal duration from T_s to $(T_s - |\tau|)$.

Neglecting the $(1 - |\tau|/T_s)$ term in (3.63), simply turn

$$|\chi(\tau, f_d)|_0 = \sin c\left[\pi BT_s\left(\frac{\tau}{T_s} + \frac{f_d}{B}\right)\right] \tag{3.64}$$

which displays a symmetrical property in τ and f_d. It is well known from the sin c property (that is, $\lim_{x\to 0} \sin(x)/x = 1$) that the peak of (3.64) occurs when

$$\frac{\tau}{T_s} + \frac{f_d}{B} = 0 \tag{3.65}$$

It can be inferred from (3.64) and (3.65), without loss of generality, that the matched-filter response in delay τ for a Doppler mismatch f_{d0} will be the response for zero Doppler translated in τ by $\tau_0 = -f_d T_s/B$. Similarly, the matched-filter response in Doppler f_d for a delay mismatch τ_0 will be a response on the f_d-axis for zero Doppler translated in f_d by $f_{d0} = -B\tau/T_s$. This demonstrates the coupling between range and range rate, which is the equivalence of translations between τ and f_d.

Putting $f_d = 0$ in (3.64),

$$|\chi(\tau, f_d)|_0 = \sin c[\pi B\tau] \tag{3.66}$$

which gives the half-power width of the central peak of the order of $1/B$. Thus the peak output is compressed from the original duration of T_s to $1/B$, which is a compression factor of BT_s, the time-bandwidth product of the chirp signal. Conversely, by letting $\tau = 0$, the half-bandwidth in Doppler is $1/T_s$ and the band compression factor is $B/(1/T_s)$, which again equals to the time-bandwidth product.

The reader might wonder if there is a lower limit of time-bandwidth product. Gabor (1946) gave this lower limit as

$$BT_s \geq \pi \tag{3.67}$$

The exact value of the time-bandwidth product is of no particular interest, as it depends on the definition of the signal duration and bandwidth. As noted by Rihaczek (1969), the important point is that the time-bandwidth product of a signal has a minimum value of the order of unity.

3.2.3 Repetition of pulsed signals

A way of generating signals with large time-bandwidth products is to repeat the input waveform. Signal repetition can be contiguous, or gaps can be left between pulses called *pulse trains* or *pulse bursts*. Like (3.52), let us allow a signal with a complex envelope $\mu_c(t)$ to be repeated coherently so that its carrier phase remains continuous from one segment to the next. Following (3.55), the signal's ambiguity function may be written as

$$\chi(\tau, f_d) = \frac{1}{N}\sum_{n=0}^{N-1}\sum_{m=0}^{N-1}\int_{-\infty}^{\infty} \mu_c(t - nT)\mu_c^*(t - mT - \tau)e^{j2\pi f_d t}dt \tag{3.68}$$

Note that T in this case is the repetition period (see Figure 3.11).

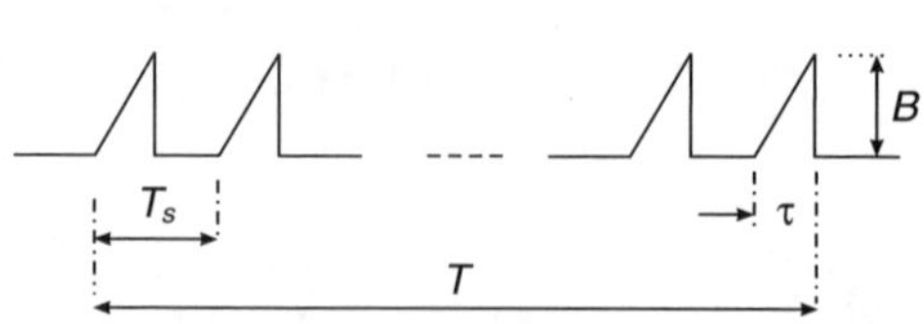

Figure 3.11 Pulse train

The expression in (3.68) can be rewritten as

$$\begin{aligned}\chi(\tau,f_d) &= \frac{1}{N}\sum_{n=0}^{N-1}\sum_{m=0}^{N-1} e^{j2\pi f_d nT}\int_{-\infty}^{\infty}\mu_c(t)\mu_c^*(t-[m-n]T-\tau)e^{j2\pi f_d t}dt \\ &= \frac{1}{N}\sum_{n=0}^{N-1}\sum_{m=0}^{N-1} e^{j2\pi f_d nT}\chi_c[\tau-(n-m)T,f_d]\end{aligned} \tag{3.69}$$

where $\chi_c(\tau, f_d) = \int_{-\infty}^{\infty}\mu_c(t)\mu_c^*(t-\tau)e^{j2\pi f_d t}dt$ is the ambiguity function of the component signal. Without further complication, it can be shown that (3.69) follows the same law as the autocorrelation function of a train of N signals. So, a solution to (3.69) is written as

$$\chi(\tau,f_d) = \frac{1}{N}\sum_{p=-(N-1)}^{N-1}\chi_c(\tau-pT,f_d)\frac{\sin[\pi f_d(N-|p|)T]}{\sin \pi f_d T}e^{j\pi f_d(N-1+p)T} \tag{3.70}$$

where its envelope is the sum of the envelopes of the individual parts. So, the overall magnitude of the pulse-train ambiguity function is

$$|\chi(\tau,f_d)| = \frac{1}{N}\sum_{p=-(N-1)}^{N-1}|\chi_c(\tau-pT,f_d)|\left|\frac{\sin[\pi f_d(N-|p|)T]}{\sin \pi f_d T}\right| \tag{3.71}$$

Hence, the gross structure is determined by the repetition of the ambiguity surface of the component pulse $|\chi_c(\tau, f_d)|$ with its magnitude decreasing by $(1-|p|/N)$. By assuming that the ambiguity function of an individual pulse has a similar simple shape, then, from (3.71), one can deduce that

- the highest peak occurs when the sine term has a value of one;
- a dependence of the mainlobe at each p surface as $(1-|p|/N)$;
- any sampling in the Doppler domain occurs in accordance with $\sin[\pi f_d(N-|p|)T]/\sin \pi f_d T$, which would have peaks at $f_d = k/T$, where k is an integer and with its ambiguity spaced out at $1/T$ being the repetition frequency;
- the half-power (−3 dB) width in Doppler is of the order of $1/NT$, the inverse duration of the pulse train.

A plot of (3.71) of the uniform pulse train gives the waveform of the matched filter shown in Figure 3.12 for 1 ms period, $N = 5$, and 50 kHz bandwidth.

One can observe, from Figures 3.11 and 3.12, that pulse repetition does not affect close-target resolvability in range, which is the same for a single pulse and a train of pulses. Close-target resolvability in range rate is improved with pulse repetition because the sampling in the Doppler domain narrows the mainlobe width in Doppler. A practical implication of using pulse repetition is that periodic signal repetition increases the time-bandwidth product at the expense of introducing pronounced range ambiguities in delay and Doppler.

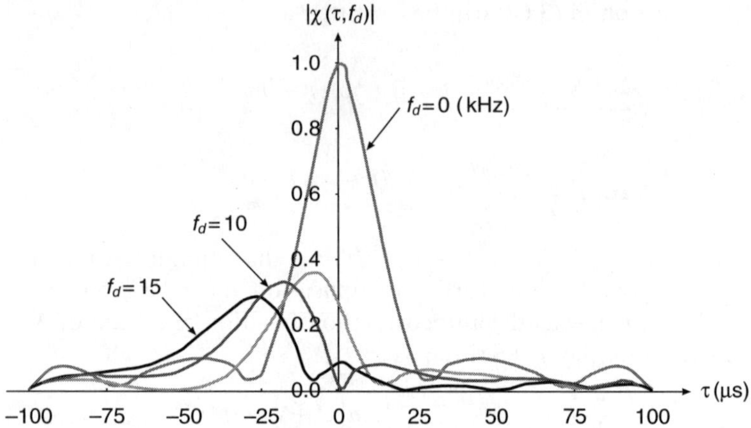

Figure 3.12 Pulse-train compression filter response, $TB = 50$, $B = 50$ kHz

3.2.4 Sidelobes suppression

The pulse compression responses for a single pulse and a train of pulses have shown that sidelobes are well pronounced in range (delay) domain. Sidelobes from any range bin are likely to appear as targets in adjacent bins. The response sidelobes may introduce significant interference even if there are relatively few targets. Also, radiation from the 'hot' ground may enter the antenna by means of the sidelobes. Consequently, the suppression of sidelobes is critical in applications expecting high target densities, extended clutter, or targets of varying reflectivity.

Sidelobes are often suppressed to an acceptable level by tapering the matched filter by weighting the transmitted waveform, the matched filter, or both in either frequency or amplitude. To simultaneously apply weighting at both the frequency and amplitude without loss of signal-to-noise ratio (S/N) is rather difficult in practice.

If Doppler spread of the targets is negligible, spectrum weighting (i.e. weighting applied only to the matched filter) suppresses the range sidelobes and hence the interference, at the cost of a small broadening of the response mainlobe. A similar advantage might be gained for more complicated target distributions. Note that spectrum weighting is the same as if a tapered spectrum has been transmitted, and true target distribution is obtained only if the range rate is constant over the entire extrapolation interval.

To suppress the Doppler sidelobes, it may be convenient to use a reference function with tapered amplitude in the correlation process, rather than to transmit the amplitude-weighted signal.

In theory, complete suppression is achievable only with signals of infinite extent in time and frequency. There are several types of spectral weighting functions, namely Dolph–Chebyschev, Taylor, Hamming, and Blackman–Harris. These weighting functions have been discussed in Chapter 1,

Table 3.1 Weighting function data (Nathanson 1969)

Weighting function	Peak sidelobe level (dB)	Pulse widening	Mismatch loss (dB)
Dolph–Chebyschev	−40.0	1.35	—
Taylor ($N = 6$)	−40.0	1.41	−1.2
Hamming	−42.8	1.47	−1.34

section 1.33. Table 3.1 shows comparative values of spectral weighting functions for a linear frequency-modulated signal with a rectangular spectrum. The Dolph–Chebyschev weighting is theoretical, with all sides equal. However, a practical approximation to the Dolph–Chebyschev is the Taylor weighting, with the number of terms, $N = 6$, meaning that the peaks of the first five sidelobes, equivalent to ($N - 1$), are equal; the sides fall off at 6 dB per octave. Weighting the received-signal spectrum to lower the sidelobes increases the mainlobe width, but reduces the peak (S/N) in comparison to the unweighted pulse compressed spectrum. If the weighting is not matched with the received-signal spectrum, a mismatch loss occurs, as shown in column 4 of Table 3.1. For example, take the case of the Hamming, reducing the sidelobes to a level of −42.8 dB of weighting results in loss in peak of 1.34 dB.

For a treatment of specific types of weighting functions, as well as some ancillary topics on sidelobe suppression, the reader is referred to Cook and Bernfield (1967).

3.2.5 Resolution

The ambiguity response, or surface $|\chi(\tau, f_d)|$, plays a central part in the analysis of resolution as well as estimating the limiting values of measurement precision. Target resolution can be analysed from the superposition of the ambiguity surfaces associated with all targets within the radar beam. Each ambiguity surface is scaled in height in accordance with the target cross-section, and it is centred at the proper delay and Doppler coordinates. As discussed by Siebert (1956) and Woodward (1953), the total volume under the ambiguity surface is invariant to, or independent of, the choice of signal. Specifically,

$$\int_{-\infty}^{\infty}\int_{-\infty}^{\infty} |\chi(\tau, f_d)|^2 d\tau df_d = |\chi(0,0)|^2 = 1 \tag{3.72}$$

This expression means that there are limits on achievable resolution performance in range and range rate. The practical implication of this expression is profound. For example, if one wants to separate closely spaced targets and for this reason chooses a waveform having an ambiguity function with a narrow mainlobe, the bulk of the fixed volume under $|\chi(\tau, f_d)|^2$ will appear elsewhere in the $\tau - f_d$ plane. There, it might introduce self-clutter, which might mask targets that are relatively far removed in range and range rate Rihaczek (1969).

Of course, the Doppler filtering offered by a DFT of N pulses could be used to separate moving targets from zero-Doppler clutter.

The width of the mainlobe of $|\chi(\tau, f_d)|$ is a measure for close-target separability, or nominal resolution, in τ and f_d, while the sidelobes and other low-level parts of the surface give an indication of the problem of self-clutter and target masking by mutual interference. Resolution in range domain corresponds to resolution in the time (range-delay) domain. For example, consider the two equal target-sin c responses shown in Figure 3.13. Each response has a bandwidth B. $\Delta\tau$ is the separation time, where the peak of one response falls directly over the null of the second. The dotted segment over which the separation time $\Delta\tau$ is the sum of the two responses.

In practice, closely spaced targets or target scatterers will appear to merge and separate as the range separation changes on the order of $\lambda/2$. This half-wavelength criterion is extremely useful in estimating the resolution capability of radar. The usefulness of this criterion is demonstrated as follows.

Consider two targets with the same range but a differential range rate of $\Delta\dot{R}$, the differential changes the differential range by $T\Delta\dot{R}$, where T is the signal duration. (Note that $T = T_s$ for a single-pulse transmission.) By setting the differential range to half-wavelength, that is $T\Delta\dot{R} = \lambda/2$, the limiting close-target resolvability, or nominal range-rate resolution, can be expressed by

$$\Delta\dot{R}_{\min} = \frac{\lambda}{2T} \tag{3.73}$$

In similar vein, if the two targets move with differential range acceleration $\Delta\ddot{R}$, the range change during duration T will be $1/2T^2\Delta\ddot{R}$. If the range change is equated to half-wavelength, then

$$\frac{1}{2}T^2\Delta\ddot{R} = \frac{\lambda}{2} \tag{3.74}$$

From this expression, the limiting close-target resolvability, or nominal resolution in range acceleration, is

$$\Delta\ddot{R}_{\min} = \frac{\lambda}{T^2} \tag{3.75}$$

For a range measurement on a stationary target, the resolution is the width in time of the mainlobe of the matched-filter response:

$$\Delta\tau = \frac{1}{B} \tag{3.76}$$

A target is considered stationary if its motion is negligible over the signal duration. Similarly, the radar deals with a constant-range-rate target not if the range rate is necessarily constant but if the effects of range acceleration are negligible over the signal duration. Any target that cannot be resolved by the radar, be it in range, range rate, or another parameter, is considered a point target (Rihaczek 1969).

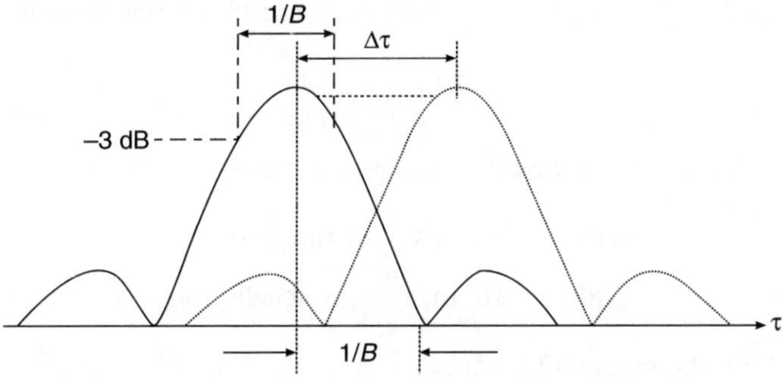

Figure 3.13 Two targets of equal matched-filter responses resolved to Rayleigh criteria

The bin width in Doppler is simply the half-power width of the matched-filter response in Doppler:

$$\Delta f_d = \frac{1}{T} \tag{3.77}$$

Angular resolution involves knowledge of radar beamwidth and how the radar aperture, D, is illuminated. The nominal angular resolution is given by

$$\Delta\phi = \frac{\lambda}{D} \quad D \gg \lambda \tag{3.78}$$

This expression is often called the *Rayleigh resolution*.

More fundamentally, equations (3.75) and (3.76) refer to the Rayleigh resolution for signals that are *windowed* by rectangular *weighting* functions of spectral width B and time duration T, respectively. Windowing and weighting are synonymous in digital signal processing: filter weighting is called windowing, already discussed in Chapter 1. Equations (3.76) and (3.77) are roughly correct for any well-matched, moderately weighted signals.

Example 3.3 A pulse width of 1 μs is to be transmitted. Two moving targets of similar range are to be resolved. If the transmitter's *duty cycle* is 0.1, at 0.1 m wavelength, estimate the nominal resolutions in range rate and in range acceleration. If the horizontal and vertical dimensions of the antenna aperture are 3 m and 0.5 m respectively, calculate the azimuth and elevation beamwidths of the antenna.

Solution

Duty cycle d_u is the ratio of the pulse width to the transmitting period, or the product of the pulse width and pulse repetition frequency (PRF); that is

$$\begin{aligned} d_u &= \frac{\tau}{T_s} \\ &= \tau PRF \end{aligned} \tag{3.79}$$

Given that $T = T_s = 1\,\mu s$ and $\lambda = 0.1\,m$, using (3.79) the duty cycle yields

$$d_u = \frac{\tau}{T} = 0.1$$

Using (3.73) and (3.75) the nominal resolutions are found as

$$\Delta\dot{R}_{\min} = 5\,\text{km/s} \quad \text{(in range rate)}$$

$$\Delta\ddot{R}_{\min} = 10^9\,\text{m/s}^2 \quad \text{(in acceleration)}$$

Using (3.78), the beamwidths are:

$$\Delta\phi = \frac{0.1}{3} = 0.033\,\text{rad} \quad (1.91^\circ) \quad \text{(azimuth)}$$

$$\Delta\phi = \frac{0.1}{0.5} = 0.2\,\text{rad} \quad (11.5^\circ) \quad \text{(elevation)}$$

3.2.6 Measurement accuracy for stationary and moving targets

Radar performance on a stationary target depends on the signal bandwidth. With Rice (1944) and Rihaczek (1969), and upon assumption of Gaussian noise, radar performance precision in the following domains is obtained:

$$\text{In range:} \quad \sigma_R = \frac{c}{2B\sqrt{\frac{S}{N}}} \quad \text{(m)} \tag{3.80}$$

$$\text{In Doppler:} \quad \sigma_{\dot{R}} = \frac{\lambda}{2T_s\sqrt{\frac{S}{N}}} \quad \text{(m/s)} \tag{3.81}$$

$$\text{In angle:} \quad \sigma_\theta = \frac{\lambda}{2D\sqrt{\frac{S}{N}}} \quad \text{(rad)} \tag{3.82}$$

where

σ_i = standard deviation of the variable i of interest
c = speed of light ($3 \times 10^8\,\text{m/s}^2$)
(S/N) = radar signal-to-noise ratio (linear unit).

However, when there is a coupling between range and range rate, for example when the target is moving and/or manoeuvring, the limiting values of measurement precision can be expressed as:

$$\text{In range:} \quad \sigma_R = \frac{c}{2B\sqrt{\frac{S}{N}}}\frac{1}{\sqrt{1-\left(\frac{\alpha_m}{BT_s}\right)^2}} \quad \text{(m)} \tag{3.83}$$

$$\text{In Doppler:} \quad \sigma_{\dot{R}} = \frac{\lambda}{2T_s\sqrt{\frac{S}{N}}} \frac{1}{\sqrt{1 - \left(\frac{\alpha_m}{BT_s}\right)^2}} \quad \text{(m/s)} \tag{3.84}$$

$$\text{In angle:} \quad \sigma_\theta = \frac{2\lambda}{\pi D\sqrt{\frac{S}{N}}} \quad \text{(rad)} \tag{3.85}$$

where α_m is the signal modulation factor

3.2.7 Effects of pulse compression on Doppler radars

Pulse compression for low- and medium-pulsed Doppler radars is subject to code sensitivity when there is a Doppler shift across the range bins. As such, the compression must be preceded by some attempt to compensate for the Doppler shift in order to minimize this effect. A constant Doppler shift produces an unwanted linear phase progression over the code length. A compensation scheme consists of rotating (or *derotating* as it is sometimes called) each complex range (*in-phase*, *quadrature* I/Q pair), by linearly changing phase angle of a range sweep. In airborne-based radar, the *derotating* rate in the range cell is calculated using Morris (1988)

$$\frac{d\phi}{dr} = 360\tau\frac{k_p V_a}{\lambda} \quad \text{(degree/range-cell)} \tag{3.86}$$

where

V_a = radar carrying platform's velocity (m/s)
k_p = radar platform dependent factor: typically $1.0 \leq k_p \leq 1.5$.

In the frequency domain, however, there is a Doppler ambiguity folding analogous to range ambiguity folding in the time domain. The maximum unambiguous Doppler, $f_{d\max}$, is

$$f_{d\max} = \text{PRF} \tag{3.87}$$

which corresponds to a maximum unambiguous relative target velocity ($V_a + V_{r\max}$). In view of (3.46), the maximum unambiguous relative target velocity is given by

$$V_a + V_{r\max} = \frac{\lambda}{2}\text{PRF} \tag{3.88}$$

Example 3.4 To have a feel for this compensation process, suppose that $\lambda = 30\,\text{cm}$, $\tau = 1\,\mu\text{s}$, $V_a = 500\,\text{m/s}$, and PRF = 10 kHz. If the mean value of k_p is taken, i.e. $k_p = 1.25$, calculate (i) the compensation velocity, (ii) the derotation rate required and (iii) the maximum unambiguous target velocity.

Solution

(i) $V_{\text{comp}} = k_p V_a = 650\,(\text{m/s})$
(ii) derotation rate, $d\phi/dr = 7.5°/\text{range-cell}$
This implies that each range bin would be rotated (or derotated) 7.5 ° more than the previous range bin. The compensation velocity and rotation process is often performed after the null pulse during data processing.
(iii) $V_t = V_a + V_{r\max} = 1.5\,\text{km/s}$.

Modern radars can detect the Doppler shift of N consecutively returning pulses as well as their potentially ambiguous range. For example, the radar assesses Doppler shift by collecting one complex sample; that is, each sample from the I and Q channels of the receiver from each of N received pulses. The radar in turn uses the N consecutive samples to form the complex fast Fourier transform (FFT). Of course, the sampling rate is PRF, or the inverse of the time interval between pulses. Amplitude detection (that is, magnitude of the I/Q phasor) is frequently used to determine the presence of a target in low-PRF search.

The use of pulse compression in the high-PRF mode of modern pulsed Doppler radar systems is obviated by duty cycle constraints and the high average powers developed. As a result, pulse compression increases the range blind zones. Range blind zones are zones where target returns cannot be received when transmitting. In practice, the receiver is off for one or two extra range gate positions after the transmitter pulse. A simple rule of thumb is used to detect the possible occurrence of the range-blind zones in a particular radar transmission. The maximum fraction d_r of the interpulse interval available for target reception may be expressed by

$$d_r = 1 - d_u \tag{3.89}$$

where d_u is the transmitter duty cycle, as defined by (3.79). The maximum fraction is also called the *clear region duty cycle*.

For example, consider a pulse width of 1 μs and PRF of 10 kHz. $d_u = 0.1$ and $d_r = 0.99$. Clearly, with this example, blind zones are not a major consideration. However, if a transmitter pulse τ of say 13 μs has been compressed to an effective pulse width of 1 μs, the maximum fraction d_r becomes 0.87. Range blind zones, in this case, are a major concern.

3.3 Summary

This chapter has looked at the antenna physics: using a simple dipole, or doublet, to formulate expressions that generate its radiation patterns. The influence of ground termination on vertical monopole antennas was also discussed.

Radiation fields were categorized into regions: reactive near-field, radiating near-field and radiating far-field, relative to the radiation source.

The principle of pulse compression which allows recognition of closely spaced targets was studied. Since the matched-filter response is of sin c shape, slowly decreasing sidelobes are present. A suppression technique that reduces sidelobes was discussed. The chapter further studied combined resolution, or close-target resolvability, in range and range rate in terms of the complete matched-filter response in delay and Doppler. The analysis presupposes resolution potential inherent in the radar.

Appendix 3A Ambiguity function of a chirp pulse

We consider a linearly swept frequency modulation pulse (chirp). The frequency is allowed to increase or decrease linearly over the pulse duration, T, so that the time phase changes quadratically. So, with the amplitude constant over T, we write

$$\mu(t) = \begin{matrix} e^{jbt^2}, & 0 < t < T \\ 0, & \text{elsewhere} \end{matrix} \tag{A3.1}$$

The ambiguity function of (A3.1) is

$$\chi(\tau, f_d) = \int_\tau^T \mu(t)\mu^*(t-\tau)e^{j2\pi f_d t}\,dt \tag{A3.2}$$

which, of course, is the combined correlation function of the pulse, where $\mu^*(t)$ is the complex conjugate of $\mu(t)$. Expanding (A3.2) and collecting terms, we have

$$\begin{aligned} \chi(\tau, f_d) &= \int_\tau^T e^{j\left[-bt^2+b(t-\tau)^2+2\pi f_d t\right]}dt \\ &= e^{jb\tau^2}\int_\tau^T e^{-2j[b\tau-\pi f_d t]}dt \\ &= \frac{e^{jb\tau^2}}{2j[b\tau-\pi f_d t]}\left\{e^{-2j[b\tau-\pi f_d t]T} - e^{-2j[b\tau-\pi f_d t]\tau}\right\} \end{aligned} \tag{A3.3}$$

The exponential terms in the curly bracket {.} can be expressed in trigonometric terms; that is, $\cos(.) + j\sin(.)$. The magnitude of the solution to (A3.3) is the magnitude of the ambiguity function of the chirp pulse. After some algebraic manipulation, we obtain

$$|\chi(\tau, f_d)| = \left|\frac{\sin(b\tau - \pi f_d)\left(\frac{1-\tau}{T}\right)T}{b\tau - \pi f_d}\right| \quad 0 < \tau < T \tag{A3.4}$$

The same result is obtained for $-T < \tau < 0$. So,

$$|\chi(\tau, f_d)| = \left| \frac{\sin(b\tau - \pi f_d)\left(\frac{1-|\tau|}{T}\right)T}{b\tau - \pi f_d} \right| \quad |\tau| < T \tag{A3.5}$$

Problems

1. Estimate the radar range Rayleigh resolution provided by a monotone pulse if its spectrum can be approximated by a rectangular spectrum of 250 kHz width.
2. A linear FM signal is expressed in terms of an arbitrary real envelope $a(t)$ as $\mu(t) = a(t)e^{j\pi kt^2}$, calculate its spectrum.
3. Since the chirp signal measures only extrapolated range, can targets that have the same extrapolated range be resolved?
4. The coupling between range and range rate for a chirp signal causes a loss in resolvability for targets that, at the instant of signal reflection, have certain combinations of range and range rate. Can targets whose differential range and range rates falling in the mainlobe of the ambiguity function be resolved from each other? Under what conditions are they resolvable?
5. A police radar operating at 10 GHz observes a Doppler shift of 1.75 kHz when the radar is pointed at an oncoming car. Estimate the radial speed of the car towards the radar.
6. A radar operates in a multiple-target environment. We observe that the interfering targets are moving at a velocity such that there is a minimum Doppler shift of f_0. Our objective is to have a signal such that $\chi(\tau, f_d) \cong 0$, $|f_d| > f_0$. Design a transmitting signal that could accomplish this task.
7. Show that the radiation resistance of a dipole whose length is $3\lambda/2$ with a sinusoidal current distribution is equal to $105.3\,\Omega$.
8. A half-wavelength dipole radiates a time-averaged power of 159.75 W in free space at a frequency of 25 MHz. Find the electric and magnetic field strengths when viewed at a radial distance 250 m from the source, elevation and azimuth angles of 85° and 30° respectively. The observation point is considered to be in the far-field region.

4

Antenna arrays

The properties of a single radiator have been discussed in Chapter 3. While it is possible to build an antenna in somewhat physically required constraints and make it look like an antenna of a different shape, it is difficult to achieve uniform current distribution and radiation pattern as well as steering the antenna in any preferred directions with such an arrangement. But, if one uses a group of similar radiators, or antennas, to produce more than a single radiating source, it is possible to obtain an antenna that has a higher gain and a radiation pattern that can be steered in any preferred directions. The collection of radiators, or antennas, is generally referred to as an array. The fields radiated from the individual antennas composing the array can add in some preferred direction and cancel in other directions.

The aim of this chapter is to discuss the basic theory of a linear array and show how the antenna array can be steered in a preferred direction, shape its radiation pattern and even feed its elements parasitically. The chapter also examines the role power and time budgets play in moulding antenna design processes.

4.1 Planar array

The antenna array radiation pattern can be derived from basic relations by considering the propagation of electric field from a set of radiating elements. The works of Schekunoff (1943) and Stratton (1941) provide the background material applicable to the linear array discussed in this section. Consider N radiating elements, equally spaced a distance d apart, each element radiating equal amplitude a_0, but with a phase progression difference between adjacent elements, as shown in Figure 4.1. The phase difference in adjacent elements may be expressed by

$$\Delta\phi = 2\pi \frac{d}{\lambda} \sin\theta \tag{4.1}$$

where θ is the look angle; that is, angle taken of the incoming wave. To accommodate for the progressive beam scanning effect, a phase progression

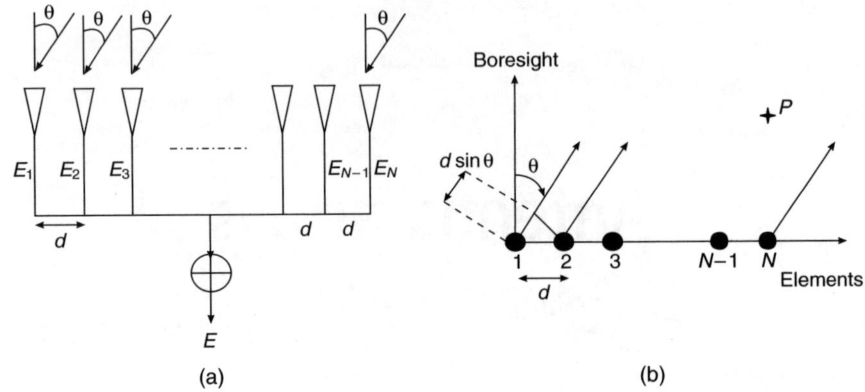

Figure 4.1 Geometry of a linear array antenna: (a) a linear array configuration; (b) line representation of radiators

$\delta\delta$ between adjacent elements is introduced. The total phase difference of the radiating fields from the adjacent elements may be expressed as

$$\psi = \Delta\phi + \delta\delta = 2\pi \frac{d}{\lambda} \sin\theta + \delta\delta \tag{4.2}$$

The outputs E_i of all the elements are summed via lines of equal length to give the sum output E, as in Figure 4.1(a). If source 1 is taken as the phase centre so that the field from source 2 is advanced by ψ, source 3 is advanced by 2ψ, and progressively onwards until the source N is advanced by $(N-1)\psi$, then the sum output E can be written as a geometric series:

$$E = a_0\left(1 + e^{j\psi} + e^{j2\psi} + \cdots + e^{j(N-1)\psi}\right) \tag{4.3}$$

For brevity, $a_0 = 1$. Multiply (4.3) by $e^{j\psi}$ and by simple geometry,

$$Ee^{j\psi} = a_0\left(e^{j\psi} + e^{j2\psi} + e^{j3\psi} + \cdots + e^{jN\psi}\right) \tag{4.4}$$

Subtracting (4.4) from (4.3) and dividing by $(1 - e^{j\psi})$,

$$E = \frac{1 - e^{jn\psi}}{1 - e^{j\psi}} = \frac{e^{\frac{jN\psi}{2}}\left[e^{\frac{-jN\psi}{2}} - e^{\frac{jN\psi}{2}}\right]}{e^{\frac{j\psi}{2}}\left[e^{\frac{-j\psi}{2}} - e^{\frac{j\psi}{2}}\right]} \tag{4.5}$$

Rearranging (4.5),

$$E = \left\{e^{\frac{j(N-1)\psi}{2}}\right\}\left[\frac{\sin\left(\frac{N\psi}{2}\right)}{\sin\left(\frac{\psi}{2}\right)}\right] \tag{4.6a}$$

noting that $e^{-jx} - e^{jx} = -2j\sin x$. Two terms emerge from (4.6a): the term $\{.\}$ in curly brackets is the phase of the field shifted $(N-1)\psi/2$; and the second term [ ] represents an amplitude factor or simply *array factor*, $f_a(\psi)$. Specifically,

$$f_a(\psi) = \frac{\sin\left(\frac{N\psi}{2}\right)}{\sin\left(\frac{\psi}{2}\right)} \tag{4.6b}$$

The array field strength is the magnitude of (4.6), and noting that ψ is directly related to the physical dimension of the antenna in the form

$$|E(\theta)| = \left|\frac{\sin\left(\frac{N\psi}{2}\right)}{\sin\left(\frac{\psi}{2}\right)}\right| = \left|\frac{\sin\left(N\left\{\frac{\pi d}{\lambda}\sin\theta + \frac{\delta\delta}{2}\right\}\right)}{\sin\left(\frac{\pi d}{\lambda}\sin\theta + \frac{\delta\delta}{2}\right)}\right| \tag{4.7}$$

Note that this expression represents voltage distribution. It can be converted to power, as the array radiation pattern, or antenna gain $G(\theta)$, by the normalized square of the amplitude:

$$G(\theta) = \frac{|E(\theta)|^2}{N^2} = \left[\frac{\sin\left(N\left\{\frac{\pi d}{\lambda}\sin\theta + \frac{\delta\delta}{2}\right\}\right)}{N\sin\left(\frac{\pi d}{\lambda}\sin\theta + \frac{\delta\delta}{2}\right)}\right]^2 \tag{4.8}$$

A normalized radiation pattern of the array for a uniformly illuminated six-element antenna array, with the inter-element distance of half-wave and two-phase progression error of 0° and 0.2°, is shown in Figure 4.2. The array field patterns have a sidelobe structure that decreases monotonically from the main beam. The effect of the phase progression error is to shift the response to the left (if $\delta\delta$ = positive), or right (if $\delta\delta$ = negative).

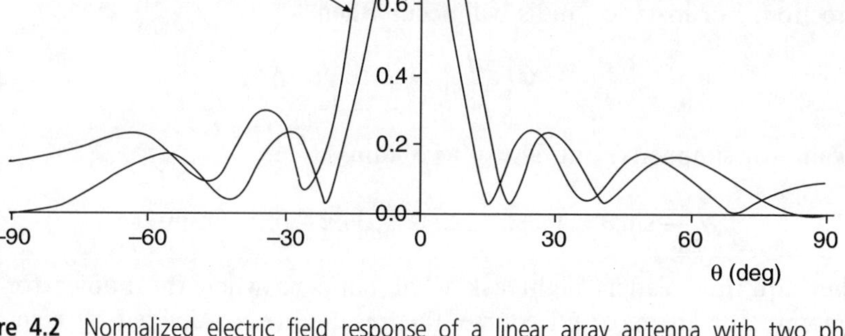

Figure 4.2 Normalized electric field response of a linear array antenna with two phase progressions of 0 and 0.2°

In essence, the array pattern could be defined as the full elevation pattern of a broadside array that substitutes (imaginary) isotropic radiators in place of the elements actually used. The broadside of the array is the direction in which maximum radiation is almost perpendicular to the plane (line) of the array.

When directive elements are used, the resultant radiation pattern is expressed as

$$G(\theta) = G_i(\theta) = \left[\frac{\sin\left(N\left\{\frac{\pi d}{\lambda}\sin\theta + \frac{\delta\delta}{2}\right\}\right)}{N\sin\left(\frac{\pi d}{\lambda}\sin\theta + \frac{\delta\delta}{2}\right)}\right]^2 \tag{4.9}$$

where $G(\theta_i)$ is the individual element factor, or radiation pattern of an individual element.

In a two-dimensional rectangular planar array, the radiation pattern may sometimes be written as the product of the two planes that contain the principal axes of the antennas. Following equation (4.8) and neglecting phase progression effect for simplicity (i.e. $\delta\delta = 0$), the product radiation pattern $G(\theta)|_{n,m}$ can be written as,

$$G(\theta)|_{n,m} = \left|\frac{\sin\left(n\frac{\pi d}{\lambda}\sin\theta_n\right)}{n\sin\left(\frac{\pi d}{\lambda}\sin\theta_n\right)}\right|^2 \left|\frac{\sin\left(m\frac{\pi a}{\lambda}\sin\theta_m\right)}{m\sin\left(\frac{\pi a}{\lambda}\sin\theta_n\right)}\right|^2 \tag{4.10}$$

where n, m = number of radiators in θ_n, θ_m dimensions with spacing d, a respectively. Note that θ_n, θ_m are not necessarily the elevation, azimuth angles normally associated with antenna beam.

An advantage of the two-dimensional array is that one can scan and shape the beam in two directions. However this type of array tends to have a rather complex and costly feed network.

4.1.1 Nulls

From (4.7) the array factor has *nulls* (zeros) whenever its numerator is zero. Also the phase of the field will remain constant whenever $E(\theta)$ has a value but changes by 180° in the direction for which $E(\theta) = 0$; that is, in the *null* direction. For instance, nulls will occur when

$$N\left\{\frac{\pi d}{\lambda}\sin\theta + \frac{\delta\delta}{2}\right\} = 0 \tag{4.11}$$

Again, for simplicity, put $\delta\delta = 0$ to obtain

$$N\frac{\pi d}{\lambda}\sin\theta = 0, \pm\pi, \pm 2\pi, \ldots, \pm n\pi \quad (n \text{ is an integer}) \tag{4.12}$$

The inquiring reader might ask what happens when the numerator and denominator are zero? Of course, this results in zero divided by zero. But, L'Hospital's rule allows for separate differentiation of numerator and denominator. By so doing it can be demonstrated that the elements' output $|E|$ is maximum when

$$\frac{d}{\lambda}\sin\theta = \pm p \tag{4.13}$$

All the maxima given by (4.13) will have the same value and will be equal to N. The first maximum (called the *main beam* maximum) will occur when $\sin\theta = 0$; that is, when $p = 0$. The other maxima define the *grating lobes*; that is, when $p \geq 1$. As expressed by (4.13), the visible range θ is real since the maximum attainable value of $\sin\theta$ is unity. Given that the first grating lobe occurs at $p = 1$, the lowest inter-element spacing at which a grating lobe will appear works out to be

$$\frac{d}{\lambda} = 1 \tag{4.14}$$

Thus, the inter-element spacing should never be allowed to reach the value of one wavelength. Therefore, one could deduce that to avoid grating lobe formation

$$\frac{d}{\lambda} < 1 \tag{4.15}$$

The case typified by (4.15) would have only one principal maximum, which is formed in the direction orthogonal to the axis of the array. Where this occurred is called the *broadside array*. As the inter-element spacing d increases to one wavelength λ, (i.e. $d \rightarrow \lambda$) grating lobes begin to appear at the *endfire* direction at angle $\theta = \pm\pi/2$, while the main beam is formed broadside. An *endfire array* has its maximum radiation parallel to the array.

As seen in Figure 4.2, secondary maxima occur at the sidelobes. The peaks of the sidelobes can be computed by differentiating $E(\theta)$ in (4.6a) with respect to ψ and setting the differential to zero:

$$\frac{dE(\theta)}{d\psi} = N\sin\left(\frac{\psi}{2}\right)\cos\left(\frac{N\psi}{2}\right) - \sin\left(\frac{N\psi}{2}\right)\cos\left(\frac{\psi}{2}\right) = 0 \tag{4.16}$$

Rearranging in terms of N to have

$$N = \frac{\sin\left(\frac{N\psi}{2}\right)\cos\left(\frac{\psi}{2}\right)}{\cos\left(\frac{N\psi}{2}\right)\sin\left(\frac{\psi}{2}\right)} = \frac{\tan\left(\frac{N\psi}{2}\right)}{\tan\left(\frac{\psi}{2}\right)} \tag{4.17}$$

This expression has its first solution at

$$N\psi = 2.8606\pi \tag{4.18}$$

By substituting (4.18) in (4.7), the sidelobe ratio for the first sidelobe is -12.06 dB. The subsequent sidelobe ratios can be estimated. The envelope of the sidelobe levels follows the $(1/\psi)$ law. In practice, the sidelobe levels of a uniformly excited array are unacceptably high and would have to be suppressed. A technique for suppressing sidelobes has been discussed in Chapter 3, section 3.2.4.

4.1.2 Beamwidth

The array *beamwidth*, θ_{BW}, can be estimated by finding out the half-power (-3 dB) points of the main beam. This is done by equating the amplitude factor of all the elements in (4.6) to $N/\sqrt{2}$:

$$\frac{\sin\left(\frac{N\psi}{2}\right)}{\sin\left(\frac{\psi}{2}\right)} = \frac{N}{\sqrt{2}} \tag{4.19}$$

The solution to (4.19) was given by Hansen (1990), within 1 per cent error margin, as

$$\sin\left(\frac{\theta_{BW}}{2}\right) = \frac{0.4429\lambda}{Nd} \tag{4.20}$$

providing that the number of array elements N is greater than 7. Rewrite (4.20) in terms of half-angle of two incoming signals θ_1 and θ_2 as

$$\begin{aligned} \sin\theta_2 &= \sin\left(\frac{\theta_{BW}}{2}\right) = \frac{0.4429\lambda}{Nd} \\ \sin\theta_1 &= \sin\left(\frac{-\theta_{BW}}{2}\right) = -\frac{0.4429\lambda}{Nd} \end{aligned} \tag{4.21}$$

Upon an application of small angle approximation, it can be shown that the beamwidth is

$$\theta_{BW} = \theta_2 - \theta_1 = \frac{0.8858\lambda}{Nd} = \frac{0.8858\lambda}{D} \tag{4.22}$$

noting that for small θ, $\sin(\theta) \approx \theta$ and D is the array aperture. Clearly, (4.22) shows that as the physical length of the array increases, the array beamwidth decreases for a given propagation wavelength.

4.2 Phase shifter

The effect of phase shifting on the array radiation pattern can be investigated via (4.8). From (4.8), it is evident that the maximum of the radiation pattern would occur when $\sin\theta = -\delta\delta/2$ corresponding to the peak of the main beam. So,

$$\delta\delta = -\frac{2\pi d}{\lambda}\sin\theta_0 \tag{4.23}$$

where θ_0 serves two definitions. First, the angle the main beam would be positioned to attain the peak if the phase shift $\delta\delta$ had been inserted at each of the elements. Second, the angle at which the main beam is steered (scanned) off the broadside direction of the antenna array, resulting in a process called *beam steering*. Equation (4.23) has been derived for a one-dimensional array antenna. It could be extended to a two-dimensional array where the radiating elements are located on a flat surface and each element is equipped with a phase shifter. By attaching n phase shifters to the output of each element, the signal generated by a single beam can be converted to an n-beam antenna. The n beams may be fixed in space, steered independently or as a group. The beams could be generated on the transmitter, or receiver, end of the antenna system. In practice, it is more convenient to generate multiple beams on the receiver only while transmitting a wide radiation pattern that gives a total coverage of the multiple receiving beams.

For a given beam position in azimuth and elevation plane, the phase shift can be computed by a digital computer and electronically inserted in each element to point the antenna in a desired spatial-beam position. In practice, phase shifters often operate with *modulo* 2π to conserve size and cost and are introduced into the feed path of each element so that the phase shift can be controlled by externally generated signal. The question is how to generate modulo 2π?

Modulo 2π is attained by either switching in a line length, or by changing the apparent impedance. If the beam is scanned as a function of time, these phase shifts also change as a function of time. A constant rate of change of phase with time is equivalent to a constant frequency. Thus a frequency difference at adjacent elements results in a scanning beam.

Increasing the array line lengths that must be switched in and out, a *time delay steering* is introduced. This process introduces path difference, which is converted to time delay difference:

$$\Delta t = \frac{d \sin \theta}{c} \tag{4.24}$$

As a result, the time delay introduces phase shift. Often both time delay units and phase shifters are employed. In such a case, the phase shifter setting corrects the phase shift introduced by the time delay units.

In general, if N phase shifters were inserted in a *series-fed* array, the signal would suffer insertion loss amounting to N times that of a single-phase shifter loss. For a two-dimensional antenna consisting $N \times N$ elements, with phase steered in both directions, would require N^2 phase shifters. This is the common form of phase array radar antenna and the most often thought of when the term *phase array* is used. If, however, the insertion were for a *parallel-fed* array, the insertion loss would amount to that of a single-phase shifter loss since the phase shifter is effectively introduced once. In order to steer the array beam in the desired direction, a single control signal is needed for a series-fed array to steer the beam while a separate control signal is needed for each of the N phase shifters in the parallel-fed array.

The prime benefit of phase array radar is its reliability, but this degrades gracefully. With it, failures can be tolerated, particularly in terms of antenna elements, providing there is no weak link in the failure chain.

The next section is intended to investigate the effect of beam steering or scanning on linear array antenna.

4.3 Beam steering

By substituting (4.23) in (4.8), a beam steered at θ_0 will have a normalized radiation pattern represented by

$$G(\theta) = \frac{\sin^2\left(N\pi \frac{d}{\lambda}[\sin\theta - \sin\theta_0]\right)}{N^2 \sin^2\left(\pi \frac{d}{\lambda}[\sin\theta - \sin\theta_0]\right)} \tag{4.25}$$

The maximum of the radiation pattern occurs when $\sin\theta = \sin\theta_0$, noting that by L'Hospital's rule, $\lim_{x\to 0}(\sin x/x) = 1$. A plot of (4.25) is shown in Figure 4.3 for six-element antenna, with a half-wave inter-element spacing (i.e. $d = 0.5\lambda$) and scanned off the boresight by 3°.

The only difference between looking at boresight ($\theta_0 = 0°$) and off boresight ($\theta_0 = 3°$) is that when steering is present the argument of the function is the difference of the *sines* of the look and steering angles. The effect of steering is simply to produce a shift either right (θ_0), or left ($-\theta_0$) of the boresight with no distortion in the electric field strength, but reduced

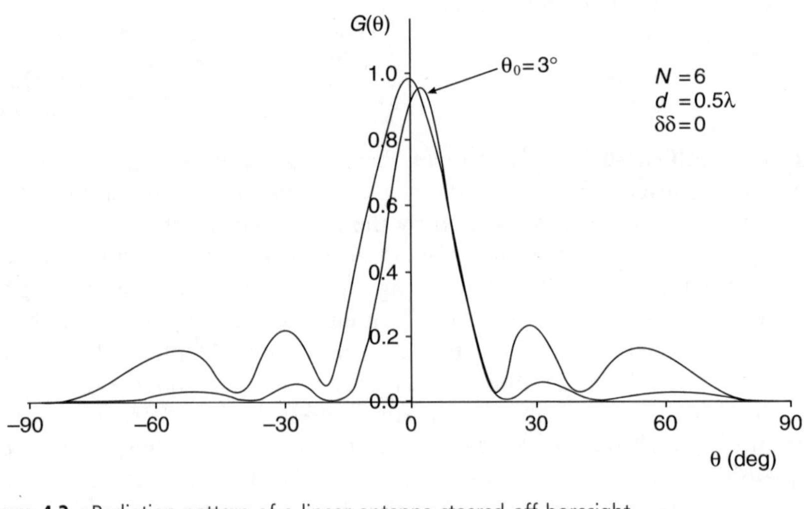

Figure 4.3 Radiation pattern of a linear antenna steered off boresight

sidelobe amplitudes. Replacing the sine in the denominator of equation (4.25) by its argument, the gain

$$G(\theta) = \left(\frac{\sin u}{u}\right)^2 \tag{4.26}$$

where

$$u = N\pi\frac{d}{\lambda}\{\sin\theta - \sin\theta_0\} \tag{4.27}$$

Equation (4.26) represents the frequently quoted sin *c* behaviour of the antenna pattern for a linear array of radiating elements. Figure 4.4 demonstrates the difference between using the *actual* (4.25) and the *approximate* (4.26) power gain expressions, for a six-element array, $d = 0.5\lambda$ and scan angle of 5°. Both responses are very close at the mainlobe where the beam is steered. Noticeable differences emerge at the sidelobes between the actual and the approximate. With this approximation typified by (4.26), the half-power beamwidth, when the spacing is half-wave, is approximately

$$\theta_{BW} \cong \frac{101.8}{N} \quad \text{(deg)} \tag{4.28}$$

The effect of changes in array bandwidth with changes in steering angle θ_0 can further be investigated using equation (4.27). The antenna power $G(\theta)$ in (4.26) will be reduced to half its maximum value when $u = \pm 0.4429\pi$. The

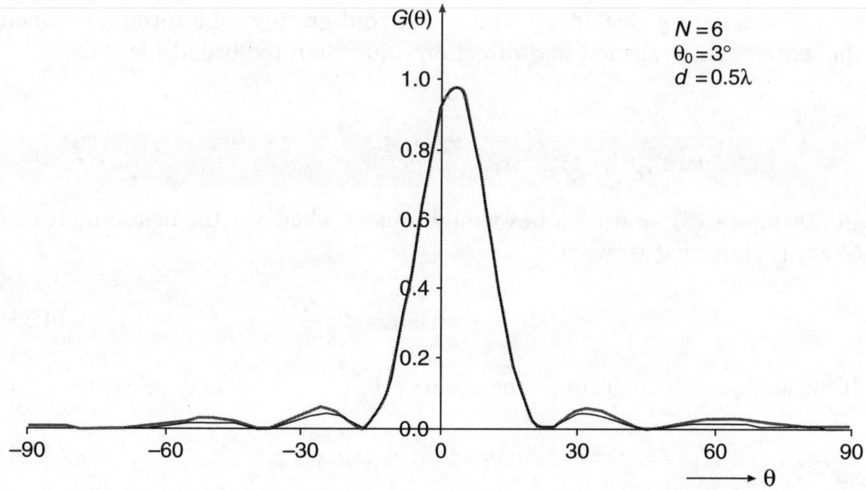

Figure 4.4 Difference between the actual and approximate expressions in the derivation of antenna array radiation patterns

positive value occurs when $\theta > \theta_0$ while the negative typifies the case when $\theta < \theta_0$. Equating $u = \pm 0.4429\pi$ to (4.27):

$$\pm 0.4429 = \frac{Nd}{\lambda}\{\sin\theta - \sin\theta_0\} \tag{4.29}$$

Following Blickmore (1958),

$$\sin\theta - \sin\theta_0 = \sin(\theta - \theta_0)\cos\theta_0 - [1 - \cos(\theta - \theta_0)]\sin\theta_0 \tag{4.30}$$

If the beam is near broadside, θ_0 is small and $\cos(\theta - \theta_0) = 1$, as such (4.30) reduces to

$$\sin\theta - \sin\theta_0 \approx \sin(\theta - \theta_0)\cos\theta_0 \tag{4.31}$$

Substituting (4.31) in (4.29), the half-power beamwidth can be approximated as

$$\theta - \theta_0 \approx \theta_{BW} = \frac{0.8858\lambda}{Nd\cos\theta_0} \tag{4.32}$$

This expression implies that, in the plane of steer, a change in beamwidth with a steer angle θ_0 off broadside, the beamwidth increases inversely as the *cosine* of the steer angle. By progressively phase shifting the array in a programmed manner, the main beam can be moved from broadside towards endfire (i.e. towards $\theta_0 = \pm\pi/2$). This is the principle of *electronic scanning*. As noted by Skolnik (1980), the expression in (4.32) is not valid for an antenna beam far removed from the broadside and also when energy is radiated in the endfire direction.

It is appropriate to investigate, at this stage, the restriction on the spacing between radiating elements in order to avoid grating lobe formation when the main beam is steered in a direction other than the broadside.

4.4 Inter-element spacing

Following (4.13), grating lobes would appear whenever the denominator of (4.25) is zero; that is, when

$$(\sin\theta - \sin\theta_0) = \pm p\frac{\lambda}{d} \tag{4.33}$$

If the angle at which grating lobe occurs is θ_g then (4.33) may be expressed as

$$\sin\theta_g = \sin\theta_0 \pm p\frac{\lambda}{d} \tag{4.34}$$

Given that the primary objective of an antenna array design is to avoid grating lobes being formed, the expression in (4.34) must be greater than

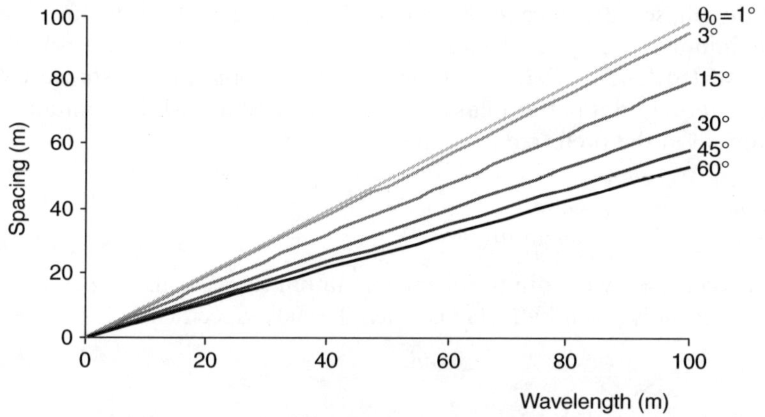

Figure 4.5 Minimum inter-element spacing required for variable beam width

unity for all values of θ_0, so that θ_0 is outside the real space. For this, two solutions would emerge:

$$|\sin\theta_0| + \frac{\lambda}{d} > 1 \tag{4.35}$$

$$|\sin\theta_0| - \frac{\lambda}{d} < -1 \tag{4.36}$$

The condition set by (4.35) can always be met even when $|\sin\theta_0| = 0$. The condition set by (4.36) is more exigent: implying that a necessary criterion for the avoidance of grating lobes in visible space must be

$$\frac{d}{\lambda} < \frac{1}{1 + |\sin\theta_0|} \tag{4.37}$$

This expression shows that by scanning to $\pm 90°$ requires a minimum element spacing of half a wavelength. Using (4.37), a plot of inter-element spacing d, where grating lobes will occur, is seen in Figure 4.5 for beam coverage between 15° and 60°. The graph demonstrates that as the beam coverage increases, the spacing between grating lobes increases for a given propagation wavelength (frequency). At shorter wavelengths ($\lambda \leq 10\,\text{m}$), wider radar coverage can be achieved with similar inter-element spacing with little consequence on radar performance. The narrower the beamwidth, the closer the separation distance is to the propagation wavelength. A knowledge of this limiting condition enables manipulation of the grating lobes.

Example 4.1 Consider an array comprising two vertical half-wave dipoles with currents of equal magnitudes being placed with the array axis along the east–west direction. (a) Determine the separation distance (in units of λ)

and the phase difference such that the horizontal pattern of the array has a maximum to the east and a null at an azimuth angle of 135° measured from the east. (b) Under what condition would the separation distance be such that the horizontal pattern has a null in any direction while maintaining the maximum in the preferred east direction?

Solution
Given $\theta = 135^\circ$ and azimuth $\phi = 0^\circ$

(a) The necessary condition for the maximum to occur in the preferred east direction is given by (4.11) is when $\theta = 90^\circ$. Specifically

$$\frac{d}{\lambda} = -\frac{\delta\delta}{2\pi}$$

For a null to occur, using (4.12) with $\theta = 135^\circ$,

$$N\frac{\pi d}{\lambda}\sin\theta = \pm\pi$$

But $N = 2$,

$$\frac{d}{\lambda} = \pm\frac{1}{2\sin 135^\circ} = \pm\frac{1}{\sqrt{2}}$$

$$\delta\delta = -135^\circ$$

(b) To maintain the maximum in the preferred direction

$$\frac{d}{\lambda} = -\frac{\delta\delta}{2\pi} \quad \text{or} \quad \delta\delta = -2\pi\frac{d}{\lambda}$$

However, if a null is to occur, (4.12) can be expressed as

$$\frac{\pi d}{\lambda}\sin\theta + \frac{1}{2}\left(-2\pi\frac{d}{\lambda}\right) = \pm\frac{\pi}{2}$$

$$d = 1.09\lambda$$

This expression indicates that there will be no null if $d/\lambda \leq 1.09$.

4.5 Pattern multiplication

The principle of pattern multiplication states that the beam pattern of an array is the product of the element pattern and the array factor. That is, the total field pattern of an array of point sources can be expressed in two parts, specifically

$$E = F_1 \times F_2 \tag{4.38}$$

where

F_1 = the pattern factor of a single point source radiator, e.g. (3.27) for a dipole
F_2 = the array factor for the n radiators, e.g. (4.6b) for a linear antenna array.

Denoting the array field by $E(\theta)$ and on application of the pattern multiplication principle, the far field of the array is expressed as

$$|E(\theta)| = \left|\frac{60I_0}{r}\left[\frac{\sin\left(\frac{N}{2}\beta d\sin\theta\right)}{\sin\left(\frac{\beta d}{2}\sin\theta\right)}\frac{\cos(\beta l\cos\theta) - \cos\beta l}{\sin\theta}\right]\right| \tag{4.39}$$

Note that $\beta = 2\pi/\lambda$.

The expression in (4.38) is a very useful because it enables the field pattern to be determined for arrays in which the elements may be other than point source radiators. The principle also shows how theorems relating to array design are independent of the particular antenna element used to form the array.

4.6 Slot antenna array

If an aperture of any shape is made in a conducting surface, and if a potential difference is applied between its two opposite sides, a radiating system is obtained called a *slot* because of the structure of most of these apertures, which take the form of an elongated rectangle, see Figure 4.6.

There exists two main relationships between the slot and the complementary dipole resulting from the Babinet's principles. These principles can be expressed as follows.

4.6.1 First property

The radiation pattern of the slot is identical to that of the complementary dipole, as long as the E- and H-fields are interchangeable. This means that the electric field radiated by a slot is polarized orthogonally to the electric

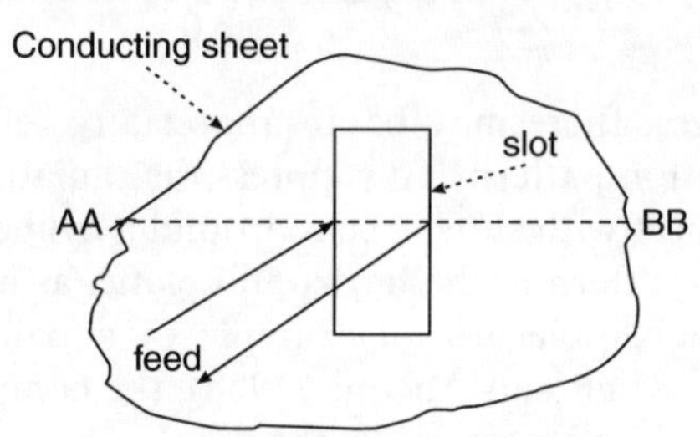

Figure 4.6 A geometry of a rectangular slot in a conducting surface

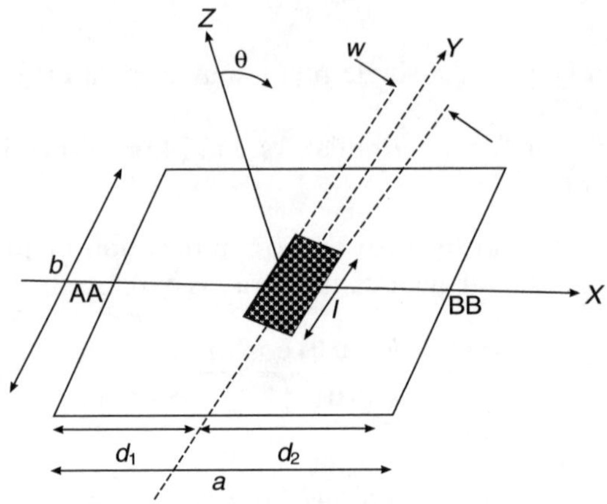

Figure 4.7 Geometry of a slot on a rectangular plate

field radiated by the complementary dipole. So for a thin slot, it is possible to state by the application of the expression (3.26) in Chapter 3 for a dipole that the magnetic field radiated is of the form

$$H = H_0 \left[\frac{\cos(\beta l \cos\theta) - \cos\beta l}{\sin\theta} \right] \tag{4.40}$$

where H_0 represents the magnetic force induced across the narrow slot of aperture, l, see Figure 4.7. AA and BB are the edges of the conducting sheet, w is the width of the narrow aperture, and a and b are the ground plane dimensions.

If the total number of dipoles in the array is $N + 1$, then the separation between dipoles is

$$d = \frac{l}{N} \tag{4.41}$$

The magnetic field radiated by the array in the far field may be expressed as

$$H = H_0 \sum_{n=1}^{N+1} \frac{\cos(N\beta d \cos\theta) - \cos(N\beta d)}{\sin\theta} \tag{4.42}$$

Because of the edges, there may be an appreciative effect, in the form of 'ripple' in the radiation pattern. To suppress the radiation from one edge of the plate, a finite sheet with slots is bent around a cylinder, thereby creating an omnidirectional pattern in the horizontal plane, as in Figure 4.8.

The slot and the dipole are an example of a pair of complementary antennas. By Drabowitch and Ancona (1988), the complementary property has an important consequence: unlike the dipole, the slot is omnidirectional in the E-plane and directive in the H-plane. This is the reason why, when

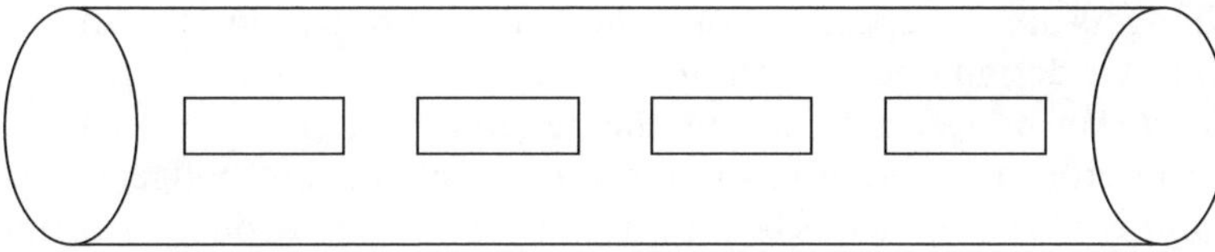

Figure 4.8 Geometry of a cylindrically bent slot array

aiming for directivity in one plane or the other, solutions based on dipoles or slots are generally adopted.

4.6.2 Second property

The impedance Z_d of the dipole is related to the impedance Z_s of the complementary slot by

$$Z_d \times Z_s = \frac{\eta_0^2}{4} \tag{4.43}$$

where η_0 is characteristic impedance of free space, which equals $120\pi\,(\Omega)$. It can be concluded from (4.43), for example, that if the impedance of a very thin half-wave slot is resistive, its impedance can be calculated approximately from (3.36):

$$Z_s = \frac{(120\pi)^2}{4 \times 30Cin(2\pi)} = \frac{(120\pi)^2}{4 \times 73.14} = 485.79\,\Omega \tag{4.44}$$

Generally, a slot antenna will radiate in two half-spaces defined by the conducting surface. However, if only one half-space is to be irradiated, the slot must be screened on the opposite side, effectively turning it into a cavity.

Slot antennas have wide applications in communications and defence industries because of low-profile conformal design. For example, slot antennas have been used for the following:

- For high-power (television) broadcasting transmitters as vertical collinear arrays. A collinear array is formed when the radiators are stacked vertically end to end with centres approximately equidistant apart.
- As radiating elements in aircraft where the vehicle skin acts as the conducting plane. To ensure directionality, the slot arrays are backed with a second conducting plane to form a plate antenna.
- As antennas providing omnidirectional coverage when wrapped around missiles, rockets and satellites.

4.7 Power and time budgets

Budgets play an important factor in moulding design processes; they set the bounds of design and operation of the radar. It is important to remember that budgets are not specifications, they only represent an allocation inherent to radar design and operation.

The power budget is bound by the design constraint of the main supply power, its efficiency and utilization by the radar. The time budget determines the number of simultaneous beams. Both time and power budgets define how well the radar can fulfil its functions under strenuous operating conditions.

Another important budget constraint is processing constraint, which is bound in terms of available power and how many targets can be tracked at any given time.

As an illustration of how power, time and processing budgets are estimated, consider a radar system in volume search mode. For this mode, the average *power* requirement relative to the transmitter average power is

$$P_{\text{av}} = n_p n_b T_f \frac{\tau}{d_u} \tag{4.45}$$

where

n_p = no of pulses per scan
n_b = total number of beams
d_u = transmitter duty cycle
τ = pulse width
T_f = frame or scan time.

Using the values in the Table 4.1 and substituting them in (4.45), it is found that 24 per cent of radar energy would be required for volume search: a relative power requirement of 0.24 is demanded.

The proportion of time spent, t_0, in a particular mode is approximately

$$t_0 = n_b n_p \frac{\text{PRI}}{T_f} = \frac{n_b n_p}{T_f \text{PRF}} \tag{4.46}$$

PRI is the pulse repetitive interval, which is the inverse of pulse repetition frequency (PRF). To have a feel for the time budget, consider the same radar system in volume search mode with other parameters in Table 4.1. A relative time occupancy of 0.0267 is obtained; that is, 2.67 per cent of the time spent in each volume.

In general, the frame time is variable for volume surveillance because it depends on the number of targets being tracked by the radar. It could be argued that the time budget for signal processing is perhaps extravagant

Table 4.1 Search data

Number of pulses per scan	2.0
Total number of beams	200
Pulse width	10 μs
PRF	5 kHz
Frame or scan time	3 s

when other functions, like tracking, require considerable time frame. The simplest solution in real time surveillance is to increase volume time frame as the number of tracks increases. This would allow other parameters that impact on detection performance to be kept constant. This would not significantly affect surveillance tracking since the surveillance frame rate is only important up to initial detection.

4.8 Summary

In this chapter, the examples of an array theory have been simplified, but the purpose has been achieved; to develop the concept of array factor and show how the antenna can be steered in a preferred direction and shape its radiation pattern by varying the excitation on the array elements. The concept demonstrates that an array can be broadband, its shape can be altered somewhat and even feed its elements parasitically.

By varying progressively phase shift of an antenna array in a programmed manner, the principle of electronic scanning is explained. The scanning criterion, for the avoidance of grating lobe formation in the desired direction, whether a progressive phase shift is introduced or not, was established. And finally, the concept and applications of slot antenna arrays were discussed.

Problems

1. A new array arrangement is envisaged to comprise half-wave dipoles oriented horizontally, all parallel to one another, to produce a single beam of width 20° ± 0.5° in the broadside direction. If the array's elements are uniformly excited, determine the necessary spacing of the antennas that lead to the fewest number of dipoles that meet the requirement.
2. Consider a 12-element array, equidistant at $d = 0.5\lambda$, centre-fed, z-directed half-wave dipoles, with centres lying symmetrically relative to the origin along the positive and negative x-axis. The feed point currents of the nth element are expressed by $I_n = I_0 e^{jn\delta}$, where the amplitude I_0 is the same for all elements. Determine (a) the width of the main beam if the phase angle δ is set to the value corresponding to an endfire array; (b) the value of the phase angle and the orientation of the receiving dipole such that the received signal is maximum if the receiving antenna is in the form of a half-wave dipole observed at 100λ away from the source and elevation angle of 75°.
3. A uniform array has 60 elements and an inter-element spacing of a quarter-wavelength. Determine the width of the mainlobe if the array is intended to operate as (a) an endfire array, (b) a broadside array, and (c) an array whose mainlobe maximum occurs at 35° relative to its axis.

4. A slot antenna is terminated by an impedance represented by $Z = 73 + j42.5\ \Omega$. Determine (a) the impedance of the antenna. (b) If the characteristic impedance is resistive and represented by a three-quarter-wavelength dipole, estimate the slot impedance value.
5. If the slot antenna is represented with an equivalent dipole of length greater than one wavelength, determine whether sidelobes will appear and when they become dominant.
6. By Babinet's principles, slots antennas can be represented by their equivalent dipoles. Will the principles hold if the dipoles are of (a) conical and (b) elliptical section? Give reasons. (c) For cylindrical dipoles, will the sinusoidal current hypothesis used in the derivation of dipole field pattern also be valid?
7. Why is a slot array easier to construct at higher frequencies than a dipole array?
8. How can the phase of a dipole array be reversed?
9. It is desired to operate a radar system that is capable of transmitting N pulses, 1 μs pulse width, PRF of 10 kHz, within a timeframe of 2.5 s. If the radar is to maintain a peak power of 1 kW when 15–20 per cent power requirement is allocated for volume search, design such a radar.
10. An isotropic radiating system means, by definition, a system where the radiated field is independent of the direction considered both in amplitude and polarization. Is such a system feasible?

5

The radar equations

The discussion on radar measurements in the previous chapter avoided the issue of noise and other losses and their effect on parameter estimation. From radar theory, the maximum range beyond where a target cannot be seen can be calculated using the radar equations, which are the main subject of this chapter. Targets have varying reflectivity: as their orientation changes so also their reflective characteristics change as a function of time. Knowledge of the target's radar cross-section, or backscattering coefficient, is essential in any radar measurement considerations. This chapter looks at models that take into consideration target descriptions of sufficient latitude to accommodate such variations in target characteristics at specific times.

The medium in which radar operates is not ideal. The environment, as well as the medium the radio waves propagate through, introduces several losses in addition to system loss. These losses will be discussed and included where appropriate in the radar equation. By making subtle changes to the basic radar equation, the laser radar and secondary radar equations will be determined.

5.1 Radar equation for conventional radar

From spherical geometry, the surface area of a sphere of radius R from the source is $4\pi R^2$. If the target is considered to be at the peak of beam, see Figure 5.1, then the power density uniformly distributed by the transmitter over the spherical surface may be expressed by

$$P_\Delta = \frac{P_t}{4\pi R^2} \tag{5.1}$$

P_t is the transmitted power in watts. Suppose a comparable antenna of gain G_t is used in place of a point source radiator, which is allowed to radiate a target of σ radar cross-section at range R on the main antenna beam axis, then the target will intercept a fraction of the radar power given by:

$$P_i = \frac{P_t G_t \sigma}{4\pi R^2} \tag{5.2}$$

The advantage of developing the radar equation through a very elementary point source radiator is that it serves as a basis for comparison of many types

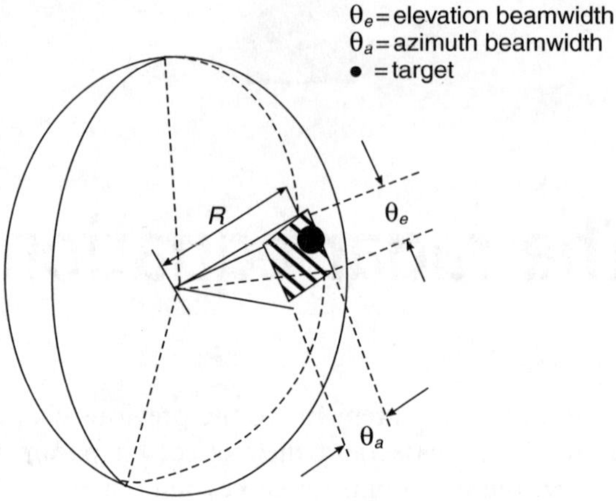

Figure 5.1 Geometry of power density of a target at peak of beam

of antennas whose performance is best expressed in terms of such a basic radiator. Before proceeding further with the development of the radar equation, it is important that the reader understands the fundamentals and their meaning.

5.1.1 Some comments on radar gain and target characteristics

5.1.1.1 Antenna gain

Gain, as applied to an antenna system, is a measure of the directivity of the antenna field pattern as compared with some standard antenna. Qualitatively, the gain is a ratio of power that must be supplied to the comparison antenna to deliver particular field strength in the desired direction, to the power that must be supplied to the directional antenna system to obtain the same strength in the same direction. The procedure of calculating the gain of an antenna system consists of assuming currents in the antenna to be investigated and in the comparison antenna, such that the field strength produced in the desired direction is the same in both cases. The total energy involved is then determined either by the Poynting vector method or in terms of the radiation resistance (Terman 1949).

The Poynting vector method models the antenna, or antenna array, in the centre of a large, imaginary sphere; thus allowing the power flowing out of the sphere to be determined in terms of density per unit of area. From this, the total power equivalent density and peak (main beam) power density can be determined. By relating the two power density values, the concentration of power in the main beam relative to the total input power can be resolved and therefore the gain of an isotropic radiator. It should be noted that this gain figure has to be corrected if a different gain reference is needed. Also, when referencing to a half-wave dipole, the gain of the dipole would need to be subtracted from any gain figure that has been referenced to an isotropic source.

5.1.1.2 Re-radiation pattern of a target

The ideal radar target is a point object sending back a spherical wave with the same polarization as the transmitted signal. Under these conditions, the orientation of the tracking antenna is such that the wave incident on its aperture is equiphasal, so this orientation corresponds to that of the target (Croft 1972). Real targets are usually of a complex structure: the wave reflected by the target's various components interfere with each other, and the target's re-radiation pattern has an irregular, lobed appearance (see Figure 5.2).

As the orientation of the target varies with time, the echo signal fluctuates according to a probability characteristic similar to Rayleigh's rules. More is said of these rules in section 5.2. In addition, adjacent lobes usually differ in phase by 180°.

5.1.1.3 Radar cross-section

The radar cross-section of most targets does not necessarily demonstrate a simple relationship to their physical area, except it is highly probable that the larger the size, the larger the cross-section. Since targets have a probabilistic distribution associated with their specular aspects, there would exist a wide fluctuation in target radar cross-section. An example of target fluctuations is ship and aircraft where their cross-sections change from moment to moment and with frequency as they change orientation. Swerling (1960) has characterized the types of fluctuation. A good albeit brief discussion is given in section 5.2.

Similar to a receiving antenna, a radar target also intercepts a portion of the power, but reflects (re-radiates) it in the direction of the radar. The amount of power reflected toward the radar is determined by the radar cross-section (RCS) of the target. RCS is a characteristic of the target that

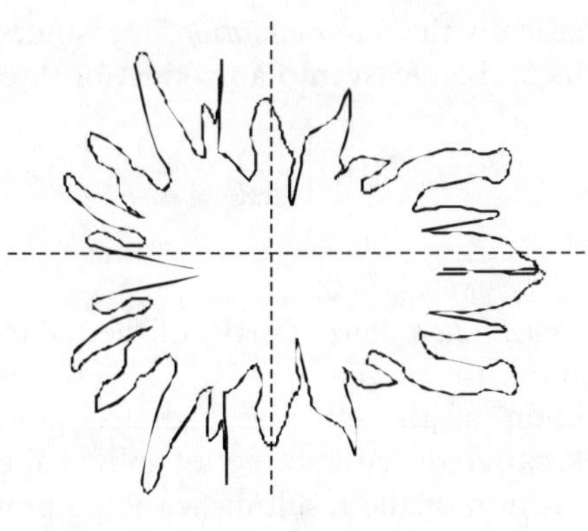

Figure 5.2 Re-radiation pattern of a real target

represents its size as seen by the radar and has the dimensions of area. RCS area is not the same as physical area. But the power re-radiated or reflected by a target, in the direction of the radar, is equivalent to the effective capture area of the receiving antenna. Therefore, the effective capture area (A_e) of the receiving antenna is replaced by the RCS. The effective capture area is also called the *aperture* of the antenna. More is said of antenna aperture in section 5.3.1.

An acceptable method of estimating radar cross-section of complex target shapes is a three-stage process:

(i) Decomposing the complex shape into a collection of simple component parts with recognizable scattering signatures.
(ii) Estimating each component's cross-section as a function of aspect angle and measuring frequency.
(iii) Arithmetically adding each component's cross-section to form the complex shape combined radar cross-section, assuming random phase between the simple component parts.

A lot of effort has been devoted to measuring radar cross-section of complex materials and objects in the literature. Two good sources for the reader are Crispin and Siegel (1968) and Ruck *et al.* (1970).

Going back to equation (5.2), a target cross-section, σ, can be replaced by an equivalent target whose geometry is symmetric, such as a sphere, which would produce at the radar a power density P_d equivalent to that of an isotropic transmitter of power P_i located at the target. Hence,

$$P_d = \frac{P_i}{4\pi R^2} = \frac{P_t G_t}{4\pi R^2}\frac{\sigma}{4\pi R^2} \tag{5.3}$$

The antenna aperture captures the re-radiated wave from the target whose power is

$$P_r = P_d A_e = \frac{P_t G_t}{4\pi R^2}\frac{\sigma}{4\pi R^2} A_e \tag{5.4}$$

This expression is basically the *radar equation*. It assumes a lossless propagation medium, which can be recast into a product of three factors:

$$P_r = (P_t G_t)\left[\frac{\sigma}{4\pi R^2}\right]\left\{\frac{A_e}{4\pi R^2}\right\} \tag{5.5}$$

where

(i) $(P_t G_t)$ = *Effective radiated power* (ERP) of the radar transmission in the direction of the target.
(ii) $[\sigma/4\pi R^2]$ = Fraction of the effective radiated power intercepted and backscattered by the target of spherical cross-section.
(iii) $\{A_e/4\pi R^2\}$ = Fraction of the resulting scattered power captured by the receiving aperture.

Equation (5.5) demonstrates that radar waves, like other forms of electromagnetic radiation, are inherently subject to the fourth power law of attenuation; that is, the echo signal will have been attenuated by a factor of $1/R^4$. This indicates that range is one of the prime considerations with radar systems and shows that for any sensor system, the energy received from a reflector decreases as the target, or object, range increases. Thus, the difficulty of target detection would increase with range. Conventional wisdom would suggest that doubling the range of the radar systems would require approximately 16 times the transmitter power. Unfortunately the design of transmitter and receiver systems is not restricted to range considerations alone but also to other factors including power losses, cost, and environmental conditions. Consequently, great care must be taken when designing radar systems to reduce losses such that radar receivers are capable of detecting target signals well below ambient noise levels. In the light of equation (5.5), it is appropriate to discuss the type of receiver–transmitter radar arrangements.

5.1.2 Receiver–transmitter arrangement

If the same antenna is used for both transmission and reception of energy, then the receiver and transmitter gains in (5.5) will be replaced with G; that is, $G_r = G_t = G$. Its subsequent expression becomes the *monostatic* radar equation. If the receiving and the transmitting antennas are not the same but are located adjacent to each other, and the separation distance d_x is far less than the distance between the receiver and the target (i.e. $d_x \ll R$), such an arrangement is called *quasi-monostatic*. In this instance, the monostatic radar equation will still be valid for quasi-monostatic arrangement. However, when the receiver and transmitter are clearly separated by a significant distance the situation is called *bistatic* and the radar equation logically follows – the bistatic radar equation. In the bistatic case, equation (5.5) will be valid with two possible changes. First, replacing the target radar cross-section σ by its bistatic value σ_b, which is expected to be functionally dependent on the angle of incidence and wavelength of the signal; that is, $\sigma_b = f(\theta, \lambda)$. And second, the target range R, will be replaced with the effective range R_b, perceived to have been measured at the mid-point between transmit and receive antennas.

5.1.3 Peak and average power

The power P_t in the radar equation (5.5) is the peak power. A distinction should be made between the powers used in radar analysis. The peak power of a sine wave is not the same as the pulse peak power. Peak power is usually equal to one-half the maximum instantaneous power. If the input power to a transmitter is pulsated, the peak pulse power is the power averaged over that carrier-frequency cycle which occurs at the maximum of the pulse power. Often a number of pulses are transmitted per cycle and the situation assessed. As such, the average power is then considered for the time the radar activity is observed. The average power P_{av}, as the name indicates, is the average transmitted power over the pulse-repetition period, T_s.

For example, if the transmitted waveform is a train of pulses of width, or length, τ and period T_s, the average power is related to the peak power in the form

$$P_{av} = \frac{P_t \tau}{T_s} \tag{5.6a}$$

As discussed in Chapter 3, equation (3.79), $d_u = \tau / T_s = \tau \text{PRF}$, which is the transmitter duty cycle. Thus, the average power can be written as

$$P_{av} = P_t d_u \tag{5.6b}$$

PRF is the pulse repetition frequency and d_u is transmit *duty cycle* of the radar. A continuous wave (CW) radar will have a unity duty cycle; that is $d_u = 1$, with $P_{av} = P_{\max}$.

Following from (5.5) the radar equation may be expressed in terms of average power and the time t_0 the radar dwells on (or observes) the target:

$$P_r = \frac{t_0 P_{av} G A_e \sigma}{(4\pi)^2 R^4} \tag{5.7}$$

The observation time t_0 can be estimated using the antenna beamwidth of θ_{BW} (deg) and scanning at the rate ω_m (rpm):

$$t_0 = \frac{\theta_{BW}}{6\omega_m} \tag{5.8}$$

Example 5.1 Consider a radar system having the following specifications:

Transmit and receive gain	33 dB
Receiver sensitivity	−110 dBm
Operating frequency	2.5 GHz
Atmospheric attenuation	0.0095 dB/km

Calculate the minimum peak transmitter power required in pulsed radar to detect a target of 15 m^2 radar cross-section at a range of 250 km.

Solution

Since the receiver sensitivity is −110 dBm, the receive power

$P_r = -110\,\text{dBm} = -140\,\text{dBW} = 10^{-14}\,\text{W}$ (Note that dBm is dB relative to 1 mW)

$R = 250\,\text{km} = 2.5 \times 10^5\,\text{m}$

$\sigma = 15\,\text{m}^2$

$\lambda = 0.12\,\text{m}$

$G_r = G_t = 33\,\text{dB} = 103.3 = 1995.3$

One-way losses = 0.0095 × 250 = 2.375 dB for one way. Hence, two-way losses = 4.75 dB or 2.985. The transmitter power must be increased at least by 2.985* calculated P_t power. In view of the above estimated values, and rearranging (5.5) in terms of P_t, the minimum peak transmitter power required is

$$P_{\text{peak}} = 2.985\frac{P_r(4\pi)^3 R^4}{G_t G_r \sigma\lambda^2} = 43\,\text{kW}$$

5.1.4 Aperture

By antenna theory, a relationship between the transmitting antenna gain and the receiving antenna area is formalized as

$$A_e = \frac{G\lambda^2}{4\pi\eta} \tag{5.9}$$

where η is the efficiency. This expression is the effective antenna area, which is also called the *aperture*. Strictly speaking, polarization and impedance mismatches reduce the effectiveness of the aperture area. Upon inclusion of these mismatches, the effective antenna aperture area can be expressed as the ratio of received power to the incident power density (Morchin 1993), specifically

$$A_e = pq\frac{\lambda^2}{4\pi}D_r(\theta, \phi) \tag{5.10}$$

where

p = polarization mismatch factor

$$q = \text{impedance mismatch factor} = 1 - |\Gamma_\Delta|^2 \tag{5.11}$$

Γ_Δ = input reflection coefficient of the antenna

$D_r(\theta, \phi)$ = directive gain (also called *directivity*) in the direction of maximum radiation intensity. It is descriptive of the antenna pattern. The directive gain is defined as

$$D_r(\theta, \phi) = \frac{4\pi P(\theta, \phi)_{\text{max}}}{P(\theta, \phi)d\theta d\phi} \tag{5.12a}$$

where $P(\theta, \phi)$ is the radiation intensity in the direction (θ, ϕ). Alternatively

$$D_r(\theta, \phi) = \frac{4\pi}{B_a} \tag{5.12b}$$

where B_a is the beam area; that is, the solid angle through which all radiated power would pass if the power per unit solid angle were equal to $P(\theta, \phi)_{\text{max}}$ over the beam area. Given that the beamwidths θ_e and θ_a in elevation θ and

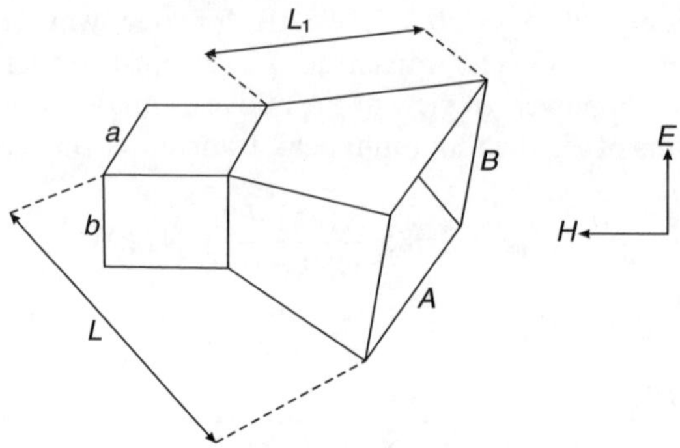

Figure 5.3 Geometry of a pyramidal horn antenna

azimuth ϕ planes respectively are orthogonal, the beam area B_a approximates to the product of the beamwidths. Consequently,

$$D_r(\theta, \phi) \approx \frac{4\pi}{\theta_e \theta_a} \quad (5.12c)$$

The expression in (5.9) can further be investigated for design purposes. The design objective is to have a sufficient beamwidth θ_{BW} to match the required vertical coverage or search sector. As such, a mathematical relationship between the aperture dimension and illumination distribution across the aperture can be established. For example, consider a pyramidal horn antenna shown in Figure 5.3.

In the figure, A and B are aperture length in the H and E planes respectively; a and b are length and breadth of matching waveguide if the horn's shortest length, L_1, possible is required; and L is the axial length to apex to the aperture.

A horn is a slightly flared end of a piece of waveguide. Instead of electrical currents, the waveguide carries a tightly focused electromagnetic wave with the electric components extending between the parallel walls (Kolawole 2002). Horns can be square in section (called pyramidal horns) but rectangular in either two orthogonal planes (called E-plane and H-plane horns).

Pyramidal horns are easily designed and often used for earth coverage antennas because of their symmetrical radiation properties. The following equations are applicable to pyramidal horns, whose length is long compared to a wavelength, λ:

$$G = 10 \log \eta \frac{4\pi}{\lambda^2} AB \quad \text{(dBi)} \quad (5.13)$$

where G is the gain and η is the pyramidal horn's efficiency, typically 50 per cent.

$$\text{3 dB beamwidth in } E \text{ plane: } \theta_E = 54 \frac{\lambda}{B} \quad \text{(deg)} \quad (5.14)$$

$$\text{3 dB beamwidth in } H \text{ plane: } \theta_H = 78 \frac{\lambda}{A} \quad \text{(deg)} \quad (5.15)$$

If it is necessary to have shortest length possible, then by scaling

$$L_1 = L\left(1 - \frac{a}{2A} - \frac{b}{2B}\right) \quad \text{(m)} \tag{5.16}$$

A horn's interior surfaces can be smooth or corrugated, depending on polarization requirements. When annular corrugations are placed on the inner wall of a circular waveguide, a hybrid-mode horn is formed. If the annular corrugations are placed in such a way that neither TE (transverse electric) nor TM (transverse magnetic) modes can be propagated, then a hybrid mode is generated. The hybrid-mode horn antennas can be used to achieve axially symmetric beamwidths, and improve cross-polarization and sidelobe performance.

Example 5.2 The earth subtends an angle of 17.3° when viewed from geostationary orbit. Estimate the dimensions and gain of a pyramidal horn antenna that will provide global coverage at 4.5 GHz.

Solution

By assuming a uniformly illuminated wave across the aperture (length and breadth) of the pyramidal antenna, the beamwidths in the E and H planes may be considered equivalent; that is,

$$\theta = 17.3^\circ = \theta_E = \theta_H.$$

Take the antenna's efficiency $\eta = 50$ per cent.
Wavelength, $\lambda = c/f = 0.3\times10^9/4.5\times10^9 = 6.67$ cm
From (5.14) and (5.15), the aperture dimensions can be computed:

$$A = 30.06 \text{ cm}$$

$$B = 20.81 \text{ cm}$$

Gain, $G = 19.46$ dB

5.1.5 Search coverage

For a uniform search, the azimuth coverage sector is expressed for a rectangular beam as

$$A_m = \frac{\Omega}{\sin\theta_u - \sin\theta_L} \tag{5.17}$$

where θ_u and θ_L correspond to upper and lower search limits, see Figure 5.4.

Antenna theory gives the relationship between the transmitting-search solid angle Ω_s and transmitter gain, G_t, as

$$\Omega_s = \frac{4\pi}{G_t L_n} \tag{5.18}$$

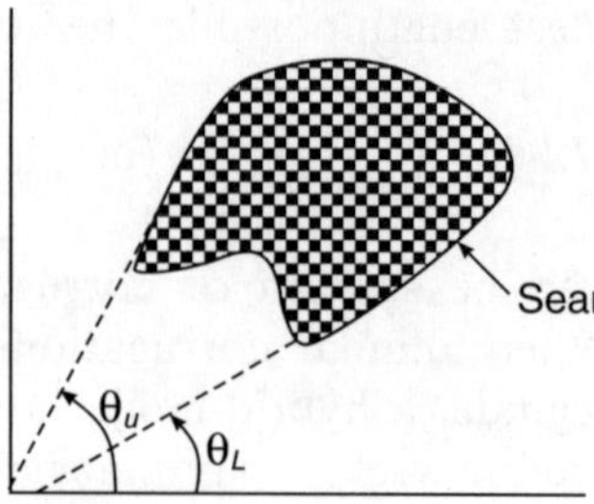

Figure 5.4 Geometry of search area

L_n is the pattern constant and accounts for power radiated outside the idealized mainlobe: typical value is between 1.2 and 1.6. The number of beam positions to be searched may be expressed by

$$n_b = \frac{\Omega_s}{\Omega} \tag{5.19}$$

The radar observation time t_0 depends on the allowable search (frame) time T_f and assigned solid angle Ω:

$$t_0 = \frac{T_f}{n_b} \tag{5.20}$$

Radar operates in a noisy environment. To utilize (5.5) or (5.7) in real life, the effect of noise in the system and environment must be accommodated.

5.1.6 Receiver bandwidth, temperature and noise

To calculate the equivalent input noise factor, the noise figures and gains of all stages must be known. The stage, in this instance, excludes the detector. Simplistically, the noise figure of passive stages, which do not contain noise sources other than thermal noise, equals their loss in decibels, or noise factor equals reciprocal of the gain:

$$F_i = 1/G_i \tag{5.21}$$

where

F_i = stage noise factor
G_i = stage gain.

Equation (5.22) is valid for situations where the passive stage's temperatures are relatively uniform. However, if a passive stage is at a higher temperature than the rest of the receiver chain, its noise figure must be adjusted for the temperature difference. To do this, assign the lowest component temperature as the system noise temperature. Thus, the noise factor for a passive (lossy) device at any temperature, T, is:

$$F = 1 + \frac{T(L-1)}{T_{amb}} \tag{5.22}$$

where

F = lossy device noise factor
L = loss of device (=1/gain)
T_{amb} = ambient temperature.

The reader might ask whether *noise figure* could be higher than the device loss. Yes, it is possible. For example, some passive stages, such as *double-balanced diode mixers*, often have a noise figure slightly higher than their loss. It should be noted also that the gains and noise factors of active stages are not correlated. As such, they must be obtained separately.

Following from (5.22), if two networks are in tandem, the networks' noise figure, F_{12}, may be expressed by

$$F_{12} = F_1 + \frac{F_2 - 1}{G_1} \tag{5.23}$$

where

F_1 = noise figure of first network
G_1 = gain of first network
F_2 = noise figure of second network.

If there are N stages comprising the receiver, the cascaded noise figure must be used as the total equivalent input noise factor for the receiver's noise figure, F_N:

$$F_N = F_1 + \frac{F_2 - 1}{G_1} + \frac{F_3 - 1}{G_1 G_2} + \cdots + \frac{F_n - 1}{G_1 G_2 \ldots G_n} \tag{5.24}$$

Concisely

$$F_N = 1 + \sum_{i=1}^{n} \frac{F_i - 1}{\prod_{j=0}^{i-1} G_j} \tag{5.25}$$

where $G_0 = 1$.

Following the noise factor derivation, the effective noise temperature T_e for N stages in cascade can be deduced as:

$$T_e = T_1 + \frac{T_2}{G_1} + \frac{T_3}{G_1 G_2} + \cdots + \frac{T_n}{G_1 G_2 \ldots G_{n-1}} \tag{5.26}$$

Concisely

$$T_e = \sum_{i=1}^{N} \frac{T_i}{\prod_{j=0}^{i-1} G_j} \qquad G_0 = 1 \tag{5.27}$$

How does the noise figure affect the radar equation?

5.1.7 Radar equation modified by noise and other losses

Noise is the primary factor limiting receiver sensitivity. Noise may originate from the receiver itself, or it may be part of the signal received via the antenna. While component segmentation could provide, as in (5.25), for the receiver's noise figure, this process might be cumbersome. As such, during the design or analysis process, the receiver's noise figure is measured as the ratio of the total noise N_O, at the output of the receiver to the thermal-noise power N_{thermal} obtained from an ideal receiver at standard temperature T_0. Specifically,

$$F_N = \frac{1}{G}\frac{N_0}{N_{\text{thermal}}} \tag{5.28}$$

G in this case is the available gain, being the ratio of the signal out, S_o, to the signal in, S_i.

The thermal noise (also called Johnson noise (Johnson 1928)) is the noise generated by the thermal motion of the conduction of electrons in the ohmic portions of the receiver-input stages. For a receiver of bandwidth B_n (in Hz) at a temperature T (in kelvin, K), this noise has been quantified as:

$$N_{\text{thermal}} = kTB_n \tag{5.29}$$

where k = Boltzmann's constant = 1.38×10^{-23} W/(Hz-K), and 1 K $\cong$ 273 + T (°C).

If the receiver circuitry were at some temperature, the thermal-noise power would be correspondingly different. The thermal-noise power N_{thermal} is primarily the input noise, N_i.

The radar receiver bandwidth, B_n, is often synonymous with the receiver's *intermediate frequency* (IF) amplification stage. B_n is an integrated bandwidth, defined as

$$B_n = \frac{\int_{-\infty}^{\infty} |H(f)|^2 df}{|H(f_0)|^2} \tag{5.30}$$

where $H(f)$ and f_0 correspond to the filter frequency response characteristic of an IF amplifier and the frequency of maximum response usually occurring at the mid-band. When $H(f)$ is normalized at the mid-band, $H(f)$ tends to unity. B_n is called *noise bandwidth*: a bandwidth equivalent to a rectangular

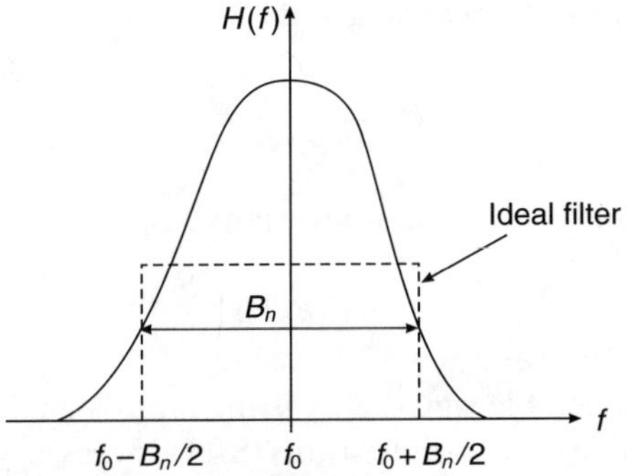

Figure 5.5 An equivalent noise bandwidth

filter whose noise-power output is similar to a filter with characteristic $H(f)$, depicted by Figure 5.5.

Note that the noise bandwidth B_n is not the half-power (3 dB). Although many receivers have noise bandwidths close enough to the 3 dB to make its use a good approximation, the measurement of noise bandwidth requires a complete knowledge of the response characteristic $H(f)$.

In low-PRF (LPRF) radar, the bandwidth of its IF stage can be very large: usually set at $B_n\tau = 1$, where τ is the pulse width. If the LPRF radar operates on time discrimination, a timing pulse is usually initiated at the start of each transmitted pulse. The target-return pulse is then matched against the timing pulse, resulting in the elapsed time between the returned and transmitted pulses. The elapsed time is proportional to the radar-target range, R. For most air search operations that require broad elevation sectors, LPRF radars are strongly favoured.

On the other hand, in high-PRF (HPRF) radars, target range rate information is primarily obtained using the Doppler principle, already discussed in Chapter 3, from the Doppler filters formed by FFT processing. The target in a given Doppler filter, or cell, competes with thermal noise that is folded into the Doppler ambiguity. In fact, the noise in any Doppler cell is a fraction of the front-end noise bandwidth, B_n. (See Appendix 5A for further discussion on the noise effect in Doppler processing and the implication on range calculation.) The Doppler filters' bandwidth is approximately, $B_d \approx 1/\tau_c$, where τ_c is the compressed pulse width. The total number of Doppler filters is calculated using the desired range of velocity coverage. Following (3.44), the number of Doppler filters, n_d, is

$$n_d = \frac{\Delta f}{B_d} = \frac{2\dot{R}}{\lambda B_d} \tag{5.31}$$

For surface search, for example sea or land vehicles, navigational or fixed structures, HPRF radars are preferred.

Equation (5.28) can be rearranged as

$$F_N = \frac{1}{\frac{S_o}{S_i}}\frac{N_0}{N_i} = \frac{\frac{S_i}{N_i}}{\frac{S_o}{N_0}} \tag{5.32}$$

In view of (5.29) and (5.32), the input signal, S_i is expressed as

$$S_i = kT_0B_nF_N\left(\frac{S}{N}\right)_0 \tag{5.33}$$

If the minimum detectable signal S_{min} is the input signal S_i that corresponds to the minimum IF signal-to-noise ratio $(S/N)_0$ necessary for detection, then (5.33) is recast as

$$S_{min} = N_{thermal}F_N\left(\frac{S}{N}\right)_0 \tag{5.34}$$

In an attempt to estimate the maximum range R_{max} beyond where a target cannot be seen, the received signal power P_r in (5.5) must equate to the *minimum detectable signal* S_{min} in (5.34). Specifically

$$P_r = \frac{P_tG_t}{4\pi R^2}\frac{\sigma}{4\pi R^2}A_e = N_{thermal}F_N\left(\frac{S}{N}\right)_0 \tag{5.35}$$

Or

$$P_r = \frac{P_tG_t}{4\pi R^2}\frac{\sigma}{4\pi R^2}\frac{A_e}{kT_0B_nF_N} = \left(\frac{S}{N}\right)_0 \tag{5.36}$$

Expressing the range R in terms of other variables

$$R_{max} = \left(\frac{t_0P_{av}G_tA_e\sigma}{(4\pi)^2kT_0B_nF_N\left(\frac{S}{N}\right)_0}\right)^{\frac{1}{4}} \tag{5.37}$$

which is in the form of a radar equation. The range for unity (or 0 dB) signal-to-noise ratio (i.e. when signal power equals the noise power) is called the *free space* range. The waveform of the transmitted signal does not enter into the radar equation. This suggests that the signal can be selected for other considerations such as range and Doppler resolution. The choice of signal, however, does play a major part of radar (or sonar) signal processing.

The reader might wonder whether an application of (5.37) is sufficient to give an accurate range estimate of the target in all conditions. Unfortunately, the answer is no because of the failure of (5.37) to include the various losses that can occur during radar operation, including an unpredictable nature of several parameters that have an effect on the radar performance. For example, the minimum detectable signal S_{min} and target radar cross-section σ are statistical in nature and must be expressed in statistical terms. Also, the environmental conditions along the propagation path(s) introduce some losses.

5.1.7.1 Other losses

Detailed discussion of all the factors that influence the prediction of radar range is beyond the scope of a single chapter. However, a summary of some of the losses is described in this section.

5.1.7.1.1 System loss

A system loss is associated with each stage of signal processing in both transmitting and receiving portions of a radar system. The losses in the transmitting portion include those from waveguide, duplexers and antenna. These losses are called plumbing losses, typically in the order of 2 to 5 dB. In the receiving portion, the losses include those from waveguide, mixers, RF and IF amplifiers, and antenna. The noise figure F_N of the receiver is an indication of its contribution to system loss.

5.1.7.1.2 Beam-shape and processing losses

One of the assumptions taken in radar analysis is that field strength is constant over the width of the beam. In actuality, as a target passes through a beam, the signal return is modulated. This causes a loss called the beam-shape loss.

Processing loss includes those due to FFT windowing, typically of the order of −2 dB.

5.1.7.1.3 Collapsing loss

When noise from different sources converges in the proximity of the true target return, collapsing loss occurs. Its effect is to increase the background noise level, thereby decreasing the detectable signal (S/N) level of the true target. A typical value of 1 dB is assigned for the range bin collapsing loss. The collapsing loss is most pronounced (less than 1 dB) in the short-range mode.

5.1.7.1.4 Propagation loss

Ducting is a form of anomalous propagation. It causes radar beams to travel in a curved line as opposed to the normal, straight line. Ducting can cause radar not to pick up objects (or targets) it would otherwise detect or that it detects objects (or targets) much further away than it normally would. It is undependable and can degrade the performance of MTI (*moving target indicator*) radar by extending the range at which ground clutter is seen.

Aside, signals propagated through the atmosphere suffer another loss, called *atmospheric attenuation*. Depending on the type of radar used, absorption of radio waves occurs differently in the lower atmosphere (called *tropospheric attenuation*), or in the upper atmosphere (called *ionospheric attenuation*). More is said about the division of the atmosphere into regions in Chapter 6.

The absorption of propagation waves in the troposphere is caused by the presence of both free molecules and suspended particles such as dust grains and water drops condensed in fog and rain. Rain attenuation is modelled as

$$A_r = aLr_r^b \tag{5.38}$$

where

(i) a and b are coefficients that are calculable theoretically from considerations of electromagnetic propagation in spherical rain drops. These coefficients are polarization and frequency dependent based on raindrop characteristics, and can be approximated to the following analytical expressions:

$$a = \begin{cases} 4.21 \times 10^{-5} f^{2.42} & f \leq 54\,\text{GHz} \\ 4.09 \times 10^{-2} f^{0.699} & 54 < f \leq 180\,\text{GHz} \end{cases} \tag{5.39}$$

$$b = \begin{cases} 1.41 f^{-0.0779} & f \leq 25\,\text{GHz} \\ 2.63 f^{-0.272} & 25 < f \leq 164\,\text{GHz} \end{cases} \tag{5.40}$$

Outside these frequency ranges, the coefficients are equated as zero. If the coefficients are linearly polarized vertically or horizontally, the coefficients for a circularly polarized wave can be calculated using:

$$a_c = 0.5(a_h + a_v) \tag{5.41}$$

$$b_c = \frac{a_h b_h + a_v b_v}{2a_c} \tag{5.42}$$

The subscripts of the constants indicate their polarization. For example, subscripts '*c*', '*h*' and '*v*' denote circular, horizontal, and vertical polarization respectively.

(ii) r_r = rain rate (mm/hr). Average values of r_r can be obtained from the Department of Meteorology (or its equivalent) of your country.

(iii) L = path length (km) of the intervening rain.

Attenuation due to absorption by electromagnetic waves, aside the rain, follows the relationship (Millan 1965)

$$A_d = 10^{-0.05\gamma L} \tag{5.43}$$

where γ is decay constant.

The decrease in signal strength, for a radio wave traversing an absorbing region, is in addition to space or inverse-square-law attenuation.

The absorption of signal energy in the ionosphere occurs when electrons, colliding with other particles, are forced to give up some of their energy to these particles. More is said about electron formation, collision and refractivity in Chapter 6. The amplitude of the signal is attenuated logarithmically with increasing distance, s (Millan 1965):

$$A_{ion} = 20 \log_e^{-\int_{s_1}^{s_2} k_a ds} \quad \text{(dB)} \tag{5.44}$$

where

A_{ion} = ionospheric attenuation
k_a = absorption coefficient of the medium
s_1 and s_2 are the limits of the path.

Since the ionosphere can be divided into distinct heights, the path differential can be expressed in terms of height differential by

$$ds = f(h)dh \tag{5.45}$$

This relationship becomes obvious in Chapter 6.

5.1.7.1.5 Polarization loss

The term polarization refers to the direction of an electric or magnetic vector in the radiated field. If the transmitting and receiving antennas were not properly polarized, because either the propagation medium changes the original polarization, or the target depolarizes the signal, polarization losses would occur. For instance, if an elliptically polarized receiver receives a linearly polarized transmitted wave, the nature of energy received would be seriously affected: either a complete loss of signal or the signal is severely distorted.

Kramer (1986) established a relationship for the polarization loss L_p between an antenna elliptically polarizing and an antenna linearly polarizing as

$$L_p = 0.5(1 + k_\chi \cos 2\chi_p) \tag{5.46a}$$

where

$$k_\chi = \frac{a_k^2 - 1}{a_k^2 + 1} \tag{5.46b}$$

a_k is the axial ratio of elliptical polarization and χ_p is the angle between linear polarization and the ellipse's major axis. Kramer further established a cumulative probability expression for estimating the polarization loss L_p for a given value, say l_p, when distributed randomly over zero to 2π angular orientation of one antenna to the other. Specifically

$$P(L_p \leq l_p) = 0.5 + \frac{1}{\pi} \sin^{-1}\left(\frac{2l_p - 1}{k_\chi}\right) \tag{5.47}$$

For brevity, L_{tot} denotes these losses. Rigorous radar system engineering involves careful evaluation of each loss term and the evolution of a design, which minimizes the losses for the intended application.

5.1.7.1.6 Multi-path reflection factor

Surface reflection is due to the modification of the free-space field that results from reflection of the waves from the surface beneath the direct path. Surface reflection causes multi-path lobing effects on target detection, and multi-path errors in tracking and radar measurement. To account for this effect, a term called *pattern-propagation factor* is included in the radar equation. The quantity F describes the ratio of a one-way field amplitude at range R to that which would have been obtained under free-space conditions in the centre of the beam. Thus the *pattern-propagation factor* can be defined as the value of F obtained with a broad antenna beam, such that the underlying surface of the earth is fully illuminated. Our discussion on the propagation factor F is restricted to a flat earth model whose geometry is shown in Figure 5.6.

Above the flat earth surface, at point A, height h_a, a radar antenna is assumed to be located. A target is located at point C distance h_t above the earth's surface and ground range (distance) R_h from the antenna. The reflected wave ABC hits the earth surface at point B and reflects to C. The direct wave is AC. Symbols θ_a, θ_t and ψ represent the antenna elevation angle, target elevation angle and grazing angle respectively. By simple geometry, ABC = ABE = R_2.

By considering the two right-angle triangles ADC and ADE, these equations are written:

$$R_1 = \left[R_h^2 + (h_t - h_a)^2\right]^{\frac{1}{2}} \approx R_h\left[1 + \frac{(h_t - h_a)^2}{2R_h^2}\right] \quad R_h \gg (h_a + h_t) \qquad (5.48a)$$

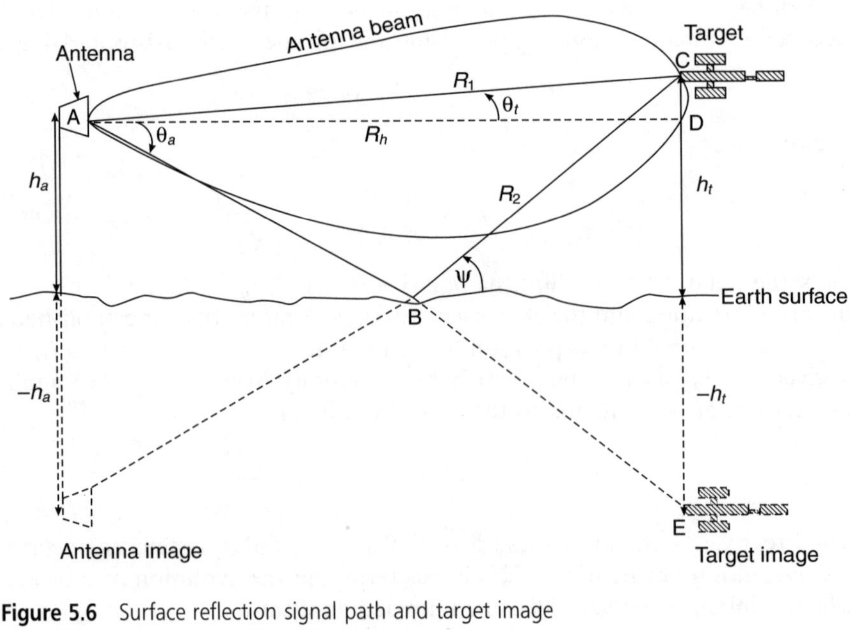

Figure 5.6 Surface reflection signal path and target image

$$R_2 = \left[R_h^2 + (h_t + h_a)^2\right]^{\frac{1}{2}} \approx R_h\left[1 + \frac{(h_t + h_a)^2}{2R_h^2}\right] \quad R_h \gg (h_a + h_t) \qquad (5.48b)$$

And the ground range:

$$R_h = (h_t + h_a)\cot\theta_a \qquad (5.49)$$

The path difference between the direct and reflected waves is

$$\Delta R_d = R_2 - R_1 \approx \frac{2h_a h_t}{R_h} \qquad (5.50)$$

And, providing that the grazing angle ψ is small, the phase caused by the path difference may be expressed by

$$\Delta\phi = \frac{2\pi}{\lambda}\Delta R_d = \frac{4\pi h_a h_t}{\lambda R_h} \qquad (5.51)$$

If the total phase difference $\Delta\phi$, in (5.51) for instance, is equal to an even multiple of π, the waves are said to be in phase. So, a signal maximum results when

$$\frac{4\pi h_a h_t}{\lambda R_h} = 2n \qquad (5.52a)$$

Conversely, the odd multiple of π gives the null or minimum when

$$\frac{4\pi h_a h_t}{\lambda R_h} = 2n + 1 \qquad (5.52b)$$

It should be noted that the expressions in (5.52) do not account for any phase change or amplitude change that might occur at the earth-reflecting surface. However, if the antenna gain does not change between the direct and the reflected rays' directions, and also that the target backscattering pattern does not change, then the direct and reflected fields reaching the target are of equal intensity but having a phase difference given by

$$\Delta\phi_0 = \frac{4\pi h_a h_t}{\lambda R_h} + \pi \qquad (5.53)$$

By allowing a complex factor F to represent the ratio of the resultant field at the target in the presence of surface reflection coefficient ρ, an expression for the factor may be written as

$$F = 1 + \rho e^{-j(\Delta\phi_0 + \gamma_f)} \qquad (5.54)$$

where γ_f is the phase angle of the reflection coefficient. The power ratio at the target is $|F^2|$. However, by assuming broad antenna pattern assumption,

$$F^2 = 1 + \rho^2 + 2\rho\cos(\gamma_f + \Delta\phi_0) \qquad (5.55)$$

Given that the same multi-path effect would occur on return signals from the target to radar, the power ratio received with and without the presence of the earth-reflecting surface equals $|F^4|$. For brevity, $\rho \approx 1$ and $\gamma_f = \pi$. Upon an application of trigonometric relationships, the resultant power ratio can be written as

$$|F^4| = (2 + 2\rho\cos\Delta\phi_0)^2 = 16\cos^4\left(\frac{\Delta\phi_0}{2}\right) \tag{5.56}$$

By substituting (5.53) in (5.56),

$$|F^4| = 16\sin^4\left(\frac{2\pi h_a h_t}{\lambda R_h}\right) \tag{5.57}$$

Using a small angle approximation technique

$$|F^4| \approx 16\left(\frac{2\pi h_a h_t}{\lambda R_h}\right)^4 \approx (\Delta\phi)^4 \tag{5.58}$$

In fact, this expression is a gain rather than a loss to the radar equation.

By including all losses and the multi-path reflection factor in (5.37), the radar equation is

$$R = \left(\frac{t_0 P_{av} G_t A_e \sigma |F^4|}{(4\pi)^2 k T_0 B_n F_N \left(\frac{S}{N}\right)_0 L_{tot}}\right)^{\frac{1}{4}} \tag{5.59}$$

This expression assumes that detected peak or echo is at the centre of the beam associated with a single PRF. In actual digital mechanization, by placing digital filters that cover the frequency span of one PRF, a particular digital filter will respond to target returns symmetrically located on either side of it. This makes the signal-to-noise ratio $(S/N)_0$ in (5.59) possible. In medium-PRF (MPRF) radars, to resolve range and/or Doppler ambiguities, two or more PRFs are used. Therefore, to calculate $(S/N)_0$, an average PRF is often used because of the closeness of the PRFs.

The detection and measurement of target reflected energy is most affected by competing clutter and thermal noise energy. To use (5.59) in a clutter environment requires knowledge of the clutter and its energy, which will be used to modify the thermally induced noise-only radar equation typified by (5.59). Clutter is unwanted echoes, typically from the ground, sea, rain or other precipitation, chaff, birds, insects and aurora. The characteristics of ground, sea and rain clutter are studied in section 5.4.

Example 5.3 Consider a transmitter with peak power of 1.5 kW with a gain of 10 dB when propagating at 3 GHz. Calculate (a) the magnitude of the signal received at room temperature by the receiver of 8 m^2 aperture if the radar is upward looking, 20.9 dB for all extraneous search losses, and the target of 1.5 m^2 cross-section is viewed at about 100 km away from the receiver. Consider a noise factor of 5 dB and noise bandwidth to match the receiver's bandwidth. (b) A mismatch between the noise and receiver bandwidth was noticed during observation. If the noise bandwidth was given as 1 kHz, calculate the magnitude of the signal received. (c) If the radar scans at 100 rpm, calculate the time frame required for the scan. (d) Calculate the antenna elevation beamwidth for an equal azimuthal beamwidth. (e) How long can the radar dwell on the target?

Solution

$$\begin{aligned}
L_{\text{tot}} &= 20.9\,\text{dB} = 10^{2.09} = 123.03 \\
F_N &= 5\,\text{dB} = 10^{0.5} = 3.16 \\
G_t &= 10\,\text{dB} = 10 \\
P_t &= 1.5\,\text{kW} = 1500\,\text{W} \\
T_0 &= 23.7^\circ\text{C} = 273 + 23.7 = 296.7\,\text{K} \\
R &= 100\,\text{km} = 10^5\,\text{m} \\
k &= 1.38 \times 10^{-23} \\
\sigma &= 1.5\,\text{m}^2 \\
A_e &= 8\,\text{m}^2
\end{aligned}$$

For an upward looking antenna, $|F^4| = 1$.

Rearranging (5.59) in terms of the receiver signal-to-noise ratio:

$$\left(\frac{S}{N}\right)_0 = \frac{P_t G_t A_e \sigma |F^4|}{(4\pi)^2 k T_0 B_n F_N R^4 L_{tot}}$$

Substituting values that correspond to the afore-listed symbols, the following numerical values are obtained:

(a) Matched filter: $B_n = 1$

$$\left(\frac{S}{N}\right)_0 = 8.55\,\text{dB}$$

(b) Mismatched filter: $B_n = 1\,\text{kHz}$

$$\left(\frac{S}{N}\right)_0 = -21.5\,\text{dB}$$

(c) If the radar scans at 100 rpm, calculate the time frame required for the scan.

$$T_f = 2\pi/100 = 3.77\,\text{s}$$

(d) Following (5.18) the solid beam angle Ω_s that corresponds to antenna gain of 10 dB is:

$$\Omega_s = \frac{4\pi}{G_t} = 1.26\,\text{rad}^2 = 72.0 \text{ degrees squared}$$

putting the pattern loss $L_n = 0\,\text{dB}$. Hence the elevation beamwidth:

$$\theta_e = \theta_a = \sqrt{\Omega_s} = 8.48^\circ$$

(e) Following (5.8), the time t_0 the radar dwells on the target is

$$t_0 = \frac{\theta_{BW}}{6\omega_m} = 0.01414\,\text{s}$$

5.2 Target fluctuation models

Basically all radar target objects produce echo signals that vary in amplitudes either in power or cross-sectional terms. These amplitudes can vary from scan to scan or between echo to echo due to aspect changes relative to the radar. This variation is often referred to as *target scintillation*.

A simple target model is shown in Figure 5.7, which could represent a reflective structure of a satellite, an aircraft, a warship, or a submarine. This figure is similar to that given in Van Trees (1971). If the direction of signal propagation is along the x-axis and the target orientation is assumed to be changing with time, then three target positions are shown in Figure 5.7(a, b, c). It is assumed that the target in Figure 5.7 is illuminated with a long pulse of duration T whose envelope is shown in Figure 5.8(a) and the received signal envelope is represented by Figure 5.8(b).

If the received envelope is distorted as shown in Figure 5.8(b), the target is changing, varying, or fluctuating, with time as the target changes its orientation. This varying-time attenuation of the received envelope is often called *time-selective fading*. On the other hand, if a short pulse is transmitted as in Figure 5.8(c), and an undistorted signal envelope is received as in Figure 5.8(d), the target can be considered to be slowly fluctuating (Van Trees 1971).

Target cross-section fluctuations are complex to quantify by a simple mathematical expression. Swerling (1960) postulated models that describe

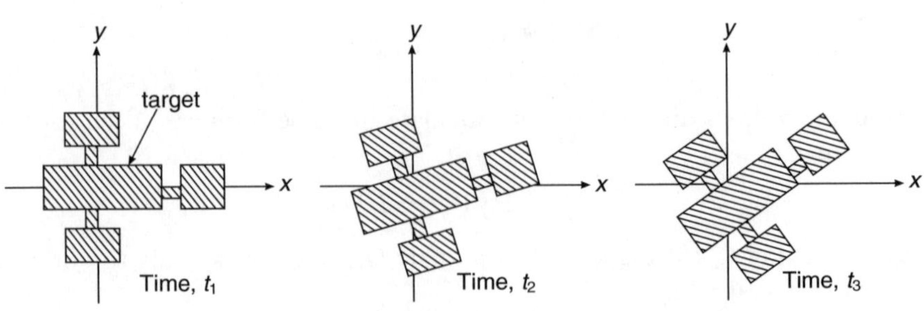

Figure 5.7 A representation of target orientations at different times

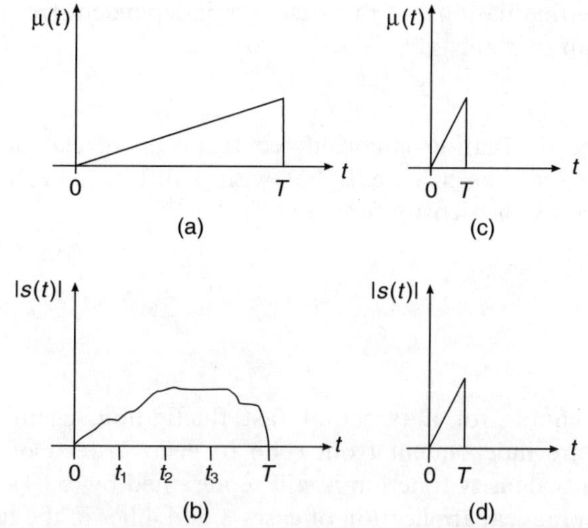

Figure 5.8 An illustration of time-selective fading of transmitted signals: (a) envelope of transmitted signal; (b) returned envelope of signal (a) with time varying attenuation; (c) short-pulse transmitted signal; (d) returned short-pulsed signal without distortion

slowly and fast varying targets. The slowly fluctuating-target model is assumed to have complete correlation, or dependence, between echo signals during a radar scan, but independent with scan to scan. The fast fluctuating-target model is assumed to have partial correlation from echo to echo instead of scan to scan. The virtue of these models lies in the fact that they are a reasonable approximation of a variety of targets. The models are briefly described as cases as follows.

Case 1

Swerling designated a target as case 1 when fluctuation is slow. For instance, when the echo signals or pulses, received from a target on any one scan, are of constant amplitude throughout the duration of the scan. These signals are uncorrelated (independent) from scan to scan. The probability density function for the cross-section σ is given by the density function:

$$p(\sigma) = \frac{e^{\frac{-\sigma}{\sigma_{av}}}}{\sigma_{av}} \qquad \sigma \geq 0 \tag{5.60}$$

where σ_{av} is the average cross-section over all target fluctuations. It must be noted that this case ignores the effect of the antenna beam shape on the amplitudes of echo signals.

Case 2

This case accounts for a target of fast fluctuation. The probability density function for the cross-section σ has a similar distribution function as that in

(5.60) but the fluctuations, in this case, are independent from echo to echo instead of scan to scan.

Case 3
In case 3, the fluctuation is considered to be uncorrelated (independent) from scan to scan, as in case 1, but with a different probability density function given by the density function:

$$p(\sigma) = \frac{4\sigma}{\sigma_{av}^2} e^{\frac{-2\sigma}{\sigma_{av}}} \tag{5.61}$$

Case 4
This case accounts for a target of fast fluctuation, as in case 2 where fluctuations are independent from echo to echo instead of scan to scan. The probability density function is still represented by (5.61).

A known practical application of cases 3 and 4 lies in the use of case 3 to represent case 1 target observed by dual-diversity radar.

The probability density function given by (5.60) and (5.61) are special cases of the chi-square, or gamma, distribution with $2n$ *degrees of freedom* ($2n$ DOF). Chi-square distribution is a general approximation of target models. Specifically

$$p(\sigma) = \frac{n}{\sigma_{av}(n-1)!}\left(\frac{n\sigma}{\sigma_{av}}\right)^{n-1} e^{\frac{-n\sigma}{\sigma_{av}}} \quad \sigma \geq 0 \tag{5.62}$$

The envelope of σ is taken as a Rayleigh random variable whose average is σ_{av}. When the unresolved target echo results from many scatterers of comparable size adding vectorially with random phases, then the echo amplitude is Rayleigh distributed (and the echo energy is thus exponentially distributed) (Heering 1977). Unlike in statistical texts where index n is only allowed to be an integer, n in this instance must be positive and can also be a real number when applying to target cross-section models.

Nakagami (1960) gave a more generalized chi-square model as

$$p(\sigma) = \frac{2n^n \sigma^{2n-1}}{\sigma_{av}^n \Gamma(n)} e^{\frac{-n\sigma^2}{\sigma_{av}}} \quad \sigma > 0 \tag{5.63}$$

where Γ (.) is the gamma function of (.) and n can be real or integer.

Another model worth mentioning is the Rician model (Rice 1944). The Rician model is suitable for the case of one dominant signal in the presence of many other small signals. Specifically

$$p(x,\overline{x}) = \begin{cases} \frac{1}{\psi_0} I_0\left[2\sqrt{\frac{sx}{\psi_0}}\right] e^{-s-\frac{x}{\psi_0}} & x > 0 \\ 0 & x \leq 0 \end{cases} \tag{5.64}$$

where

s = ratio of steady reflector's radar cross-section to the combined average cross-section of Rayleigh scatterers

$$\bar{x} = \psi_0(1 + s) \tag{5.65a}$$

$$\sigma = \psi_0\sqrt{1 + 2s} \tag{5.65b}$$

ψ_0 = mean value of Rayleigh component of x
I_0 = modified Bessel function of the first kind of zero order.

It must be acknowledged that little, if any, real targets fit a mathematical model with any precision. Targets have complex geometry. As such, the various mathematical models cannot be expected to produce precise predictions of system performance. In effect, the use of constant (non-fluctuating) cross-section in radar equation is a very attractive alternative when prior information about the target is minimal.

5.3 Detection probability

The signal-to-noise ratio $(S/N)_0$ required to achieve target detection is statistical. It depends on probabilities of detection and false alarm, and other additional factors that enter into target detection. The minimum $(S/N)_0$ that is required at achieving a specific detection probability without exceeding a specified false-alarm probability could be calculated. An expression that connects $(S/N)_0$ with the specific probabilities as well as with target scintillation was developed by Neuvy (1970) as

$$\left(\frac{S}{N}\right)_0 = \frac{\alpha_n \log\left(\frac{\ln 2}{P_{fa}}\right)}{n_p^{\frac{2}{3}}\left[\log\left(\frac{1}{P_d}\right)\right]^{\beta_n}} \tag{5.66}$$

where n_p is the number of signal pulses transmitted. This expression has an inverse and behaved reasonably well in real-life scenarios. The symbols α_n and β_n are coefficients, each assumes a specific value as per Swerling case, shown in Table 5.1.

By using the Neuvy expression given by (5.66), a family of curves was plotted, as shown in Figures 5.9 to 5.13, for a single pulse and different Swerling

Table 5.1 Neuvy's coefficients

Swerling case	α_n	β_n
0	$1 + 2e^{\frac{-n_p}{3}}$	$\frac{1}{6}$
1	$\frac{2}{3}\left(1 + \frac{2}{3}e^{\frac{-n_p}{3}}\right)$	1
2	1	$\frac{1}{6} + e^{\frac{-n_p}{3}}$
3	$\frac{3}{4}\left(1 + \frac{2}{3}e^{\frac{-n_p}{3}}\right)$	$\frac{2}{3}$
4	1	$\frac{1}{6}\left(1 + 2e^{\frac{-n_p}{3}}\right)$

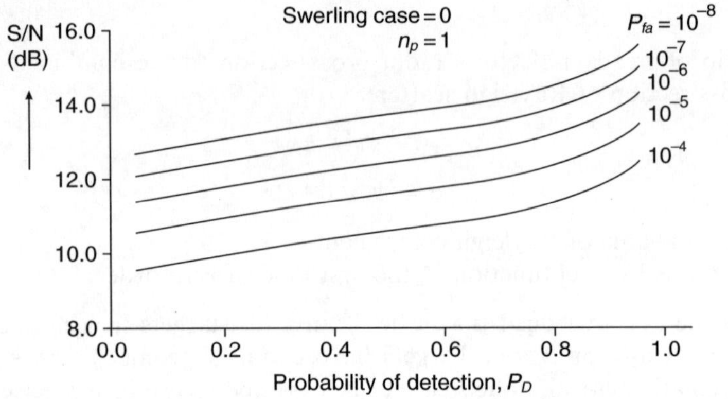

Figure 5.9 Curves of minimum signal-to-noise ratio versus probability of detection for various probability of false alarm

cases and probability of false alarms. If the number of pulses transmitted increases, the magnitude of the expected minimum signal-to-noise ratio to detect a fluctuating target decreases. Thus, for a specific probability of detection, P_D, and probability of false alarms, P_{fa}, the minimum signal-to-noise ratio required to achieve detection of target with variable reflectivity can be estimated.

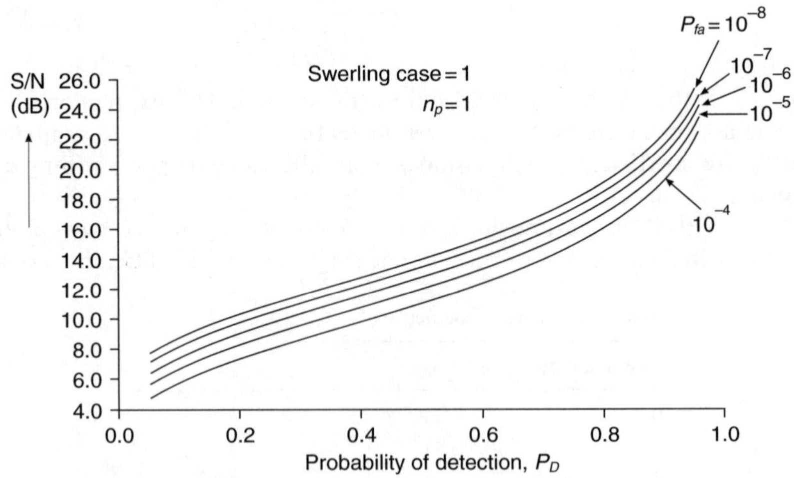

Figure 5.10 Curves of minimum signal-to-noise ratio versus probability of detection for various probability of false alarm

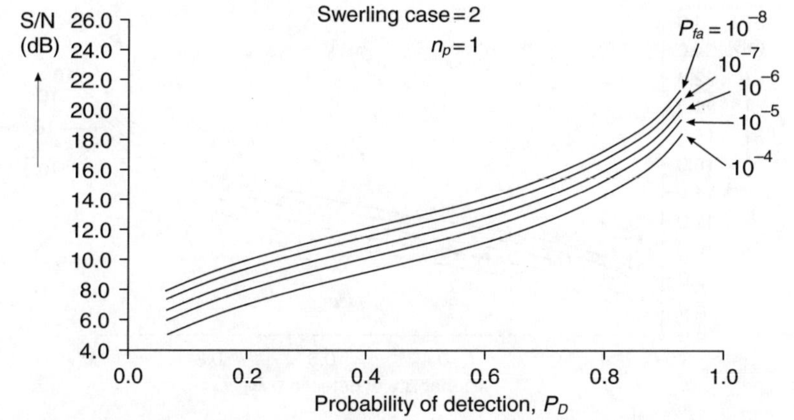

Figure 5.11 Curves of minimum signal-to-noise ratio versus probability of detection for various probability of false alarm

The minimum detectable signal has also been described by detectability factor, D_x defined as the energy ratio necessary to achieve detection. If the target fluctuating density function is described by gamma distribution, with $2n$ degrees of freedom, then the expression for the detectability factor for n_p transmitted pulses is defined by Barton (1988):

$$D_x(n_p) = \frac{L_f^{\frac{1}{ne}}}{k_e n_e}\left[\frac{\log(P_{fa})}{\log(P_d)} - 1\right] \tag{5.67}$$

where

L_f = the steady-state apparent fluctuation loss

k_e = number of degrees of freedom describing the target function. This is equivalent to half the number of independent gaussian components added together to form a target signal

n_e = number of independent signals or pulses integrated during N-pulse transmission.

If k_e and n_e are large, (5.67) then describes a steady target (i.e. case 0), thus:

$$D_x(n_p) = \frac{L_f}{n_p}\left[\frac{\log(P_{fa})}{\log(P_d)} - 1\right] \tag{5.68}$$

For other Swerling cases, the parameters k_e and n_e are as defined in Table 5.2, when applied to the generalized expression of (5.67). The fluctuation loss in (5.67) may be considered as a *diversity gain*, G_d, for a system taking

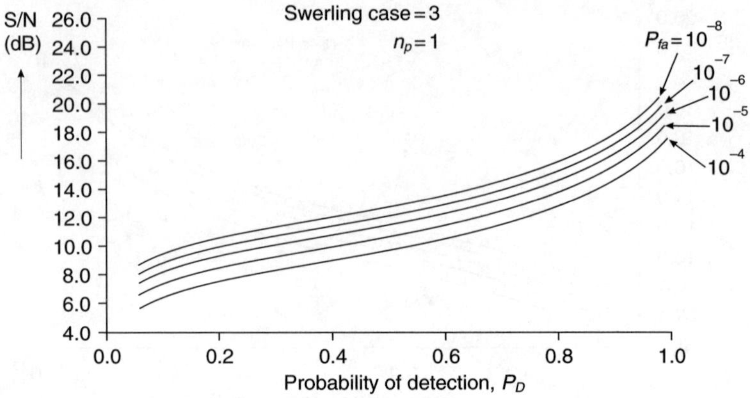

Figure 5.12 Curves of minimum signal-to-noise ratio versus probability of detection for various probability of false alarm

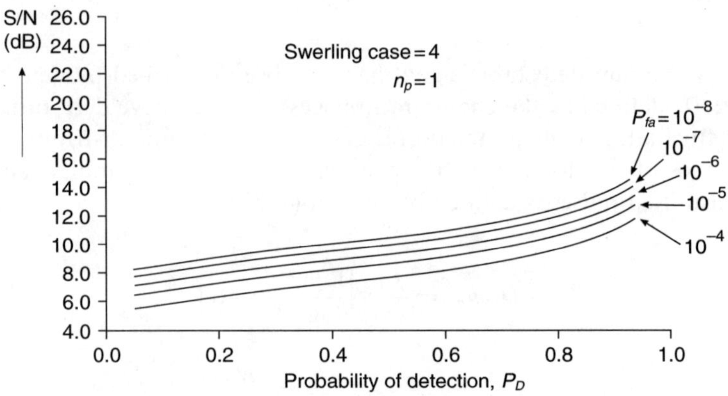

Figure 5.13 Curves of minimum signal-to-noise ratio versus probability of detection for various probability of false alarm

samples over intervals in time or frequency. The diversity gain may be defined as:

$$G_d(n_e) = (L_f)^{1-\frac{1}{n_e}} \tag{5.69}$$

Table 5.2 Independent parameters

	Parameters	
Swerling case	k_e	n_e
1	1	1
2	1	n_p
3	2	2
4	2	$2n_p$

Diversity is only possible if a non-diverse system has a fluctuation loss. Strictly speaking, two cases (time and frequency) can be distinguished for diversity, with the third being a combination of the two. These diversity cases are discussed briefly as follows.

5.3.1 Time diversity

Time diversity is when n_e independent samples are obtained at intervals equal to the correlation time of the target. The requirement of time diversity requires the signal observation (or integration) time t_0 exceeding the target correlation time t_c. Target correlation time approximates to

$$t_c \approx \frac{\lambda}{2\omega_{ma} l_t} \tag{5.70}$$

where ω_{ma} and l_t correspond to rate of rotation of the radar (rad/s) and target length or target broadest part (in metres). In fact, the length should be the section measured normal to the radar axis of rotation. When the surveillance of long dwells is observed, the correlation time must be much less than the *pulse repetition interval* (PRI), i.e. $t_c < 1/\text{PRF}$. In real life, integration is carried out over several scans. But if targets move between scans, integration within a narrow range of cells might be difficult and, when this situation arises, integration is performed cumulatively. The number of independent samples may be expressed as

$$n_e = 1 + \frac{t_0}{t_c} \tag{5.71}$$

Note that n_c may not necessarily equal the number of pulses transmitted, n_p.

5.3.2 Frequency diversity

Frequency agility is a situation in which n_e independent samples are received rapidly by changing transmitter frequency from pulse to pulse. Frequency agile radar can approach Swerling case 2 classification. In the frequency diversity case, the number of independent samples is estimated using

$$n_e = 1 + \frac{B_{na}}{f_c} \tag{5.72}$$

B_{na} and f_c are available bandwidth for integration and target correlation frequency respectively. Similar to *time diversity* analysis, target correlation frequency is related to the target radial length l_r and speed of light, c. Specifically

$$f_c = \frac{c}{2l_r} \tag{5.73}$$

In a cluttered environment, a fractional change in frequency between pulses would decorrelate the clutter, thereby permitting an increase in target-to-clutter ratio when the decorrelated pulses are integrated. However, clutter statistics are non-Rayleigh, particularly sea clutter, where clutter spikes persistently appear – spikes that tend to correlate over a relatively long duration, which may reduce the benefit of frequency agility. Conversely, when a radar is hoisted on a moving platform, the clutter might also decorrelate as the radar resolution cell looks at a different patch of clutter.

5.3.3 Time and frequency diversity

The third diversity case, the combined time and frequency diversity, is a case where time and frequency effects are used to increase the number of independent samples. Specifically

$$n_e = \left(1 + \frac{t_0}{t_c}\right)\left(1 + \frac{B_{na}}{f_c}\right) \tag{5.74}$$

With this scheme, it is essential to ensure that the transmissions are uniformly distributed over the time-frequency space to avoid correlation between pulses, which invariably reduces n_e.

In essence, equation (5.66) or (5.67) corresponds to the desired value of detection probability P_d and false-alarm probability P_{fa}, which can be fed into the radar equation (5.59).

Example 5.4 Consider a transmitter with a peak power of 100 kW with a gain of 50 dB. The transmitter sends three pulses of equal width of 1 μs at every second. It is desired to have a low probability of false alarm at 10^{-6} and detection probability of 0.9. The receiver is matched to receive the 1 μs-width pulses. It also has an aperture of 8 m^2 and noise factor of 5 dB. The total propagation losses envisaged are not more than 18.3 dB. Calculate the maximum range required detecting a type III target of 3.2 m^2 radar cross-section at a temperature of 32.8 °C.

Solution

$$\begin{aligned}
T_s &= 1\ \text{s} \\
n_p &= 3 \\
P_D &= 0.9 \\
G_t &= 50\ \text{dB} = 10^5 \\
k &= 1.38 \times 10^{-23} \\
\tau &= 1\ \mu\text{s} \\
T_0 &= 273 + 32.8 = 305.8\ \text{K} \\
P_{fa} &= 10^{-6} \\
P_t &= 100\ \text{kW} = 10^5\ \text{W} \\
\sigma &= 3.2\ \text{m}^2 \\
B_n &= 1/\tau = 1\ \text{MHz} \\
F_n &= 5\ \text{dB} = 10^{0.5} = 3.162 \\
L_{\text{tot}} &= 18.3\ \text{dB} = 10^{1.83} = 67.608
\end{aligned}$$

Using the Neuvy expression of (5.66) as well as Table 5.1, the signal-to-noise ratio $(S/N)_0$ can be determined.

For type III target:

$$\alpha_n = \frac{3}{4}\left(1 + \frac{2}{3}e^{\frac{-n_p}{3}}\right) = 0.9339 \quad \beta_n = \frac{2}{3}$$

$$\left(\frac{S}{N}\right)_0 = \frac{\alpha_n \log\left(\frac{\ln 2}{P_{fa}}\right)}{n_p^{\frac{2}{3}}\left[\log\left(\frac{1}{P_d}\right)\right]^{\beta_n}} = 13.12\,\text{dB}\,(20.5)$$

This expression indicates that for the target to be detectable, the received signal must be at least 13.12 dB. With this information, the maximum detectable range can be estimated. For brevity, $|F^4| = 1$ for an upward looking antenna. Using (5.59) while writing 'τP_t' instead of '$t_0 P_{av}$' and substituting values, the detectable range

$$R = \left(\frac{\tau P_t G_t A_e \sigma |F^4|}{(4\pi)^2 k T_0 B_n F_N \left(\frac{S}{N}\right)_0 L_{\text{tot}}}\right)^{\frac{1}{4}} = 3.06\,\text{km}$$

5.4 Target detection range in clutter

To derive the radar equation required to evaluate the target detection range in a background of clutter requires knowledge of the reflectivity of clutter sources. Instead of the signal-to-noise ratio (S/N) concept previously used, the signal-to-clutter ratio (S/C) is used. Interference is defined as the combination of system noise and clutter, which is assumed to add incoherently. The clutter discussed in this section includes rain clutter, land and sea clutters.

Regardless of the purpose for which radar is intended, clutter is very harmful because it always appears to accompany the useful target signal. It is thus imperative to provide a mechanism for rejecting clutter by radar designers and signal processing professionals; a summary of how the clutter rejection issue is approached is discussed in section 5.4.3. For a sample of the background material applicable to the clutter models discussed in this section see Barton (1988), Beckmann and Spizzichino (1987), Guinard and Daley (1970), Katzin (1957), Kerr (1951), Keydel (1976), Rice (1944), Sinnott (1989), Trunk (1972), Ulaby *et al.* (1986), Vizmuller (1995) and Ward (1982).

5.4.1 Land and sea clutter

Clutter from land, or sea, surfaces can be treated as a target that produces a radar cross-section, σ_c. To quantify σ_c requires knowledge of many factors such as surface composition, measurement wavelength, roughness, polarization, look (depression) angle, wind velocity (for sea), etc. Land and sea clutter cross-section σ_c is proportional to the product of land reflectivity, σ^0, or sea reflectivity $\overline{\chi}$ (to distinguish it from land reflectivity) and the illuminated surface area A_c (in m^2) within a radar resolution cell. Reflectivity is dimensionless. So, the clutter radar cross-section is written as

$$\sigma_c = A_c \sigma^0 \quad \text{m}^2 \tag{5.75}$$

This equation is the clutter radar cross-section of a single unambiguous range within a cell. However, in a given ambiguity range, all the contributions from all range cells that map on to the cell that is being resolved must be added. Knowing the clutter radar cross-section, the clutter power, P_c, in a given range ambiguity can be quantified.

Before exploring further, it is necessary to have a close look at the nature of clutter in a typical radar antenna pattern. Although radar attempts to concentrate its energy in a tight beam, in fact, it transmits and receives energy to some extent from all directions: mainlobe and sidelobes – comprising the near sidelobes (those closest to the mainlobe) and the far sidelobes of different intensities, see Figure 5.14. However, for analytical purposes, these sidelobes are lumped as the same.

Whenever the mainlobe and sidelobes illuminate a target, surface clutter is returned with the signal, see Figure 5.15. Clutter received from the mainlobe is called *mainlobe clutter* (MC) and that via the sidelobe is called *sidelobe clutter* (SC). In addition, in the mainlobe, there is another clutter called *residual mainlobe clutter* (RMC). RMC is present in all detection cells, its power is more important than the MC since detection is not attempted in detection cells containing MC. The residual clutter power is the same as that in MC but modified by the MC rejection factor, K_{rej}. For completeness, if the radar antenna is hoisted above the surface at altitude h_a (m), *altitude clutter* could be received directly below the radar platform. Since radar platform motion is relatively stable and constant, the altitude clutter is centred on zero

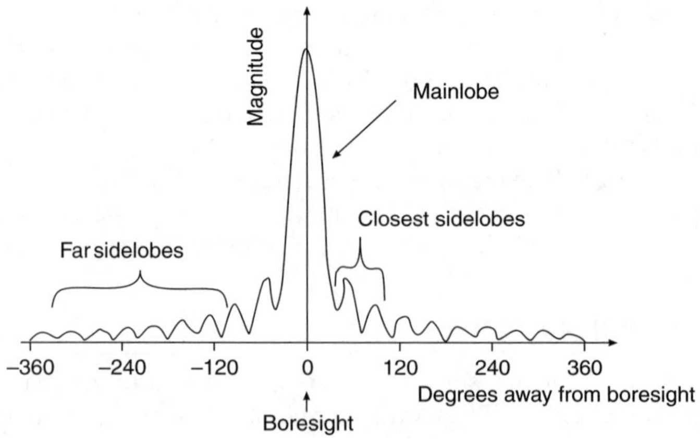

Figure 5.14 Typical radar antenna pattern

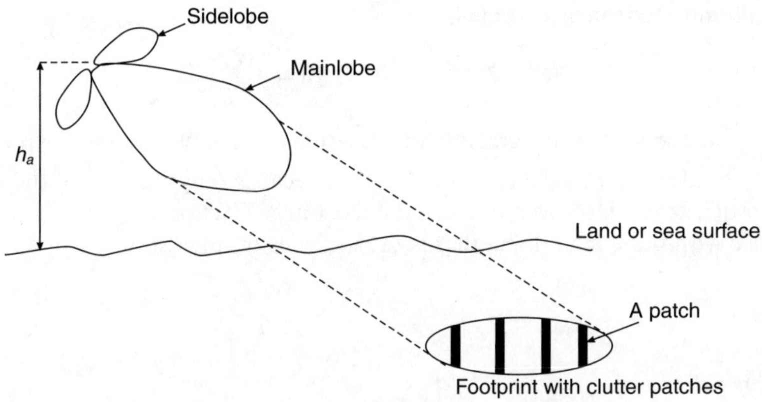

Figure 5.15 An example of footprint clutter patches in a range cell

Doppler, hence neglected. Consequently, the primary clutter powers of concern are that of the SC and RMC, denoted as $P_{c(\mathrm{SC})}$ and $P_{c(\mathrm{RMC})}$ respectively.

A cross-sectional view of a footprint shows a number of clutter-ring patches. For analytical purpose, the ith clutter patch is considered, as in Figure 5.16, where R_i, Δx_i, A_i and ψ_i are the range to the clutter patch, the elemental extent, area and grazing angle of the ith clutter patch respectively.

Using Figure 5.16(b),

$$\sin \psi_i = \frac{h_a}{R_i} \tag{5.76a}$$

$$\Delta x_i = \frac{c\tau}{2} \tag{5.76b}$$

$$x_i = R_i \cos \psi_i \tag{5.76c}$$

$$A_i = 2\pi x_i \Delta x_i \tag{5.76d}$$

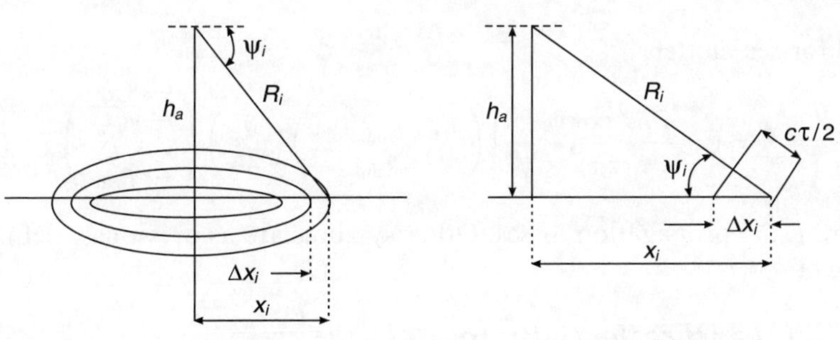

Figure 5.16 Geometry of ith ring of clutter patch: (a) clutter rings; (b) line representation of ith ring

The illuminated surface area A_c is

$$A_c = \sum_i A_i = 2\pi \cos\psi\left(\frac{c\tau}{2}\right)\sum_i R_i \tag{5.77}$$

Note that the sum of all grazing angles equals ψ: i.e., $\psi = \sum_i \psi_i$. Expression (5.77) is valid for the sidelobe consideration because only the area of the entire ring is of interest. However, in the mainlobe, only a fraction of the ring's circumference within the main lobe, that is, $\Delta y_i/2\pi x_i$, is of interest. Consequently,

$$A_{c(MC)} = \sum_i A_i \frac{\Delta y_i}{2\pi x_i} \tag{5.78}$$

where

$$\Delta y_i = \theta_g x_i \tag{5.79a}$$

θ_g = angular extent of a ring's circumference, in radians. It can also be expressed in terms of angular extent of the *i*th ring patch in the mainlobe footprint and its gain relative to the mainlobe gain. Specifically,

$$\theta_g = \sum_i g_i^2 \theta_{i(\text{mainlobe})} \tag{5.79b}$$

By substituting (5.57) in (5.56),

$$A_{c(\text{MC})} = \theta_g \cos\psi\left(\frac{c\tau}{2}\right)\sum_i R_i = \left(\frac{c\tau}{2}\right)\cos\psi \sum_i g_i^2 R_i \theta_{i(\text{mainlobe})} \tag{5.80}$$

Having defined the mainlobe and sidelobe illuminated area given respectively by (5.80) and (5.77), the next task is to define the reflectivity for land surface, σ^0, and sea surface, $\overline{\chi}$, for the associated clutter radar cross-section to be determined. After this, the clutter power P_c, analogous to (5.5), can be quantified. Specifically, for land clutter

$$\begin{pmatrix} P_{c(\text{RMC})} \\ P_{c(\text{SC})} \end{pmatrix} = \frac{P_t \lambda^2 \cos\psi}{\eta(4\pi)^3}\left[\frac{c\tau}{2}\right]\left(\frac{K_{rej} G^2_{\text{mainlobe}}}{G^2_{\text{sidelobe}}}\right)\left(\frac{\theta_g}{2\pi}\right)\begin{pmatrix} \sum_i \frac{R_i}{R_i^4 L_{pi}} \\ \sum_i \frac{R_i}{R_i^4 L_{pi}} \end{pmatrix}\sigma^0 \tag{5.81}$$

And for sea clutter,

$$\begin{pmatrix} P_{c(\text{RMC})} \\ P_{c(\text{SC})} \end{pmatrix} = \frac{P_t \lambda^2 \cos\psi}{\eta(4\pi)^3}\left[\frac{c\tau}{2}\right]\left(\frac{K_{rej} G^2_{\text{mainlobe}}}{G^2_{\text{sidelobe}}}\right)\left(\frac{\theta_g}{2\pi}\right)\begin{pmatrix} \sum_i \frac{R_i}{R_i^4 L_{pi}} \\ \sum_i \frac{R_i}{R_i^4 L_{pi}} \end{pmatrix}\overline{\chi} \tag{5.82}$$

where L_{pi} = propagation losses. Other symbols are as previously defined in the text.

5.4.1.1 Land reflectivity model

A simple model for land reflectivity, at grazing angle ψ, is

$$\sigma^0 = \gamma \sin\psi \tag{5.83}$$

where γ has values between 0.03 to 0.15, characterizing different terrain types. For instance (Barton 1988; Levanon 1988):

(i) $0.03 \leq \gamma \leq 0.1$ land covered by crops, bushes and trees;
(ii) $\gamma \approx 0.01$ desert, grassland and marshy terrain; and
(iii) $\gamma \approx 0.32$ urban, or mountainous regions.

At low grazing angles, as applied to ground-based radar, propagation considerations become dominant.

With (5.54) and (5.61) in (5.59), the clutter power from land surface is written as

$$\begin{pmatrix} P_{c(\text{RMC})} \\ P_{c(\text{SC})} \end{pmatrix} = \frac{\gamma P_t \lambda^2}{\eta(4\pi)^3} h_a \cos\psi \left[\frac{c\tau}{2}\right] \left(\frac{K_{rej} G^2_{\text{mainlobe}}}{G^2_{\text{sidelobe}}}\right) \left(\frac{\theta_g}{2\pi}\right) \left(\sum_i \frac{1}{R_i^4 L_{pi}}\right) \tag{5.84}$$

For practical purpose, $R_c = \sum_i R_i$ is replaced by R_c; the clutter range situated at the centre of the clutter in any given resolution cell, and $L_T = \sum_i L_{pi}$ being the effective propagation loss.

5.4.1.2 Sea reflectivity model

Sea clutter reflectivity is a complex mix because it requires several parameters to realistically develop it. The parameters include frequency, grazing angle, sea state, polarization, wind direction and surface roughness. In the current form, the expression (5.61) does not encompass realistic environmental features.

The sea reflectivity $\overline{\chi}$ (to distinguish it from land reflectivity, σ^0) can vary from one radar resolution cell to another. Clutter in each of the radar beams, be it narrow or broad beams, will be seen by the radar as the same. The wind is assumed to be blowing in a way that allows propagation and detection. While wave swells make reflectivity measurement accuracy difficult, an approximate value is often settled for. As such, the mean value of the reflectivity is expressed in (5.63), with appropriate adjustments, and makes it as real as possible:

$$\overline{\chi} = \overline{x}_{\text{ref}} + k_g + k_s + k_p + k_d \tag{5.85}$$

The terms comprising (5.85) are adjustment factors that are defined as follows.

(i) Sea state adjustment factor, k_s, is defined by

$$k_s = \overline{x}_{\text{ref}}(S - \overline{x}_{\text{ref}}) \tag{5.86}$$

where the reference reflectivity $\overline{x}_{\text{ref}}$, which applies to all sea states, is constant and taken as 5. The sea state S is an integer, see Table 5.3, column 1.

Table 5.3 Description of state of sea

Code figure of sea state, S	Description of sea state	Significant wave height (m)	Average period of maximum wave (s)
0	Calm (glassy)	0	—
1	Calm (rippled)	0–0.1	—
2	Smooth (wavelets)	0.1–0.5	—
3	Slight	0.5–1.25	—
4	Moderate	1.25–2.5	7.0
5	Rough	2.5–4.0	7.7
6	Very rough	4.0–6.0	8.5
7	High	6.0–9.0	9.0
8	Very high	9.0–14.0	10.0
9	Phenomenal	over 14.0	10.0

(ii) The grazing angle adjustment factor, k_g, consists of three regions:

(a) For small grazing angles ($\psi < 0.1°$), $k_g = 0$.
(b) For grazing angles less than the transitional angle ψ_t, i.e. ($0.1° \leq \psi \leq \psi_t$), reflectivity $\overline{\chi}$ increases by $20 \log \psi$. The transitional angle, ψ_t, is defined as

$$\psi_t = \sin^{-1}\left(\frac{0.066\lambda}{\sigma_z}\right) \tag{5.87}$$

where σ_z = root-mean-square of wave height (m).
(c) For grazing angles beyond ψ_t, $\overline{\chi}$ increases as $10 \log \psi$. To estimate the grazing angle adjustment factor, k_g, two conditions have to be met: when $\psi_t \geq 0.1°$ and when $\psi_t < 0.1°$. The dependent of k_g on the grazing angle and transitional angle for these conditions are:

(a) For $\psi \geq 0.1°$:

$$k_g = \begin{cases} 0 & \psi < 0.1° \\ 20\log(10\psi) & 0.1° \leq \psi \leq \psi_t \\ 20\log(10\psi_t) + 10\log\left(\frac{\psi}{\psi_t}\right) & \psi_t < \psi < 30° \end{cases} \tag{5.88}$$

(b) For $\psi_t < 0.1°$:

$$k_g = \begin{cases} 0 & \psi \leq 0.1° \\ 10\log\left(\frac{\psi}{\psi_t}\right) & \psi > 0.1° \end{cases} \tag{5.89}$$

(iii) Polarization adjustment k_p

The depolarization component of k_p is zero. Also, with vertical polarization, the adjustment k_p is also zero. So, the adjustment factor for horizontal polarization may be written as

$$k_p = \begin{cases} 1.7\ln(w_h + 0.015) - 3.8\ln(\lambda) - 2.5\ln\left(0.0001 + \frac{\psi}{57.3}\right) - 22.2 & f < 3 \\ 1.1\ln(w_h + 0.015) - 1.1\ln(\lambda) - 1.3\ln\left(0.0001 + \frac{\psi}{57.3}\right) - 9.7 & 3 \leq f \leq 10 \\ 1.4\ln(w_h) - 3.4\ln(\lambda) - 1.3\ln\left(\frac{\psi}{57.3}\right) - 18.6 & f \geq 10 \end{cases} \tag{5.90}$$

where f and w_h correspond to the propagation frequency (in GHz) and the mean wave height (m), see Table 5.3, column 3. Note that $\ln = \log_e$.

(iv) For downward looking radar, the wind direction adjustment, k_d, is defined by

$$k_d = -2\left(2 + 1.7\log\left\{\frac{1}{10\lambda}\right\}\right)\sin^2\left(\frac{\theta_e}{2}\right) \tag{5.91}$$

For an upwind looking radar, $k_d = 0$.

In essence, with the knowledge of parameters denoted by (5.86) through (5.91), and upon their substitution in (5.85) and (5.84), the sea clutter power can be evaluated.

The task now is to account for clutter by calculating the signal-to-clutter ratio (S/C). If the major clutter contributor is from a land, or sea, surface, then replace N_i in (5.32) with $P_{c(\text{RMC})}$, $P_{c(\text{SC})}$ from (5.84). However, if the combined noise-plus-clutter power is considered, assuming both effects occur incoherently, then the clutter is the sum of input noise power – that is, N_i from (5.33) – and surface (land or sea) – that is, $P_{c(\text{RMC})}$, $P_{c(\text{SC})}$ from (5.84).

Knowing the (S/C) required to achieve the desired detection performance (either extrapolating from performance curves, or using Neuvy's expressions, in section 5.3 in conjunction with an appropriate probability of detection, P_d, and acceptable probability of false alarm, P_{fa}) the range where target detection is possible can be estimated.

The preceding development assumes that there are no additional *clutters* in the 'look-path' of the radar. If another clutter is present, for instance rain, the previous equations will need to be modified – discussed in the next section.

5.4.2 Rain clutter

For the rain clutter to be meaningful, rain rate is taken to be the average over a widespread 'stratiform' rainfall. Rain rate, r_r, and hence mean reflectivity, η_v are assumed to vary spatially within any typical storm. The cross-section of precipitation, σ_r, is proportional to the product of rain reflectivity η_v(m^2/m^3), and the volume V_c(m^3) within a radar resolution cell:

$$\sigma_r = \eta_v V_c \quad (m^2) \tag{5.92}$$

Similar to land and sea clutter, backscattered power is directly proportional to reflectivity with its proportionality constant being the volume the rain occupied in a cell.

5.4.2.1 Volume resolution cell

Consider the clutter range R_c to be situated at the centre of the clutter in the resolution cell. The geometry of volume clutter is shown in Figure 5.17.

The volume resolution cell V_c is defined as

$$V_c = \Delta w \Delta H \Delta R \quad (m^3) \tag{5.93}$$

The vertical extent of the beam in the rain or height of the radar resolution cells (whichever is lesser) is ΔH, which is defined by

$$\Delta H = \theta_v R_c \quad (m) \tag{5.94}$$

In the cross-range direction, the width Δw of the illuminated area is determined by the horizontal antenna beamwidth θ_H, defined by

$$\Delta w = \theta_H R_c \tag{5.95}$$

The difference between the leading edge of the pulse and the end of the pulse being reflected from the surface at a given time delay ΔR is defined by

$$\Delta R = \frac{1}{2} c\tau \tag{5.96}$$

This expression is valid for simple uncoded pulses, where c is the speed of light. For pulse compression radar the time-bandwidth product of the transmitted pulse equals the pulse compression ratio so that τ in (5.96) can

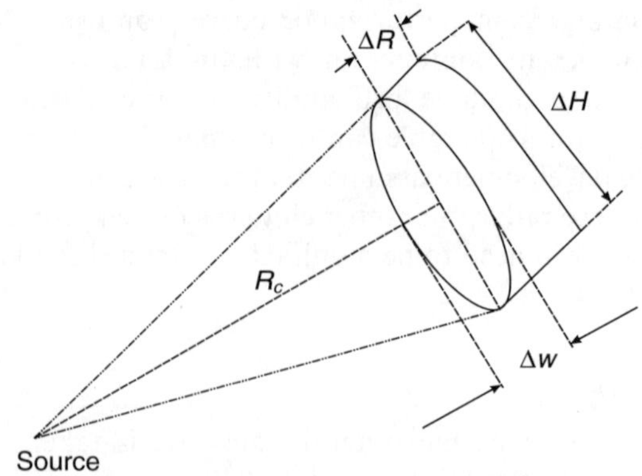

Figure 5.17 Geometry of volume clutter

be interpreted as the compressed pulse width τ_c. However, for a matched-filter receiver with rectangular spectral envelope,

$$\Delta R = \frac{c}{2B_n} \tag{5.97}$$

where B_n is the receiver beamwidth in Hz.

5.4.2.2 Rain reflectivity model

Rain reflectivity, η_v, fluctuates with time within each radar volume resolution cell. The fluctuation in η_v within each cell is governed by the exponential probability density function

$$p(\eta|\overline{\eta_v}) = \frac{1}{\overline{\eta_v}} e^{-\left(\frac{\eta}{\overline{\eta_v}}\right)} \tag{5.98}$$

where $\overline{\eta_v}$ is the mean reflectivity for each cell. Mean reflectivity and rain rate r_r (mm/hr) are assumed approximately constant within each resolution cell and are functionally dependent of propagation frequency f (GHz):

$$\overline{\eta_v} = \kappa f^4 r_r^{1.6} \quad (\text{m}^2/\text{m}^3) \tag{5.99}$$

where κ is the proportionality constant, defined as

$$\kappa = \begin{cases} 7 \times 10^{-48} & f \le 6\,\text{GHz} \\ 13 \times 10^{-48} & f = 35\,\text{GHz} \end{cases} \tag{5.100}$$

Values of κ in between the specified frequencies are obtained by linear interpolation thus:

$$\kappa = [7 + 0.206897(f - 6)] \times 10^{-48} \tag{5.101}$$

Figure 5.18 shows the variability of mean rain reflectivity against frequency.

In view of the preceding expressions, the rain clutter radar cross-section is expressed as

$$\sigma_c = \kappa f^4 r_r^{1.6} R_c^2 \theta_H \theta_v \frac{c\tau}{2} \tag{5.102}$$

This relationship holds when the radar range has no ambiguities in which clutter is present. Like the land and sea surfaces' power derivation, the rain clutter power can be expressed as

$$P_{c(\text{rain})} = \frac{P_t G_t^2 \lambda^2}{\eta (4\pi)^3 R_c^4} \kappa f^4 r_r^{1.6} R_c^2 \{\theta_H \theta_v\} \left(\frac{c\tau}{2}\right) \tag{5.103}$$

Since f is in GHz and the propagation wavelength $\lambda = 0.3/f$, then $\theta_H \theta_v = 4\pi/G_t$. So,

$$P_{c(\text{rain})} = \kappa r_r^{1.6} \left(\frac{P_t G_t}{\eta}\right) \left(\frac{0.3 f}{4\pi R_c}\right)^2 \left(\frac{c\tau}{2}\right) \tag{5.104}$$

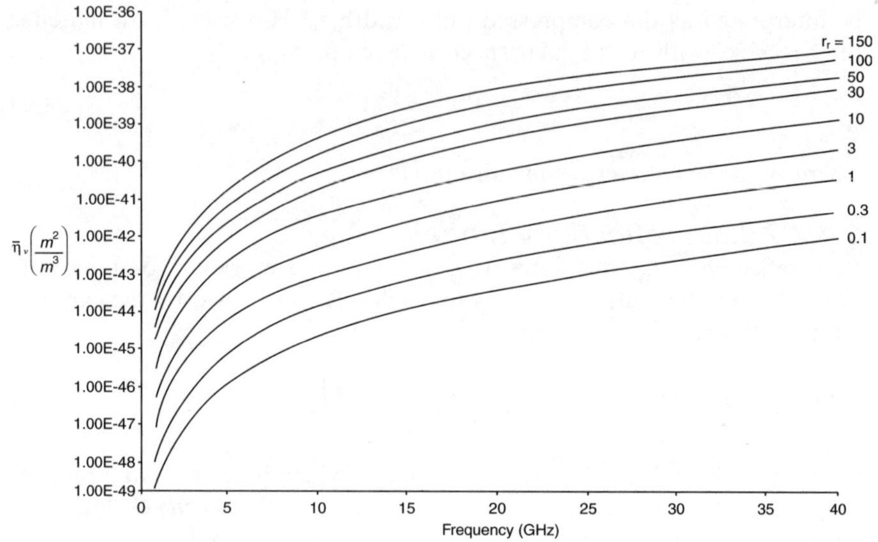

Figure 5.18 Mean reflectivity of rain against propagation frequency. Note that the unit of rain rate (r_r) is in mm/hr

To account for clutter in the radar equation, replace the input noise power N_i in (5.33), with the rain contribution in (5.104), if and only if rain is the major contributor. However, if the combined noise-plus-clutter power is considered, assuming both effects occur incoherently, then the clutter is the sum of noise N_i in (5.61) and rain (5.104); that is, $C = N_i + P_{c(\text{rain})}$.

Knowing the (S/C) required to achieve the desired detection performance (either extrapolating from performance curves, or using Neuvy's expressions, in section 5.3 in conjunction with an appropriate probability of detection, P_d, and acceptable probability of false alarm, P_{fa}) the range where target detection is possible can be estimated.

In summarizing therefore that since rain clutter, for a defined rain rate and propagation frequency, has both mean reflectivity and Doppler components that are statistically distributed, the mean reflectivity will vary in space between rain cells and temporally (in time) as a consequence of variation in rain rate. The fluctuation in reflectivity over short periods of time does not invalidate the reflectivity expressions.

5.4.3 A summary of clutter rejection techniques

There are many ways to reject, or at least reduce, clutter. Each of these techniques has received much attention in the literature of which the section could not do substantial justice to its description and implementation. However, a summary of these techniques is described (Mao 1993):

1. Preventing the clutter energy from entering the radar antenna by

   (a) installing the radars in high mountains,
   (b) tilting the radar antenna to higher elevation angles, and
   (c) surrounding the radar antenna with a 'clutter shelter fence'.

   All these methods can easily be applied to existing radars.
3. Shaping the beam pattern of the radar antenna to enhance its signal-to-clutter ratio. A typical illustration is the use of dual beam antennas for airport surveillance, where a high receiving beam is used to increase the signal strength of neighbouring aircraft.
4. Adopting the polarization technique to enhance its signal-to-clutter ratio. For instance, circular polarization can reduce the raindrop radar cross-section by $15 \sim 30$ dB while the cross-polarization technique would reduce the target-to-precipitation echo by $15 \sim 25$ dB.
5. Reducing the clutter energy by decreasing the size of radar's resolution cell. Narrowing the pulse width, narrowing the beamwidth (though limited by the antenna size), or adopting pulse compression can achieve this. This method is particularly relevant to sea clutter rejection in ship-borne radars.
6. Preventing the receiver from saturation.
7. Suppressing the clutter in the time domain with the *constant false alarm ratio* (CFAR) detector or adaptive threshold or clutter map. More is said of CFAR in Chapter 10. However, these models only can obtain super-clutter visibility (S_uCV).
8. Suppressing the clutter in the frequency domain with *moving target indication* (MTI) or *moving target detection* (MTD) techniques. These techniques can obtain sub-clutter visibility (SCV).

5.5 Radar equation for laser radar

Laser radars constitute a direct extension of conventional radar techniques to very short wavelengths. Like the acronym derived for conventional radar, laser radar is called either *ladar* (laser detection and ranging) or *lidar* (light detection and ranging). Laser radar systems are active devices that operate similarly to microwave radars but at a much higher frequency (Hovanessian 1988). This higher frequency has a beneficial effect because of smaller components and remarkable angular resolution, but suffers considerable atmospheric attenuation losses at higher frequencies if built to operate on the ground. Laser radars built for ground operations are range limited (about 10 km). However, space-borne laser radars have larger ranges (i.e. $R \gg 100$ km) because they suffer very little, if any, atmospheric attenuation losses.

5.5.1 Laser performance calculations

The design of laser radar follows the same general principles as other radars, but with subtle differences. For example, when the target is in the far field of the laser radar, and if the laser beam is greater than the target's width, (5.4) applies. However, when the laser beam is less than the target's width, (5.4) would still hold with certain modification. Also, if the laser radar is operated in the *near-field* situation, its beamwidth expression will modify the radar equation. These conditions are discussed in this section.

For the case of *far-field operations* and the *beamwidth* greater than the target's width, instead of the antenna gain G_t, the laser beamwidth is usually measured. As such the gain can be expressed as

$$G_t = \left(\frac{\pi}{\theta_{BW}}\right)^2 \tag{5.105}$$

Upon substitution in (5.4), yielding

$$P_r = \frac{P_t \sigma A_e}{16 \theta_{BW}^2 R^4} \tag{5.106}$$

Beamwidth is expressed as a function of lens diameter, D_L (m), and wavelength, λ (m):

$$\theta_{BW} = k_\theta \frac{\lambda}{D_L} \tag{5.107}$$

where k_θ is the aperture constant determined by the aperture illumination function. For example, if the aperture is uniformly illuminated

$$0.84 \leq k_\theta \leq \frac{4}{\pi} \tag{5.108a}$$

And for a Gaussianly illuminated aperture

$$k_\theta = 2.44 \tag{5.108b}$$

It is appropriate at this junction to make a distinction between the beamwidth measurements in conventional microwave and laser radars.

In conventional microwave radars, the one-half power (3 dB) point is usually applied. For instance, the 3 dB value for a sin *c* functioned beam structure is equal to the bandwidth, expressed by

$$\theta_{BW} = 0.886 \frac{\lambda}{D_r} \tag{5.109}$$

where D_r is the radar's aperture diameter (m).

In the case of laser radar, as in optical systems, e^{-1}(=0.36788) is used. So,

$$\theta_{BW} = 1.05 \frac{\lambda}{D_L} \tag{5.110}$$

Following a similar procedure in obtaining (5.34), the minimum detectable signal for laser radar can be developed as follows. Unlike the microwaves where the receiver sensitivity is determined by thermal noise, quantum effects determine the sensitivity of laser receivers. The equivalent input noise power is given by,

$$N_i = hfB_n \tag{5.111}$$

where h is the Planck's constant ($=6.6256 \times 10^{-34}$ W-s^2), and B_n is the noise bandwidth.

Quantum-limited receivers are analogous to superheterodyne (heterodyne or coherent) receivers in microwave radars. Laser radar of this type is also called a photomixer. In general, when the background noise is low, and for short-pulse modulation, the laser detector operates as a quantum limited device and gives the same detectivity (meaning, inverse of equivalent noise power) as heterodyne detectors (Skolnik 1980).

For laser radar with a *video* receiver,

$$N_i = 2hfB_n \tag{5.112}$$

Video receivers, when employed in microwave radars, are far less sensitive. Video receivers are also called incoherent (envelope) receivers or direct photodetection. Photodetection receivers are less complicated than the photomixing type. As such, photomixing receivers require local oscillators and stable transmitters.

Example 5.5 Compare the thermal noise power and quantum noise power of the microwave and laser radars if the propagation frequency and noise bandwidth equal 1 GHz and at room temperature (≈27 °C).

Solution

From (5.29), the microwave thermal noise, $N_{\text{thermal}} = kTB_n = -114\,\text{dB}$
From (5.111), the laser (quantum) noise, $N_i = hfB_n = -152\,\text{dB}$
Frequency controls primarily the level of noise in laser radar while temperature primarily influences that of the noise in the microwave radar.

The equivalent noise power expressed by (5.111) and (5.112) assumes that the sum of the residual powers (i.e. contributions from the dark current power, local oscillator power and background power) is far less than the received power and their effect on the minimum detectable signal is negligible. For more discussion on the selection of design components and their responsiveness, the reader is advised to read Jelalian (1992).

By setting (5.111) or (5.112) to thermal noise as in (5.34) (i.e. $N_{\text{thermal}} = N_i$), an equivalent noise power (or temperature) can be estimated for the laser receiver. Laser receivers are generally of greater effective temperature (or noise figure, F_N) than the contemporary microwave receivers. Subsequently, for n_p photoelectron emissions (analogous to the number of pulsating signals received by microwave radar), the minimum detectable signal $S_{\min}$ for quantum-limited detection is

$$S_{\min} = n_p \frac{N_i}{\eta_0} = n_p \frac{hfB_n}{\eta_0} \quad (5.113)$$

where η_0 is the detector quantum (or optical) efficiency. Considerable care must be taken to compensate for large Doppler frequency shift. For instance, when a target is in motion relative to laser radar, a large frequency shift occurs which can place the echo signal outside the receiver passband. To arrest this large shift, a rapidly tuning laser local oscillator and/or a bank of IF filters are necessary in the laser radar circuitry.

Like in microwave radar, the laser radar received signal power P_r in (5.90) equates to the minimum detectable signal $S_{\min}$ in (5.113). Specifically,

$$P_r = \frac{P_t \sigma A_e}{16\theta_{BW}^2 R^4} = n_p \frac{hfB_n}{\eta_0} \quad (5.114a)$$

From this expression, the maximum target range is written as:

$$R = \frac{1}{2}\left(\frac{\eta_0 P_t \sigma A_e}{n_p \theta_{BW}^2 hfB_n}\right)^{\frac{1}{4}} \quad (5.114b)$$

This expression is the laser radar equation, where A_e is the effective aperture area (m^2).

If, however, the laser beam is less than the target's width, the effect of its surface is generally included in the target's radar cross-section estimation. If the surface is a diffuse (i.e. Lambertian) scatterer, of reflectivity ρ, then the target's radar cross-section may be expressed as:

$$\sigma = \rho A_t \quad (5.115a)$$

where the target area is

$$A_t = \pi \left(\frac{R\theta_{BW}}{2}\right)^2 \cos\phi \quad (5.115b)$$

ϕ is the angle between the surface normal and incident radar signal. If the target is normal to radar beam, $\phi = 0$. By substituting (5.115) in (5.114), the coherent laser radar equation can be written for a Lambertian scatterer as:

$$R = \frac{1}{8}\sqrt{\frac{\pi\rho\eta_0 P_t A_e}{n_p hfB_n}\cos\phi} \quad (5.116)$$

For an extended diffuse radar target, scattering is often restricted to a half-sphere. In that a case, the target radar cross-section would be expressed as

$$\sigma = 2\rho A_t \quad (5.117)$$

And consequently, substituting (5.115) and (5.117) in (5.114), the laser radar equation:

$$R = \frac{1}{4}\sqrt{\frac{\pi\rho\eta_0 P_t A_e}{2n_p hfB_n}\cos\phi} \tag{5.118}$$

5.5.2 Near-field operation

It is not unusual for laser radar to operate in the near field of the optical systems. If that situation arises, the near-field beamwidth must be modified. Instead of (5.91), a near-field beamwidth is formed (Jelalian 1992):

$$\theta_{BW} = k_\theta\left(\frac{\lambda^2}{D_L^2} + \frac{D_L^2}{R^2}\right)^{\frac{1}{2}} \tag{5.119}$$

where R is the range to target and beamwidth constants.

By substituting (5.119) in (5.89), and following procedures for obtaining equations (5.114) and (5.116), the maximum detectable laser range when operating in the *near field* can be expressed as

$$R = \left(\frac{\eta_0 k_\theta^2 P_t D_L^2 f\sigma A_e}{1.44n_p hB_n}\right)^{\frac{1}{4}} \tag{5.120}$$

providing that $R \gg D_L^2/\lambda$, a condition that satisfies that stipulated by (3.38a) for a radiating near-field region.

5.5.3 Search field

The objective of a search radar is to detect and locate a target within a defined volume of space during a specified time interval. An ideal search radar will consist of the following:

- a matched-filter receiver; that is, where the receiver is matched to the signal spectrum so that the product of the pulse width and bandwidth is unity, if a rectangular pulse is used;
- the radar beams are uniformly shaped and abut perfectly; that is, the beams do not overlap or establish gaps; and
- the search pattern is uniform with 100 per cent antenna efficiency or at least the delivery transmitted energy uniformly over the designated search area.

As in the microwave radar search parameters, specified by (5.18), the laser radar search solid angle Ω_s can be defined:

$$\Omega_s = \frac{A_s}{R^2} \tag{5.121}$$

where A_s corresponds to the area to be searched. If the laser radar diffraction-limited transmitting aperture solid angle is denoted by $\Omega = (k_\theta(\lambda/D_L))^2$, the number of cells n_c to be searched can be determined as the ratio of search solid angle to the aperture solid angle. Specifically,

$$n_c = \frac{\Omega_s}{\Omega} = A_s\left(\frac{D_L}{k_\theta \lambda R}\right)^2 \tag{5.122}$$

The frame time required to search a field by the laser radar is expressed by

$$T_f = t_0 n_c = t_0 A_s\left(\frac{D_L}{k_\theta \lambda R}\right)^2 \tag{5.123}$$

Like the conventional radar, t_0 is measurement-interval time or time dwelled on the target. It can be recognized from (5.123) that a laser radar would require high repetition rates, or long acquisition time, for it to perform a target-search function unless multiple beams are utilized. Would this be a handicap for operational reasons? Not necessarily so because laser radar angular resolution, combined with modulation capability, allows substantial target measurement capability during a single measurement (Jelalian 1992).

5.6 Search figure of merit

Figure of merit (FOM) is an aspect of performance analysis of any radar systems. From an analysis of propagation condition, FOM can be related to radar availability. In operational cases, FOM is used in conjunction with propagation estimates to predict radar detection performance.

Equations (5.17) and (5.18) establish a relationship between solid angle, area of search and transmitter gain as

$$G_t = \frac{4\pi}{A_m(\sin\theta_u - \sin\theta_L)} \tag{5.124}$$

for a unity pattern constant $L_n = 1$. In view of (5.124) the microwave radar equation (5.37) is recast in terms of the received power as

$$P_r = \underbrace{\left\langle \frac{1}{4\pi} \right\rangle}_{constant} \underbrace{\left(\frac{1}{L_{tot}}\right)}_{\substack{total\\losses}} \underbrace{\left\{ \frac{t_s P_{av} A_e}{n_b L_n k T_0 B_n F_N A_m(\sin\theta_u - \sin\theta_L)} \right\}}_{\substack{radar\\capability}} \underbrace{\left[\frac{\sigma |F^4|}{R^4} \right]}_{\substack{target\\characteristics}} \tag{5.125}$$

where P_r is the target signal collected at the radar receiver. Reading after the equality sign from left to right of (5.125), the following terms are described. The first term is the proportionality constant. The second term (.) represents the losses due to the environment. The third term {.} is the radar capability. This term is called the radar *figure of merit* (FOM). The radar FOM involves the power-aperture area product. The larger the FOM the more capable is the radar system to scan a larger field in a given time frame, t_s. Radar wavelength λ is not particularly obvious in (5.125) and could be said to be not particularly associated with any of the terms. However, all of the terms change with frequency. The fourth term [.] is the target characteristics.

A similar expression to (5.125) for the case of laser radar can be written. In view of (5.106) and (5.121), a similar expression to (5.125) for the case of laser radar can be written.

$$P_r = \underbrace{\left\langle \frac{1}{16} \right\rangle}_{constant} \underbrace{\left\{ P_t \frac{\lambda^2}{\theta_{BW}^2} \right\}}_{\substack{radar \\ capability}} \underbrace{\left(\frac{\sigma}{R^2} \right)}_{\substack{target \\ characteristics}} \tag{5.126}$$

The radar capability term {.} demonstrates the wavelength-beamwidth dependence of the laser radar search FOM unlike the microwave, which involves the radar power-aperture area product.

5.6.1 Summary

The preceding discussion on radar and the subsequent development of the radar equations are concerned with primary radar in which the target acts as a passive reflector. The inverse fourth power relationship between the reflected signal power and range presents a major problem when long-range detection is envisaged. It also presents a problem when attempting to estimate the size of a moving target. Another type of radar, called secondary radar, helps to overcome these difficulties by actively interrogating the target. The well-known secondary radars are beacon and transponder. As will be seen in the next section the power requirements of secondary radars are modest in comparison to the previous because transmission is only one way.

5.7 Radar equation for secondary radars

A secondary radar system is a radio visualization system based on the comparison of reference signals with radio signals retransmitted from the position to be determined. Examples of secondary radar are beacons, which can be land based or mobile on ship, and the transponder-based surveillance on aircraft.

A radar beacon system is a passive device until a suitably coded signal triggers it, which in turn emits a series of pulses back to the transmitting radar. The process by which the transmitting signal triggers the beacon is called interrogation. So, it can be said that when a beacon is interrogated, it emits a series of pulses, which are received by the transmitting radar (the interrogator). The beacon's response is a reply to the interrogator.

Three principal system requirements (Johnson and Jasik 1984) frequently imposed on beacon antennas have the ability to:

- support each one-way link from a power budget as well as time-on-target viewpoint;
- facilitate extraction of echo responses only from main-beam interrogations and process returns received only in the main beam; and
- estimate target bearings from the responses.

These requirements are based on one-way transmission.

The power and frequency of the return signal are fixed by the beacon transmitter and are not dependent of the target cross-section, or on the received signal power, providing the triggering signal is at least at the required threshold. Since there are two distinct events, interrogation and response, two radar equations would be required depicting these events.

(i) Interrogation

$$R^2 = \frac{P_t G_t A_{eb}}{4\pi S_{b\,\min}} \tag{5.127}$$

(ii) Response

$$R^2 = \frac{P_b G_b A_e}{4\pi S_{\min}} \tag{5.128}$$

where

(i) P_t and G_t are, respectively, transmit power and antenna gain of the interrogating radar.
(ii) P_b and G_b are the power output and gain of the beacon antenna respectively. This gain has been found to be approximately π even for a small airborne antenna (Barton 1988).
(iii) $S_{b\min}$ and $S_{\min}$ correspond to the minimum detectable signal of the beacon receiver and radar receiver.
(iv) A_{eb} and A_e correspond to effective aperture area of the beacon antenna and radar receiving antenna.

In practice, R in (5.127) and (5.128) are approximately equal. However, if the estimated ranges in (5.127) and (5.128) are different, the lower value applies.

Example 5.6 Estimate the power received by a radar beacon that pumps out 100 W power, with a gain of 30 dB when transmitting at 3 GHz if a target is 100 km away.

Solution

Rewriting (5.127) in terms of the received power as well as substituting (5.9) in place of aperture area,

$$S_{\min} = P_t \left(\frac{0.3 G_t}{4\pi f R}\right)^2 = 100\left(\frac{0.3 \times 1000}{4 \times \pi \times 3 \times 10^5}\right)^2 = 0.633\,\mu\text{W}$$

5.7.1 Application of beacon radar systems

Beacon radar systems are used for different applications. Examples include instances where there is a need:

1. To enhance the target return signals with respect to their strength and/or information contents (Johnson and Jasik 1984).
2. To provide useful information on the capability of observation data link.
3. To assist in the surveillance of moving targets or provide information on surveyed points for self-location (e.g. distress signal picked up by satellite or other sensors).
4. To serve as a position reference for over-the-horizon radar.
5. To maintain accurate target tracking. For instance, when a target of interest is at a distance far from the radar, the signal reflected from the target might be too weak to be received. Under such circumstances, accurate tracking can be maintained by placing a beacon on the target.
6. To identify a friend or foe (IFF) target. It has been used, and is still used, extensively for identifying night fighters by conveying aircraft altitude and position coordinates to the ground controller as collision avoidance systems.
7. To assist aircraft homing in to their bases or making rendezvous with ocean-based convoys.
8. To navigate a ship within horizon range of land with very good precision.

5.8 Summary

This chapter has derived radar equations for three radar types, namely conventional, laser and secondary. Included in these radar equations were system and atmospheric losses as well as surface effects. The equations enabled us to estimate the radar's detectable range in benign and clutter environments. The figure of merit for a specific radar time frame is also studied.

Appendix 5A Noise in Doppler processing

Noise in a Doppler filter can be obtained as follows. It is known in Chapter 3 that Doppler bandwidth is inversely proportional to the compressed pulse width by

$$B_d \approx \frac{1}{\tau_c} \tag{5A.1}$$

Alternately

$$B_d = \frac{\text{PRF}}{N_p} \tag{5A.2}$$

where N_p = number of samples coherently processed.

It is also known in Chapter 1, from the Nyquist theorem, that a foldover, or an aliasing, occurs at twice the sampling frequency ($2f_0$). So, the number of ambiguities, N_{amb}, that can be folded, or mapped, into a given cell is

$$N_{amb} = \frac{B_n}{\text{PRF}} \tag{5A.3}$$

Consequently the noise in a Doppler is a fraction of the front-end noise bandwidth. Specifically,

$$\begin{aligned} N_d &= C_r \underbrace{F_n k T_0 B_n}_{\textit{front end noise}} \left[\frac{B_n}{\text{PRF}}\right]\left(\frac{B_d}{B_n}\right) \\ &= C_r \frac{F_n k T_0 B_d}{\tau \text{PRF}} \end{aligned} \tag{5A.4}$$

Note that

$B_n = 1/\tau$, where τ = transmitted (uncompressed) pulse width
C_r = compression ratio
$\tau\text{PRF} = d_u$; the duty cycle.

If the noise power in the minimum detectable signal of (5.33b) is replaced with that in the Doppler (5A.4), the input signal can be written as

$$S_i = \frac{C_r F_N k T_0 B_d}{\tau \text{PRF}}\left(\frac{S_0}{N_0}\right) \tag{5A.5}$$

Equating this expression to the received signal power in (5.5), the resulting equation is in the form of a radar equation:

$$R_{\max} = \left(\frac{P_{av} G_t A_e \sigma}{(4\pi)^2 k T_0 B_d F_N \left(\frac{S_0}{N_0}\right)}\right)^{\frac{1}{4}} \tag{5A.6}$$

where $P_{av} = P_t \tau \text{PRF} = P_t d_u$ and $C_r = 1$.

The only modification to the radar equations developed in range-cell processing when Doppler processing is that the noise bandwidth, B_n, is replaced with that of the Doppler, B_d.

Problems

1. Why is the figure of merit important in the design of a radar system?
2. Assume that you are tasked to design a radar system, what are the salient questions to ask?
3. Examples 5.1 and 5.2 demonstrate the applicable range limits. How will you obtain significant target detection beyond the limits?

4. Design a computer program that evaluates the significant target detection in the face of combined clutter and noise presence during radar surveillance.
5. Is it possible to combine the microwave and laser technologies to overcome the inherent problems in radar applications? What steps would you take to overcome mutual interference from both systems?

Part II

Ionosphere and HF Skywave Radar

This part comprises two chapters: 6 and 7. When a wave traverses the regions comprising the atmosphere it results in the degradation of signal-target information due to spatial inhomogeneities that exist and vary continuously with time in the atmosphere. The spatial variations produce statistical bias errors, which are an important consideration that must be accounted for when formulating and designing a high-frequency (HF) skywave radar system. Chapter 6 explains how these errors are quantified including the polarization rotational effect on the propagation wave. Chapter 7 explains the design consideration and performance of the skywave radar.

6

The ionosphere and its effect on HF skywave propagation

This chapter explains the structural composition of the atmosphere and the propagation errors introduced into the skywave radar measurements as a result of atmospheric anomalies. Propagation errors manifest themselves as refractive bending, time delays, Doppler errors, rotation of the phase of polarization (called Faraday effect), dispersion effects, and attenuation. Atmospheric anomalies brought about by man-made devices are ignored.

6.1 The atmosphere

The structure of the Earth's atmosphere is shown in Figure 6.1. The Earth's atmosphere varies in density and composition as the altitude increases above the surface. The lowest part of the atmosphere is called the troposphere and it extends from the surface up to about 10 km. The gases in this region are predominantly molecular oxygen (O_2) and molecular nitrogen (N_2). The Earth's weather is confined to this lower region (troposphere) containing 90 per cent of the Earth's atmosphere and 99 per cent of the water vapour. All of our normal day-to-day activities occur within this lower region. The high altitude jet stream is found near the tropopause at the upper end of this region. The atmosphere above 10 km is called the stratosphere. In this region, the gas composition changes slightly as the altitude increases while the air thins rapidly. Within the stratosphere, incoming solar radiation at wavelengths below 240 nm is able to break up, or dissociate, molecular oxygen, O_2, into individual oxygen atoms, each of which, in turn, may combine with an oxygen molecule to form ozone, a molecule of oxygen consisting of three oxygen atoms (O_3). This gas reaches a peak density of a few parts per million at an altitude of about 25 km becoming increasingly rarefied at higher altitudes. At heights of 80 km, the gas is so thin that free electrons can only exist for short periods of time before they are captured by a nearby positive ion. The existence of charged particles at this altitude and above signals the beginning of the ionosphere: a region having the properties of a gas and of plasma. The upper atmosphere is collectively called the ionized atmosphere

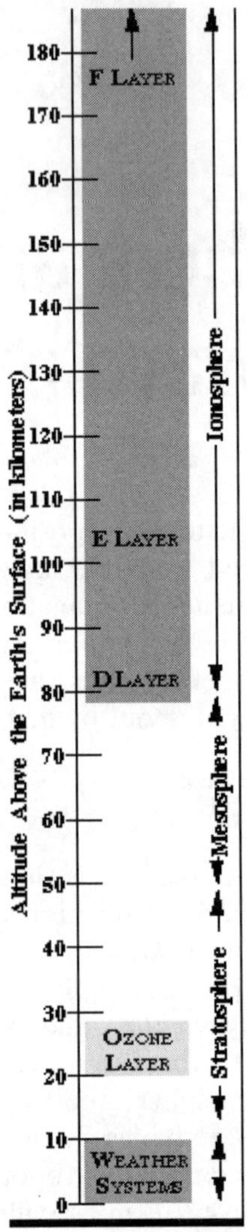

Figure 6.1 Structure of the atmosphere (courtesy: NASA)

(simply the *ionosphere*), comprising D, E and F layers. It is the ionized layers that constitute the principal factors in radiowave propagation. The composition of the ionized atmosphere is discussed in detail later under each layer's heading. It will be instructive to look at how the ionosphere is formed.

6.2 The ionosphere

At the outer reaches of the Earth's environment, solar radiation strikes the atmosphere with an average power density of 1.37 kW/m^2, a value known as the 'solar constant'. This intense level of radiation is spread over a broad spectrum ranging from radio frequencies (RF) through infrared (IR) radiation and visible light to X-rays. Solar radiation at ultraviolet (UV) and shorter wavelengths – in the 30 nm and 120 nm range – is considered to be 'ionizing' since photons of energy at these frequencies are capable of dislodging an electron from a neutral gas atom, or molecule, during a collision.

When an incoming solar-radiation incident on a molecule, or gas atom, occurs the molecule absorbs part of this radiation and a free electron and a positively charged ion are produced. Of course, cosmic rays and solar wind particles also play a role in this process but their effect is minor compared with that due to the Sun's electromagnetic radiation.

At the Earth's outer atmosphere (i.e. thermosphere and protonosphere, the highest levels), solar radiation is very strong but there are few atoms to interact with, so ionization is small. As the altitude decreases, more molecules are present so the ionization process increases. At the same time, however, an opposing process called recombination begins to take place in which a free electron is 'captured' by a positive ion if it moves close enough to it. As the gas density increases at lower altitudes, the recombination process accelerates since the gas molecules and ions are closer together. A point of balance between these two processes determines the degree of 'ionisation' present at any given time.

The number of molecules increases further even at lower altitudes thereby creating more opportunity for absorption of energy from a photon of UV solar radiation albeit at reduced radiation intensity because some of it was absorbed at the higher levels. The radiation profile through the atmosphere is neither constant nor monotonic with height. A point is reached, however, where lower radiation, greater gas density and greater recombination rates balance out and the ionization rate begins to decrease with decreasing altitude. This leads to the formation of ionization peaks or layers. Since the composition of the atmosphere changes with height, the ionization rate also changes and this leads to the formation of several distinct ionization layers called the 'D', 'E', 'F1', and 'F2' layers or regions.

The altitude of the D layer is between 70 and 90 km above the Earth's surface, the E layer is between 90 and 130 km, and the F1 layer is between 130 and 200 km. The F2 layer is above 200 km and its upper limit varies with the latitudes; namely, at the mid-latitudes, F2 altitude is between 250 and 350 km, while at the equatorial latitude it is between 350 and 500 km (Rush 1986).

The solar radiation that comes from the hotter regions is closely linked with sunspot groups on the surface of the sun. The activity of the sunspot groups varies markedly from month to month and from year to year. Solar activity also varies, on the average, with an 11-year cycle. The fact that the ionosphere is created by the sun suggests that its structures and electron-peak densities will vary greatly with time of day (diurnal variation), season of year (seasonal variation), the 11-year sunspot cycle, and geographical location (latitudinal variation).

As seen in Figure 6.1, the ionosphere envelops the Earth at varying heights from the D layer to F2 layers. During the day, all the various layers are present and each layer has its critical frequency (more is said about critical frequency later in the text). At nighttime, there is no ionizing radiation and the electrons and ions recombine to form neutral atoms or molecules, thereby causing the low layers to disappear very quickly and leaving only the F2 layer existing, although at a reduced electron density. The F2 layer is the most important for HF propagation because

- it is present all day long,
- it allows the longest hop lengths to be achieved due to its high altitude, and
- the highest frequencies in the HF band may be reflected.

Each of the ionospheric layers features different chemical and physical composition, which is briefly discussed in the next few paragraphs.

6.2.1 Composition

The D layer corresponds to a sparse layer of polyatomic ion 'clusters' with electron density (N_e) between 10^8 and $10^{10}\,m^{-3}$. N_e has a mathematical functional relationship with altitude, temperature, zenith angle and molecular composition – more is said about this in the next section. The D layer plays an important part in low-frequency/very low-frequency (LF/VLF) propagation. This layer is important also at HF because of its absorbing properties, which stem from the relatively high air density and consequent large collision frequency between electrons and neutral molecules (Rishbeth 1988). Because of the absorption property, the D layer is not used as a reflecting medium for HF skywave radar signals.

The E layer corresponds to a moderately electron dense layer ($10^9 \leq N_e \leq 10^{11}\,m^{-3}$) of molecular NO+ ions and atomic O_2+ ions, occasionally 'peaking' in the so-called *sporadic E* (Es) phenomenon. As the name suggests, the sporadic E layers are often patchy in nature and occur sporadically in the E layer. A typical patch may extend horizontally for about 10 km. At times, they may be continuous over large distances. The Es layer is important in practice because when it is dense it affects radio propagation quite seriously; causing fading and preventing any echoes reaching the upper layers, but when it is patchy it displays near perfect mirror characteristic creating near perfect reflection when continuous over large distances.

The F region corresponds to an electron dense layer ($10^{11} \leq N_e \leq 10^{12}\,m^{-3}$) of atomic O_2+ ions. One still finds subdivision into the F1 region – the transition between molecular and atomic ions – and the F2 region – the 'peak' of atomic O_2+ ions.

Drukarev (1946) seems to have been the first in foreseeing that photoionization would produce an electron gas with excess energy, and that its temperature *T* should greatly exceed that of the neutral gas when the rate of ion production

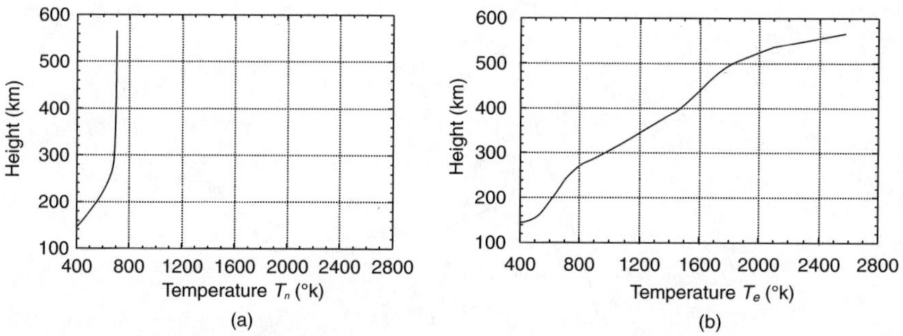

Figure 6.2 Vertical profiles of neutral and electron temperatures in daytime middle latitude: (a) neutral gas temperature, T_n; (b) electron gas temperature, T_e

is high. Typical daytime and nighttime temperature curves at mid-latitude for the E and F regions of the ionosphere are depicted in Figure 6.2. In the figure, T_n and T_e are the vertical profiles of the temperatures of neutral gas temperature and electron gas temperature respectively. At night, however, thermal equilibrium is restored because photoionization has stopped and the electron temperature T_e collapsed to T_n. Good fit approximations for daytime and nighttime temperatures for mid-latitude may be expressed as

$$T_n(h) \cong \begin{cases} 50.801e^{2.5225\times10^{-3}h} & h < 300 \\ 700 & \text{otherwise} \end{cases} \tag{6.1}$$

$$T_e(h) \cong 125.04e^{5.7052\times10^{-4}h} \tag{6.2}$$

Although thermalization of the electron gas and ion gas proceed much more rapidly than the mutual thermalization of the electrons and ions, there occurs a situation when both electrons and ions belong to approximately thermalized populations (Giraud and Petit 1978). This process does not translate to equal temperatures for electron and ion temperatures. By thermalization process the description of the ionosphere changes to a whole medium consisting of not just the ionized component but charged particles embedded in the neutral gas and permeated by the magnetic field of the Earth. The reader can consult Giraud and Petit (1978, Chapter VIII) if more information is required on the thermalization process. In addition, the Earth's magnetic field has some of the propagation waves traversing the ionosphere. This influence becomes clearer to the reader in section 6.2.2.6.

6.2.2 Ray tracing and propagation errors

6.2.2.1 Refraction and reflection

The level in the atmosphere to which any frequency penetrates depends on its absorption hardness and the gases it can ionize. For this reason as a signal is beamed from a transmitter, it undergoes refraction or bending; the extent

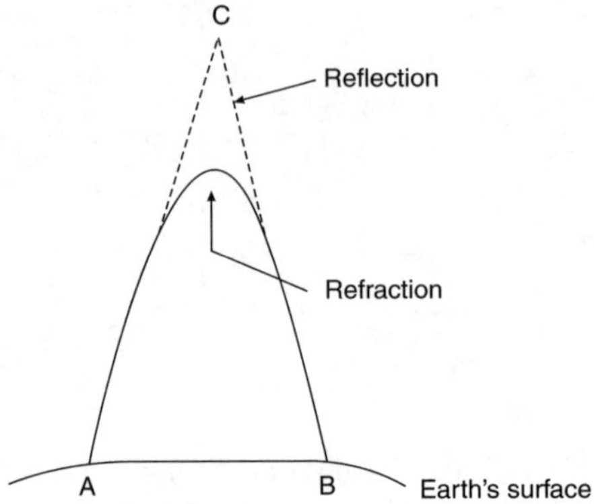

Figure 6.3 An illustration of refraction and reflection

of bending depends on the propagation wavelength. When the signals undergo sufficient refraction, they return to the Earth's surface. Reflection and refraction are sometimes difficult to separate. As an illustration, consider a radio wave being received at point B as shown in Figure 6.3.

The radio wave could equally have been refracted by the ionosphere as it travelled from point A. Also it could also have been reflected by an *apparent* layer at point C. It is apparent therefore that there would be a critical frequency, f_c, at which only partial reflection will occur. Conversely, only frequencies above this critical frequency can traverse the ionosphere. The critical frequency has a mathematical physical explanation, to be discussed later in this section.

The ionosphere, as a medium, is composed of dielectric materials with variable dielectric constants, or refractive indices. It is logical to suggest that the refractive indices are a varying function of the propagation path. From elementary physics it is known that when radio waves are subjected to refraction they undergo a change in direction, or refractive bending, and retardation in the velocity of propagation. The change in direction causes errors, which are introduced in the radar angular and range measurements of target position. To quantify the propagation errors, caused by the ionosphere, knowledge of the height variation of each layer's dielectric constant or refractive index is required.

6.2.2.2 Refractive index

By using the transmission line theory, one can generally define a plane wave that propagates along the x-axis in the ionosphere as having field strength formalized by

$$E = E_0 e^{\gamma x} \sin \omega t \tag{6.3}$$

where

E_0 = amplitude of the plane wave
γ = propagation coefficient, which is a complex number, definable as

$$\gamma = \alpha + j\beta \tag{6.4a}$$

The reference part, α, of the propagation coefficient represents an attenuation, which is called the attenuation coefficient. The quadrature component, β, of the propagation coefficient represents a change of phase down the ionospheric medium, and so β is called the phase-change coefficient. So, the travelling wave would move at a velocity, v, in the direction of decreasing x defined as

$$v = \frac{\omega}{\beta} \tag{6.4b}$$

This expression is called retarded function. If the absorption coefficient is zero, the quadrature component, β, can be neglected. This situation occurs for higher frequencies, or smaller concentrations of electron densities.

The force, F, exerted on an electron of charge e_e, in the direction of the electric field E, may be defined as

$$F = e_e E \tag{6.5}$$

The value of an electron charge is known; that is, $e_e = 1.602 \times 10^{-19}$ (coulomb). This force on the accelerating electrons equates to

$$F = am_e = m_e \frac{d^2x}{dt^2} \tag{6.6}$$

where 'a' is the acceleration of the electron and m_e is the electron mass whose value is known; that is, $m_e = 9.1 \times 10^{-28}$ (gm). Since the mass of an ion is far greater than that of an electron, the motion of an ion in the field is considered negligible. Hence, by equating (6.5) to (6.6), and neglecting the propagation coefficient (exponential) term in (6.3), the differential equation of the electron motion in the x-plane can be expressed as

$$m_e \frac{d^2x}{dt^2} = e_e E_0 \sin \omega t \tag{6.7}$$

Integrating this expression, the velocity of the electron may be defined by

$$\frac{dx}{dt} = \frac{e_e E_0 \sin \omega t}{m_e \omega} \tag{6.8}$$

The motion of the electrons produces a convectional-current density i_c defined by

$$i_c = e_e N_e \frac{dx}{dt} \tag{6.9}$$

where N_e = the electron density (m^{-3}): (more is said about this quantity later in section 6.2.2.4). By substituting (6.8) in (6.9),

$$i_c = \frac{e_e^2 N_e E_0 \cos \omega t}{m_e \omega} \tag{6.10}$$

The electric field gives rise to a displacement current density i_D. By Maxwell theory, this current may be thought of as due to the rate of electric flux in the dielectric medium and defined by

$$i_D = \varepsilon \frac{\delta E}{\delta t} \tag{6.11}$$

where $\varepsilon =$ the medium permittivity, or dielectric constant, with no electrons present. The displacement current simplifies to

$$i_D = \varepsilon \omega E_0 \cos \omega t \tag{6.12}$$

The total current density, i, is simply the sum of the convectional and displacement current densities:

$$i = \left(\varepsilon - \frac{N_e e_e^2}{m_e \omega^2} \right) \omega E_0 \cos \omega t \tag{6.13a}$$

where

$$\varepsilon = \varepsilon_r \varepsilon_0 \tag{6.13b}$$

$\varepsilon_r =$ relative permittivity, which is unity for air at standard temperature and pressure $\varepsilon_0 = 8.84194 \times 10^{-12}$; the permittivity of free space.

Rearranging (6.13a) in view of (6.13b),

$$i = \left(1 - \frac{N_e e_e^2}{\varepsilon_0 m_e \omega^2} \right) \varepsilon_0 \omega E_0 \cos \omega t \tag{6.14}$$

If electrons are present in the medium, their presence will reduce the dielectric constant from ε to

$$\left(1 - \frac{N_e e_e^2}{\varepsilon_0 m_e \omega^2} \right) \tag{6.15}$$

It is understood from elementary physics that a transmission medium with zero conductivity will have its refractive index, n, measured by simply the square root of its dielectric constant. Hence, the presence of electrons in the ionosphere causes a decrease in the dielectric constant to

$$n = \sqrt{\left(1 - \frac{e_e^2 N_e}{\varepsilon_0 m_e \omega^2} \right)} \tag{6.16}$$

If the propagation wave moves at a constant phase, at any point in the propagation medium, its phase velocity, V_p, may be defined as

$$V_p = \frac{c}{n} = \frac{c}{\sqrt{\left(1 - \frac{e_e^2 N_e}{\varepsilon_0 m_e \omega^2} \right)}} \tag{6.17}$$

where $c =$ speed of light (m/s). It is interesting to note from this expression that if there are no electrons present in the medium, $V_p = c$; that is, the velocity of propagation in free space. Also, the phase velocity approaches infinity $V_p \to \infty$ when $n \to 0$: this condition represents a situation when wave propagation is impossible.

The maximum electron density of an ionized layer can be determined by transmitting radio waves vertically incident to the ionosphere. Reflection will occur up to the frequency for which the refraction index equals to zero. Specifically, equating (6.16) to zero:

$$1 - \frac{e_e^2 N_e}{\varepsilon_0 m_e \omega^2} = 0 \tag{6.18}$$

If the frequency is still increased, the radio waves will penetrate the layer resulting in no reflection. The limiting frequency f_c at which the reflections begin to disappear is called the critical frequency of the layer, which from (6.18) is given by

$$\omega_c^2 = \frac{e_e^2 N_e}{\varepsilon_0 m_e} \tag{6.19a}$$

Noting that $f_c = \omega_c / 2\pi$, the critical frequency f_c can be written as

$$f_c = \frac{1}{2\pi}\sqrt{\frac{e_e^2 N_e}{\varepsilon_0 m_e}} \tag{6.19b}$$

This expression is also known as the electronic *plasma frequency*, f_p. Plasma occurs when an atom has been stripped of its electron resulting in a net positive electrically charged gas. Evidently, an alternative definition for the index of refraction, n, can be written as

$$n = \sqrt{1 - \frac{\omega_c^2}{\omega^2}} = \sqrt{1 - \frac{f_c^2}{f^2}} \tag{6.20}$$

The critical frequency f_c, of each of the reflecting layers E, F1 and F2, is denoted on ionograms by foE, foF1 and foF2 respectively, see Figure 6.4. Also, h'E, h'F1 and h'F2 correspond to each layer's virtual height of reflection (more is said about virtual heights in section 6.2.3). *Ionograms* are recorded tracings of reflected HF radio pulses generated by a sounder or ionosonde (more is said about ionograms in section 6.3). The subscripts '*o*' and '*x*' denote 'ordinary' and 'extraordinary' wave trace. The ordinary and extraordinary are components associated with a characteristic wave that propagates through the ionosphere having a polarization property. How an ionogram is interpreted is explained fully in section 6.2.3.1.

6.2.2.3 Modelling critical frequencies

Some good fit approximations that consider the problem of seasonal variations have been given for estimating the critical frequencies foE and foF1 (in MHz). But models of the critical frequency of the F2 region, foF2, are available in the form of numerical coefficients.

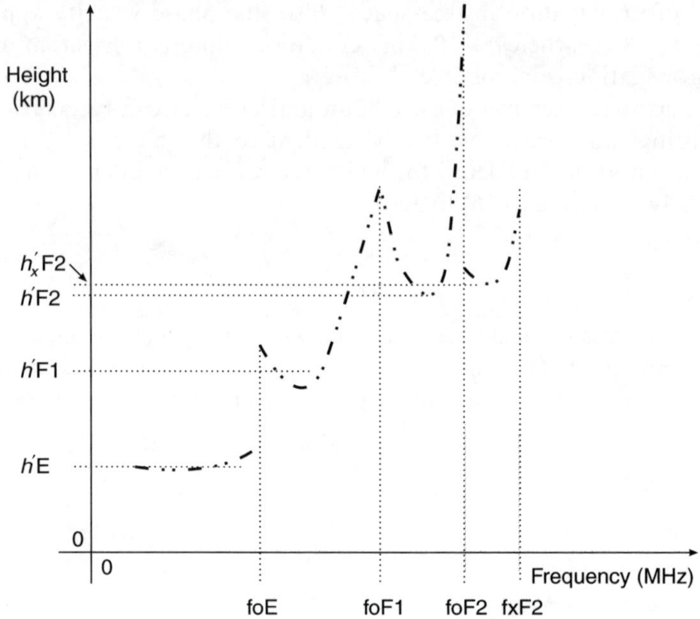

Figure 6.4 An ionogram showing critical frequencies and virtual heights

6.2.2.3.1 E layer

$$\text{foE} = 0.9[(180 + 1.44R_{12})\cos\zeta]^{0.25} \tag{6.21}$$

The notations ξ and R_{12} are the solar zenith angle and yearly (12-monthly) smoothed relative sunspot number defined by

$$R_{12} = \frac{1}{12}\sum_{n-5}^{n+5} R_k + 0.15(R_{n+6} + R_{n-6}) \tag{6.22}$$

This expression is the most widely used index in ionospheric studies, and as in (6.22) it depicts the smoothed index for the month represented by $k = n$ and where R_k is the mean of R_n for a single month k.

R_n is the sunspots' occurrence measured by the Wolf, or Zurich,[1] sunspot number, given by

$$R_n = K(10g_\Delta + s) \tag{6.23}$$

where g_Δ and s are the number of sunspot group and number of observed individual spots respectively. The scale or correction factor K (usually less than unity) depends on the observer and is intended to effect the conversion to the scale originated by Wolf.

[1] Records contain the Zurich number through 31 December, 1980, and the International Brussels number thereafter.

Sunspots are dark spots that appear and disappear with time. They appear dark because their surface temperature is low (about 3000 K) compared to 6000 K of the ambient photosphere. The activity of the sunspot groups varies markedly from month to month and from year to year: some last for a few days whereas a few survive for four or five solar rotations (of about 27 days each). Sunspots tend to cluster or group together. A group may contain a single spot or several tens. The most notable feature of sunspots is that they occur, on the average, with an 11-year cycle.

The solar zenith angle ζ is an angle measured at the Earth's surface between the Sun and the zenith (in degrees). This angle can be determined from

$$\cos\zeta = \sin\Delta_{lat}\sin\delta + \cos\Delta_{lat}\cos\delta\cos\Delta_h \tag{6.24}$$

where

Δ_h = hour angle of the sun measured westward from apparent noon expressed by (Schutte 1940)

$$\Delta_h = \cos^{-1}\left(\frac{\tan\delta}{\tan\Delta_{lat}}\right) \quad \text{(deg)} \tag{6.25}$$

for an azimuth up to 90°

Δ_{lat} = geographic latitude (deg). Geographic latitude is measured from 0° at the Earth's equator up to 90° at its pole, positive to the north, negative to the south

δ = solar declination (deg). For monthly averages, solar declination has been formalized to a sufficient accuracy by (Davies 1990)

$$\delta = 23.44\sin[0.9856(Y_n - 80.7)] \tag{6.26}$$

where Y_n = day number starting on 1 January.

Leftin (1976) gave different expressions for midnight and sunrise and sunset as follows:

$$\text{foE(midnight)} = 0.36[1 + 0.0098R_{12}]^{0.5} \tag{6.27}$$

$$\text{foE(sunrise, sunset)} = 1.05[1 + 0.008\text{R}_{12}]^{0.5} \tag{6.28}$$

Equations (6.27) and (6.28) do not hold in high latitudes; that is, above 70° latitude. Above this latitude (>70°), which is in the auroral zone, the night-time ionization is produced by particles from the magnetosphere.

6.2.2.3.2 F1 layer

$$\text{foF1} = [4.3 + 0.01R_{12}]\cos^{0.2}\zeta \quad \text{(MHz)} \tag{6.29}$$

Ducharme *et al.* (1971) gave a more detailed expression:

$$\text{foF1} = f_{00}(f_{100} - f_{00})\frac{I_{\text{F2}}}{100}\cos^x\zeta \quad \text{(MHz)} \tag{6.30}$$

where

$$f_{00} = 4.408 + 0.0076\Delta_{glat} - 0.00015\Delta_{glat}^2 \quad (6.31)$$

$$f_{100} = 5.365 + 0.0129\Delta_{glat} - 0.000248\Delta_{glat}^2 \quad (6.32)$$

$$x = 0.11 + 0.0038\Delta_{glat} - 0.000045\Delta_{glat}^2 + 0.0003I_{F2} \quad (6.33)$$

Δ_{glat} = geomagnetic latitude (rad)
I_{F2} = ionospheric index.

The expressions (6.21) through (6.33) hold for values of $\xi < 40°$, with $I_{F2} = 100$. Rosich and Jones (1973) gave similar expressions to that of Ducharme *et al.* (1971) but arrived at a peak value of foF1 $\approx$ 6 MHz for $I_{F2} = 150$ and $\Delta_{glat} \approx 45°$.

6.2.2.3.3 F2 layer

Unlike the expressions for the foE and foF1, the critical frequency of the F2 layer, that is, foF2, does not follow the cosine rule either diurnally or seasonally but exhibits a marked longitudinal effect due to geomagnetic control. Using spherical harmonics, world maps have been developed for the F2 peak critical frequency foF2. Similar maps have been established for the propagation factor M(3000)F2, which is related to the height of the F2 peak (Rush *et al.* 1984). Models for estimating foF2 are available in the form of numerical coefficients. CCIR provided an atlas of these coefficients (CCIR) with subsequent updates, enabling regional centres to produce their ionogram predicting periodic foF2. An example is Figure 6.5 produced by IPS for the city of Brisbane, Australia.

Other sources of data are the *international reference ionosphere* (IRI) and the Chinese Reference Ionosphere (CRI) (Tiehan and Peihan 1996). IRI is an international project sponsored by the Committee on Space Research (COSPAR) and the International Union of Radio Science (URSI). The IRI build-up, and what and how the formulas are derived, are detailed in Bilitza *et al.* (1979). Care must be taken while using and interpreting data produced by any of these agencies ensuring that a common reference is adopted. Some of the composite models are discussed under *composite parameter model* in the next section.

6.2.2.4 Models for electron density

The previous expressions have shown the linkage of the critical frequency with the electron density, N_e. Numerous models have been reported in the literature that attempt to chart the electron-density profile. These models provide analytical expressions that are amendable to mathematical manipulation including the following, which are used extensively in radio wave propagation work, and to some extent in estimating the virtual heights of reflection.

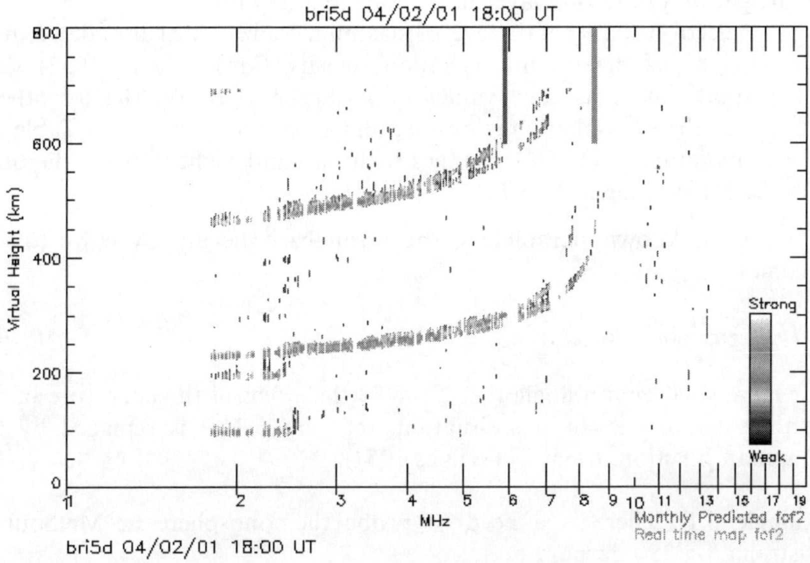

Figure 6.5 Real time map of foF2 for the city of Brisbane, Australia. (Courtesy: IPS, Australia)

6.2.2.4.1 Chapman model

The Chapman model (Chapman 1931) is the simplest type of ionized layer that can be predicted theoretically even though the model is formed under highly idealized conditions, namely, the atmosphere is isothermal, the ionizing radiation from the sun is monochromatic, and the recombination coefficient, or ion decaying, is constant with height. If the distribution of electron density with height is quasi-stable and homogeneous, then any layer's electron density can be defined by Chapman (1931)

$$N_e(h) = N_0(h)e^{\frac{1}{2}\left(1-\hat{H}-\sec\zeta e^{-\hat{H}}\right)} \quad (\mathrm{m}^{-3}) \tag{6.34a}$$

and the maximum electron density is

$$N_0(h) = N_m(h)\sec^{\frac{1}{2}}\zeta \quad (\mathrm{m}^{-3}) \tag{6.34b}$$

The scale, or normalized height, $\hat{H}$, is defined for a homogeneous ionosphere at height h (in km) and temperature T (in degree-kelvin, °K) by

$$\hat{H} = \frac{gm_m}{kT}(h - h_{\max}) \tag{6.34c}$$

where

m_m = mean molecular mass of air = 4.8×10^{-23} (gm)
k = Boltzmann's constant = 1.38×10^{-23} (joule/°K)
g = gravitational constant, $g = 9.807\,\mathrm{m/s^2}$

h = height of a reflecting layer in the ionosphere (km)
$N_m(h)$ = electron density at the level of maximum ionization at altitude h (cm^{-3})
h_{max} = height of maximum ionization density (km). Mitra (1952) gave approximate average values of h_{max} per layer, on the hypothesis that the ionospheric regions are all Chapman, as shown in Table 6.1
T = temperature (°K). This changes with day and night. During daytime, $T = T_e$, while at night $T = T_n$.

In terms of known parameters, the normalized height given by (6.34c) becomes

$$\hat{H} = \frac{34.11}{T}(h - h_{max}) \tag{6.34d}$$

For large solar zenith angle (i.e. $\zeta > 80°$) the effect of the curvature of the Earth is important. In this situation, $\sec\zeta$ in (6.34a) is replaced by the Chapman function, $Ch(x, \xi)$ (Wilkes 1954).

Example 6.1 There is a need to probe the ionosphere at Melbourne, Australia, on 25 February at

(a) 3.00 pm local time at 122 km, 256 km and 335 km,
(b) 9:12 pm local time at 132 km and 276 km.

Calculate for each layer of the ionosphere the electron density, critical frequency and refractive index when the ionosphere is probed at 1.2 MHz.

Solution
Inserting numerical values, appropriate values to the following notations are obtained.
From the Atlas World map, Melbourne geographic latitude, $\Delta_{lat} = 37.45$ °S
Day number starting on 1 January, $Y_n = 56$
From (6.26), $\delta = 23.44 \sin(-24.34) = -9.6624°$
From (6.25), $\Delta_h = 77.16°$
From (6.24), calculate the solar zenith angle $\zeta = 73.98°$

Table 6.1 Electron density at maximum ionization

Layer	h_{max} (km)	T (°K)	$N_m(m^{-3})$
Daytime			
E	100	341	1.5×10^{11}
F1	200	1360	3.0×10^{11}
F2	300	1710	12.5×10^{11}
Nighttime			
E	120	341	0.8×1010
F	250	1540	4.0×1011

Table 6.2 Computed values for ionospheric layers functions

	Daytime			Nighttime	
Parameters	E	F1	F2	E	F
$N_e^* 10^{11}(\mathrm{m}^{-3})$	1.282	2.990	11.236	0.0798	3.399
f_c (MHz)	0.102	0.155	0.301	0.025	0.166
n	0.996	0.992	0.968	1.0	0.99

Using relevant values of temperature, h_m and N_m appropriate to each layer from Table 6.1, computed values of electron density, critical frequency and refractive index for each layer and time of the day are tabulated in Table 6.2.

General comment on the Chapman layer
Diffusion or scattering has been suggested to affect the ionospheric layer profile particularly in F2 layer (Kato 1980). Even when diffusion was included in the electron-density analytical expressions, the shape of the F2 layer is approximately the same as that produced using the Chapman model. Of course, the Chapman model has its limitations because of the underlying assumption used in developing the model, namely, isothermal, single species, single ionizing radiation, which do not apply to the upper atmosphere. The model, however, provides an invaluable guide to analysing data and as a useful reference.

6.2.2.4.2 Linear model

$$N_e = \alpha^0 (h - h_0) \tag{6.35}$$

where α^0 is the electron density gradient and h_0 the layer base.

6.2.2.4.3 Exponential model

$$N_e = N_r e^{\frac{(h-h_r)}{\hat{H}}} \tag{6.36}$$

where N_r is the electron density at a reference height h_r and $\hat{H}$ is a scale height that is negative in the topside of the ionosphere.

6.2.2.4.4 Sec *h*-squared model

$$N_e = N_m \sec h^2 \left(\frac{h - h_m}{a} \right) \tag{6.37}$$

where N_m is the maximum electron density at a height h_m and a is the layer's thickness.

6.2.2.4.5 Quasi-parabolic model

$$N_e = N_m \left[1 - \left(\frac{r_0 (r - r_m)}{r (r_m - r_0)} \right)^2 \right] \tag{6.38}$$

where r is the radial distance from the centre of the Earth, r_m is the radial distance of the peak electron density N_m and r_0 is the radial distance to the bottom of the layer.

6.2.2.4.6 Composite model

One of the composite models available is a two-parabola model: one parabola representing the E layer and the other representing the F2 layer. The two parabolas could overlap or be distinct. An example of such parabolas is shown in Figure 6.6, developed by Bradley and Dudeney (1973).

The height of the peak density of the F2 layer is derived from

$$h_{mF2} = \frac{1490}{M(K) + D} - 176 \tag{6.39}$$

and its semi-thickness found as

$$y_{mF2} = h_{mF2}\left(1 + \left[\frac{0.618}{Q_x - 1.33}\right]^{0.86}\right) - h_{\min F2} - 104\left[\frac{0.618}{Q_x - 1.33}\right]^{0.86} \tag{6.40}$$

where

$h_{\min F2}$ = minimum height of the F2 layer (km)

$$D = \frac{0.18}{Q_x - 1.4} \tag{6.41a}$$

$$Q_x = \frac{\text{foF2}}{\text{foE}} \tag{6.41b}$$

The parameters foE, foF2, and $h_{\min F2}$ can be obtained from ionograms.

$$M(K) = \frac{\text{MUF}(K)}{f_c} \tag{6.42}$$

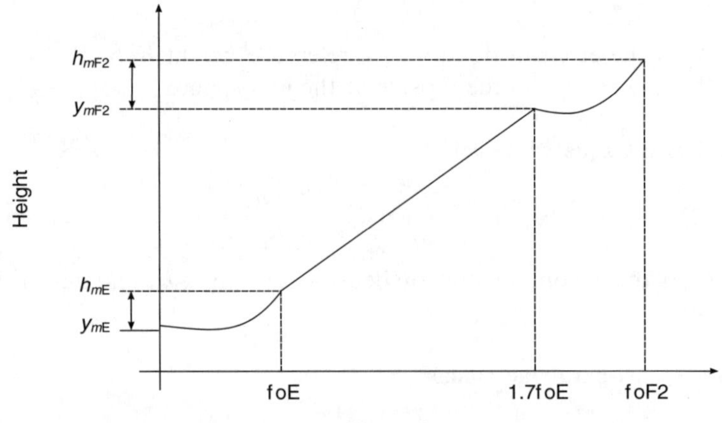

Figure 6.6 Electron density profile using two-parabola model

The percentage of dependence of the *MUF* on *K* for a short-term prediction is given by (Barghausen *et al.* 1969)

$$\mathrm{MUF}(K) = \left(\frac{p_0 + bK}{100}\right)\mathrm{MUF}(0) \tag{6.43}$$

where

p_0 = an intercept ≈ 100
b = constant that varies between −13 and 20. The negative value represents evening and in the low geographic latitude, while the positive value indicates morning and in the high latitudes. Both b and p_0 depend on season, sunspot number R_n, geographic latitude and local time.

MUF(K) = the maximum-usable-frequency for the magnetic index K. That is, the upper frequency limit that can be used for transmission between two points at a specified time. It is also defined as a median frequency applicable to 50 per cent of the days of a month, as opposed to 90 per cent cited for the *lowest usable high frequency* and *optimum working frequency* (FOT) – designated from French initials.

The magnetic index K is a 3-hour range designed to measure the irregular variations associated with magnetic field disturbance. Each observatory assigns an integer from 0 to 9 to each of the 3-hour UT (universal time) intervals: (000–0300, 0300–0600, . . . , 2100–2400). The magnetic K indices range in 28 steps from 0 (quite disturbed) to 9 (greatly disturbed) with fractional parts expressed in thirds of a unit. For example:

- K-value equal to 27 means 2 and 2/3 or 3−;
- K-value equal to 30 means 3 and 0/3 or 3 exactly; and
- K-value equal to 33 means 3 and 1/3 or 3+.

Since the K-value varies from one observatory to another, the arithmetic mean of the K-values from 13 observatories gives K_p.

A word of caution! The 'short-term prediction model' cannot be used as a precursor for long-term prediction, particularly for the F2 layer because of the departures of the *f*oF2 spatial correlation coefficient of the day-to-day from the median value.

Bent *et al.* (1978) developed the ionosphere electron density profile with particular emphasis on the topside. The model does not include the lower layers (D, E and F1) and uses a simple quadratic relationship between CCIR[2]'s M(3000)F2 factor and the height of the F2 peak. In their model, the bottomside is described by a bi-parabola:

$$N_e(h) = N_{m\mathrm{F2}}\left[1 - \left(\frac{h - h_{m\mathrm{F2}}}{y_{m\mathrm{F2}}}\right)^2\right]^2 \tag{6.44}$$

[2] CCIR stands for International Radio Consultative Committee.

The topside profile below 1000 km was subdivided into four intervals: the upper three covered equal height intervals and assumed a constant logarithmic decrement, which depended on the 10.7 cm solar noise flux. While the fourth interval just above the peak used a parabolic shape, which met the lowest of the exponential intervals in a way that the gradient was continuous at the junction height h_0, defined by

$$h_0 = h_{m\mathrm{F2}} + \frac{y_{m\mathrm{F2}}}{4} \quad \text{(m)} \tag{6.45}$$

N_0 is the electron density at height h_0, given by

$$N_0 = 0.864\, N_{m\mathrm{F2}} \quad (\mathrm{m}^{-3}) \tag{6.46a}$$

M(3000) F2 (=MUF(3000)/foF2) is a propagation factor closely related to the height of the F2 peak (Bilitza 1990; Bilitza *et al.* 1979). MUF(3000) is the highest frequency that, refracted in the ionosphere, can be received at a distance of 3000 km. As earlier defined, foF2 is the critical frequency of the F2 layer, or F2 peak plasma frequency, which is related to the F2 peak density $N_{m\mathrm{F2}}$ by

$$N_{m\mathrm{F2}} = 1.24 \times 10^{10}\ \mathrm{foF2} \quad (\mathrm{m}^{-3}) \tag{6.46b}$$

where the unit of foF2 is in MHz. Both parameters foF2 and M(3000)F2 are routinely scaled from the ionograms.

For a propagation to be possible on a particular circuit, the operating frequency f must be less than MUF. That is, at a given altitude h,

$$\mathrm{MUF}(h) = f_p(h) \sec\theta_{inc} \tag{6.47}$$

where θ_{inc} and f_p denote, respectively, the angle at which the propagation wave incidents the layer and the electronic *plasma frequency* of the layer – the same as (6.19b). At higher frequencies, the wave will penetrate the ionosphere and the reusable frequency may be expressed as

$$\mathrm{MUF}(h) = Q f_p(h) \tag{6.48}$$

where Q is called the *obliquity factor*. In the simplest form, $Q = \sec\theta_{inc}$. (This concept is revisited in section 6.2.4 to discuss 'skip zone'.) The important thing is that Q must be greater than or equal to the ratio of the operating frequency to plasma frequency. In view of (6.20),

$$Q \geq \frac{f}{f_p} = \sqrt{\frac{1}{1-n^2}} \tag{6.49}$$

The number of hops, the magneto-ionic component of characteristic wave, and the distance involved may modify the basic MUF. For example, the 1F2(4000)MUF(o) path via the F2 layer by the ordinary wave. The transmission curve for a distance of 3000 km is often used as a reference, given by

$$M(3000) = \frac{\mathrm{MUF}(3000)}{f_c} = \frac{67.6542 - 0.014938 h_v}{\sqrt{h_v}} \tag{6.50}$$

where h_v = virtual height (km); the method of measuring the virtual height is discussed in section 6.2.3.

The several models given in the literature, particularly those by CCIR, IRI and CRI, have demonstrated the complexity of the F2 layer and its variation in measurements. It may be inaccurate to predict the state of the ionosphere at one point of the data and region for another location. This suggests that more observatories are needed to spread representatively across regions and latitudes. The current observatory locations are skewed and their data are far from being globally representative.

6.2.2.5 Refraction errors by ray tracing

Due to the propagation anomaly of bending of radio waves traversing the ionosphere, measurement errors are introduced, namely, refraction angle error, range error, Doppler error for moving targets and polarization error. These errors are investigated in this section under the appropriate headings.

For simplicity, a spherical model is employed to explain the concept of ray tracing and to quantify the propagation errors caused by refraction. Though simple, the spherical method is capable of rendering theoretical estimates of propagation errors to a rather high degree of accuracy. The basic assumption considered in the ray tracing method is that the ionosphere can be stratified into spherical layers of thicknesses h_i and refractive indices n_i. For brevity, the analysis in this text is restricted to the three regions of interest for radio wave propagation, namely E, F1 and F2, as shown in Figure 6.7, where $i = 0, 1, 2$ corresponding to E, F1, F2. Of course, the same geometry can be used for N layers and variable thicknesses and indices in the troposphere.

Let us start by tracing a ray from point a to point m as it propagates through the E to the F2 layer, as shown in Figure 6.7. There are two possible paths to reach point m from a: namely the direct line-of-sight path, am, and the apparent ray path, abem. Point o is the centre of the Earth.

Following (6.16), each layer's refractive index can be estimated. Specifically, for ith layer

$$n_i = \sqrt{\left(1 - \frac{e_e^2 N_{e(i)}}{4\pi^2 \varepsilon_i m_e f^2}\right)} \tag{6.51}$$

where $N_{e(i)}$ and ε_i denote the ith layer electron density and permittivity respectively. From Figure 6.6, the ionosphere's apparent elevation angle is α_0 (angle bam) and its true elevation angle is α_{0t} (angle bad). The angle between apparent path direction and the direct line-of-sight path is called the ionosphere's elevation angle error, or refraction angle error, $\Delta\alpha_{ref}$ expressed by

$$\Delta\alpha_{ref} = \alpha_0 - \alpha_{0t} \tag{6.52}$$

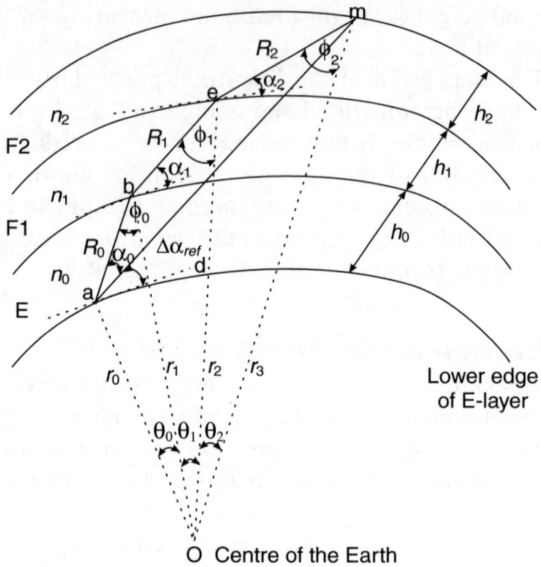

Figure 6.7 Ray path geometry by layer stratification

Using the sine's law,

$$\frac{\sin \phi_0}{r_0} = \frac{\sin\left(\alpha_0 + \frac{\pi}{2}\right)}{r_1} = \frac{\cos \alpha_0}{r_1} \tag{6.53a}$$

Alternatively,

$$\phi_0 = \sin\left(\frac{r_0}{r_1} \cos \alpha_0\right) \tag{6.53b}$$

where

$$r_0 = r_e + h_e \tag{6.54a}$$

h_e = the altitude of the lower edge of the ionosphere above the Earth's surface (km)

r_e = radius of the Earth at the equator (km) ≈ 6378.4 km. If the observation point is not at the equator, the elliptical distance r' of the point as a function of the geographic latitude Δ_{lat} and equatorial radius of the earth r_e can be calculated by (Schutte 1940)

$$r' = r_e[0.99832 + 0.001684 \cos(2\Delta_{lat}) - 0.000004 \cos(4\Delta_{lat}) \cdots] \tag{6.54b}$$

The angle that the ray makes with the horizon at the E layer is obtained, using Snell's law for symmetrical surface, as

$$\begin{aligned} n_0 r_0 \cos \alpha_0 &= n_1 r_1 \cos \alpha_1 \\ n_2 r_2 \cos \alpha_2 &= n_1 r_1 \cos \alpha_1 \end{aligned} \tag{6.55a}$$

where

$$r_1 = r_0 + h_0$$
$$r_2 = r_1 + h_1 \qquad (6.55b)$$
$$r_3 = r_2 + h_2$$

Subsequently,

$$\alpha_0 = \cos^{-1}\left[\frac{n_1(r_0 + h_0)}{n_0 r_0}\cos\alpha_1\right] \qquad (6.56)$$

Generalizing as

$$\alpha_i = \cos^{-1}\left[\frac{n_{i+1}(r_i + h_i)}{n_i r_i}\cos\alpha_{i+1}\right] \qquad (6.57)$$

$$\phi_j = \sin\left(\frac{r_j}{r_{j+1}}\cos\alpha_j\right) \qquad (6.58)$$

where $i = 0, 1, 2$.

It is easier to measure the apparent ground elevation angle α_g than the apparent ionospheric elevation angle α_0. It follows therefore that, by Snell's law, the relationship between the apparent (ground and ionosphere) elevation angles is established as

$$\alpha_0 = \cos^{-1}\left(\frac{r'}{r' + h_e}\cos\alpha_g\right) = \cos^{-1}\left(\frac{r'}{r_0}\cos\alpha_g\right) \qquad (6.59)$$

Applying sine law again to the direct path, the true elevation angle is obtained as

$$\alpha_{0t} = \cos^{-1}\left[\frac{r_3}{R_{012}}\sin\left(\sum_{j=0}^{2}\theta_j\right)\right] \qquad (6.60)$$

where $\theta_j = (\pi/2) - \alpha_j - \phi_j$.

And using the cosine law the direct radar range, R_{012}, (i.e. path am), is expressed as

$$R_{012} = \sqrt{r_0^2 + r_3^2 - 2r_0 r_3 \cos\left\{\sum_{j=0}^{2}\theta_j\right\}} \quad \text{(km)} \qquad (6.61)$$

The apparent paths ab $= R_0$, be $= R_1$, and em $= R_2$ (concisely as R_i, where $i = 0, 1, 2$) can be expressed as

$$R_i = r_{i+1}\frac{\sin\theta_i}{\cos\alpha_i} \quad \text{(km)} \qquad (6.62)$$

In view of (6.57) through (6.61), the measurement elevation angle error, or refraction angle error, is readily obtained:

$$\Delta\alpha_{ref} = \cos^{-1}\left[\frac{r'}{r_0}\cos\alpha_g\right] - \cos^{-1}\left[\frac{r_3}{R_{012}}\sin\left\{\sum_{j=0}^{2}\theta_j\right\}\right] \quad \text{(deg)} \tag{6.63}$$

6.2.2.5.1 Range, or time delay, error

If an imaginary observer were placed at the same point on the envelope of an advancing wave in the ionosphere, he/she will observe the group velocity, V_g, of the wave, which may be expressed as

$$V_g = \frac{d\omega}{d\gamma} \tag{6.64a}$$

where the wave's phase constant γ may be defined as

$$\gamma = \frac{\omega}{V_p} \tag{6.64b}$$

The phase velocity, V_p, has already been defined in (6.17). In view of (6.64) and (6.17), the group velocity can be readily shown:

$$V_g = \frac{V_p}{1 - \left(\frac{\omega}{V_p}\right)\left(\frac{dn}{d\omega}\right)} \tag{6.65}$$

Differentiating (6.20) with respect to ω, and substituting the result in (6.65),

$$V_g = cn \tag{6.66}$$

Observing the time of travel of the ray path layer by layer, the time to reach the E layer will be

$$t_0 = \frac{R_0}{V_{g(0)}} \tag{6.67a}$$

where $V_{g(0)}$ is the phase velocity at the E layer. Since the generalized group velocity is already defined by (6.66), the group velocity at the E layer may be written as

$$V_{g(0)} = cn_0 \tag{6.67b}$$

Equations for the ray path travel time to F1 and F2 can similarly be written. The total time of travel[3] of the beam in the stratified layers can be written as

$$t_{tot} = \frac{1}{c}\sum_{i=0}^{2}\frac{R_i}{n_i} \tag{6.68}$$

[3] If in the troposphere, the total time travel would be calculated from the phase velocity approach; that is,

$$t_{tot} = \frac{1}{c}\sum_{i=0}^{m} n_i R_i$$

where m = total number of stratified layers in the troposphere.

Since the product of speed and total time of travel (i.e. ct_{tot}) measures the radar range in the deviating medium, the range error, ΔR, is difference between the refracted path and the direct path, which may be expressed as

$$\Delta R = ct_{tot} - R_{012} = \sum_{i=0}^{2} \frac{R_i}{n_i} - R_{012} \tag{6.69}$$

Using (6.61) and (6.62), the range error is easily evaluated.

In summary, the expressions for time delay or range error (6.69) and refraction error (6.63) demonstrate that the measurement errors are cumulative.

Example 6.2 A 15 MHz wave is used to probe the ionosphere. The frequency at which reflection occurs is taken to be 3.04, 4.38 and 5.86 MHz respectively for the E, F1 and F2 layers. Estimate the refractive index of each layer and the likely measurement refraction angle and range errors if the sensor is located at latitude 5°S, longitude 132°E and apparent elevation angle of 9°. Each layer is approximately 100 km thick. The upper limit of the D layer is about 115 km above the surface of the Earth.

Solution

Geographic latitude $\Delta_{lat} = -5°$ (south of the equator)
Apparent elevation angle, $\alpha_g = 9°$
Height of the lowest edge of the E layer, $h_e = 115$ km
Equatorial radius of the earth $r_e = 6378.4$ km
Using (6.20) that is, $n = \sqrt{1 - f_c^2/f^2}$, calculate each layer's refractive index:

$$n_0 = n_E = 0.9792$$

$$n_1 = n_{F1} = 0.9564$$

$$n_2 = n_{F2} = 0.9255$$

Since the observation point is not at the equator, then from (6.54b) the elliptical distance to the edge of the Earth's surface $r' = 6378.24$ km. Hence,

$$r_0 = r' + h_e = 6493.24 \text{ km}$$

From (6.59), the ionosphere's apparent elevation angle is obtained as $\alpha_0 \approx 14.02°$.
Knowing α_0 and r_0, solve other apparent angles and distances iteratively:

$$R_0 = 371.9 \text{ km} \quad R_1 = 421.2 \text{ km} \quad R_2 = 701.4 \text{ km}$$

So from (6.63), the refraction angle error, $\Delta\alpha_{ref} \approx 8.72°$.
And, from (6.69), the range error, $\Delta R \approx 93.45$ km

6.2.2.5.2 Doppler effect

Doppler effect introduces an error, which is localized. This is different to the refraction and range errors, which are cumulative. The deviating medium only acts as a refractive medium. Figure 6.8 shows the ray path to target position. This figure shall be used to explain how errors introduced in the measurement of the target Doppler velocity can be quantified.

Let us assume that a target at point 'a' is travelling with a velocity V_t in an arbitrary direction in any part of the ionosphere of refractive index, n_T. The target velocity can be resolved into various orientations: ray path direction, V_r, direct path direction, V_d, and apparent path direction, V_a.

$$V_r = V_t \cos(\psi + \Delta\alpha_T) \tag{6.70a}$$

$$V_d = V_t \cos\psi \tag{6.70b}$$

$$V_r = V_t \cos\psi \cos\Delta\alpha \tag{6.70c}$$

where

$\Delta\alpha$ = refraction error angle
$\Delta\alpha_T$ = angle between the ray path and the direct part at the target location
ψ = orientation of the target velocity.

The error introduced in the target Doppler velocity may be expressed by

$$\Delta V = V_a - V_r \tag{6.71}$$

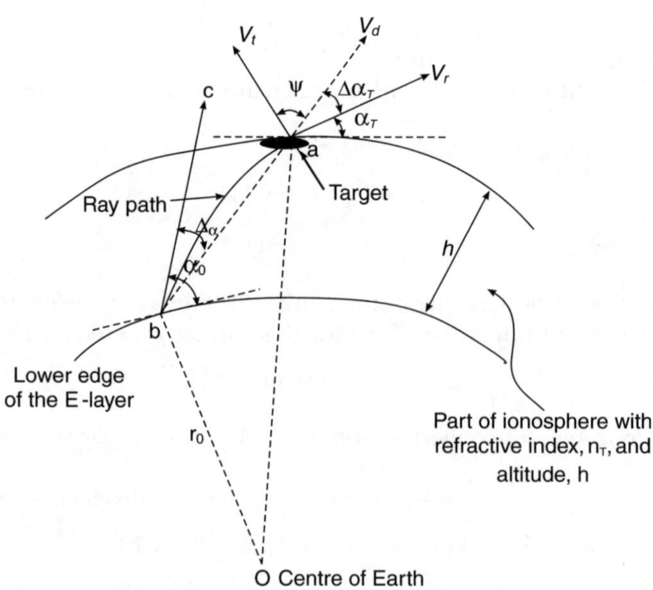

Figure 6.8 Deviation of ray path trace at target position

In view of (6.70), this error is simplified further as

$$\begin{aligned}\Delta V &= V_t[\cos\psi\cos\Delta\alpha - \cos(\psi + \Delta\alpha_T)] \\ &= V_t\cos\psi[\cos\Delta\alpha - \cos\Delta\alpha_T] + V_t\sin\psi\sin\Delta\alpha_T\end{aligned} \tag{6.72}$$

Note that

$$\cos(a \pm b) = \cos a\cos b \mp \sin a\sin b \tag{6.73}$$

Since $\Delta\alpha$, $\Delta\alpha_T$ and are very small angles, their trigonometric functions may be written in a Maclaurin series, noting that

$$\begin{aligned}\cos x &= 1 - \frac{x^2}{2!} + \frac{x^4}{4!} - \cdots \\ \sin x &= x - \frac{x^3}{3!} + \frac{x^5}{5!} - \cdots\end{aligned} \tag{6.74}$$

In view of this series expansion, the error in (6.72) is rewritten as

$$\begin{aligned}\Delta V = {} & V_t\cos\psi\left[\frac{(\Delta\alpha_T)^2 - (\Delta\alpha)^2}{2!} + \frac{(\Delta\alpha_T)^4 - (\Delta\alpha)^4}{4!}\cdots\right] \\ & + V_t\sin\psi\left[\Delta\alpha_T - \frac{(\Delta\alpha_T)^3}{3!} + \cdots\right]\end{aligned} \tag{6.75}$$

The higher-order terms can be neglected because in practice the values of $\Delta\alpha$ and $\Delta\alpha_T$ are in the order of one millionth of a radian. As such, the cosine term in (6.75), which is the target radial component, is neglected, reducing the error, ΔV, to

$$\Delta V = V_t\sin\psi[\Delta\alpha_T] \tag{6.76}$$

This expression shows that the target Doppler velocity error in the radial direction attributed to refraction is composed only of the tangential velocity component of the target velocity. The error is a maximum when the velocity vector is perpendicular to the direction of the direct path but a minimum when the target travels along the direct path. The error encountered in the measurement of the Doppler (frequency) shift Δf_d is easily determined for a target with approaching radial velocity, using the definition of (3.46) and in view of (6.76), as

$$\Delta f_d = -2\frac{\Delta V}{\lambda} = -\frac{2f}{c}V_t\sin\psi(\Delta\alpha_T) \tag{6.77}$$

This expression suggests that the Doppler shift Δf_d will be positive if the target is inbound (approaching), or negative if the target is outbound (receding). Expression (6.77) also implicitly suggests that any magnitude of target speed can be measured. Next task is to express $\Delta\alpha_T$ in terms of apparent ground elevation angle and distance from the point of observation on the Earth's surface.

By Snell's law, one can write

$$n_0 r_0 \cos \alpha_0 = n_T r_1 \cos \Delta\alpha_T \tag{6.78a}$$

which, in turn, gives

$$\Delta\alpha_T = \cos^{-1}\left(\frac{n_0 r_0}{n_T r_1}\cos \alpha_0\right) \tag{6.78b}$$

where

n_T = refractive index in the medium the target is traversing
n_0 = refractive index in the layer prior to the ionosphere, normally taken as unity.

As expressed in (6.59), there is a relationship between apparent ionospheric elevation angle α_0 and the apparent ground elevation angle α_g. So, rewrite (6.78b) as

$$\Delta\alpha_T = \cos^{-1}\left(\frac{r'}{n_T(r_0 + h)}\cos \alpha_g\right) \tag{6.79}$$

Example 6.3 A sensor operating at 12 MHz frequency is situated at an elevation angle of 7.6°, latitude 15°S, and longitude 132°E, indicating during surveillance that signal returns are from an inbound target. These signals, when analysed, suggested that they are reflected off refractive layer at about 5.6 MHz, 150 km above the Earth's surface. The target is estimated to be travelling at 85 km/s, bearing 15° east of the zenith. Estimate the Doppler frequency error.

Solution
It is obvious that the target is traversing in the E region, and $r_0 + h_0 = r' + 150$.
From (6.54b), calculate $r' \approx 6377$, so $r_0 = 6527$ km
From (6.20), calculate the refractive index, $n = 0.8844$
From (6.77) the error introduced in the Doppler frequency measurement by an inbound target, at speed $V_t(= 85\,\text{km/s})$, may be expressed by

$$\Delta f_d = -2\frac{\Delta V}{\lambda} = -\frac{2f}{c}V_t \sin\psi(\Delta\alpha_T) = 1.54\,\text{kHz}$$

Noting that $c = 3 \times 10^8$ m/s, $(\Delta\alpha_T) \approx 1.095$, and $\alpha_g = 7.6°$.

Before proceeding to the discussion on the polarization effect of radio wave propagation in stratified layers, it is appropriate to examine the effect of collisions of electrons with other particles including the effect of the Earth's magnetic field on them. This effect has been ignored in the previous analyses.

6.2.2.5.3 Effect of Earth's magnetic field on electron collisions

The polarization properties of electromagnetic waves in magnetized plasma have been studied extensively in the literature. Radio wave propagation through the ionosphere is a complex mix of interactions between the ionized constituents, the Earth's magnetic field, and the parameters of the propagating signal (such as frequency, polarization, strength or amplitude, direction, etc.). The direction of propagation can be resolved into two orthogonal directions, namely, parallel and perpendicular to the magnetic field, and the characteristic waves. A *characteristic wave* is defined as a wave that propagates through the ionosphere without any change in the polarization state. The characteristic wave that propagates perpendicularly to the magnetic field is further divided into two independently acting waves: the ordinary '*o*' wave and the extraordinary '*x*' wave. The ordinary wave has its electric vector aligned along with the magnetic field, meaning that the electrons move in the same direction as the constant-force lines in the magnetic field and no interactions occur. A snapshot of the characteristic wave would produce two distinct traces. Thus, on each ionogram two traces of '*o*' and '*x*' waves are present – more is said of ionograms in section 6.2.3.

An inquiring mind might immediately ask: How does this division of characteristic waves into two magneto-ionic components '*o*' and '*x*' affect the refractive indices of the stratified layers? The next subsections will shed some light on the question.

6.2.2.5.3.1 No Earth's magnetic field present during electron collision

Collision of vibrating electrons with the ions and neutral particles frequently occurs. When electrons collide with other particles they give up some of their energy to these particles, and in the absence of a magnetic field, some will be absorbed and converted into thermal energy. The thermal speed of electrons v_e has some mathematical function, given by

$$v_e = \sqrt{\frac{3kT_e}{m_e}} \tag{6.80}$$

To the plasma frequency, there corresponds a characteristic length λ_D, called the *Debye length*, which may be defined by

$$\lambda_D = \frac{v_e}{\sqrt{3}(2\pi f_p)} \tag{6.81a}$$

Note that f_p denotes plasma frequency, which is the same as the critical frequency f_c expressed by (6.19b). In terms of known parameters, the Debye length is

$$\lambda_D \approx 69\sqrt{\frac{T_e}{N_e}} \tag{6.81b}$$

Symbols are as defined previously in the text. The Debye length is basically the distance covered by an electron during one cycle of a plasma oscillation, representing the distance over which potential differences find themselves naturally *shielded* by their effect on the charged particles' distribution. This means that fluctuations in electron concentration can exist independently of ions only at scales smaller than this Debye shielding length. Also, plasma oscillations can develop for wavelengths greater than the Debye length only.

Example 6.4
Calculate the Debye length for Example 6.1.

Solution
By using the relevant variable values in Table 6.2 and (6.81), the Debye length for each layer is estimated as shown in Table 6.3. Note that daytime temperature $T_e = T$ of daytime in Table 6.1.

By treating each stratified layer of the ionosphere as homogeneous, the vibrating electrons may have an effective angular collision frequency, ν. It is reasonable to suggest that the angular collision frequency, ν, obeys the exponential law:

$$\nu = \nu' e^{\hat{H}} = \nu' e^{\frac{3.411\times10^{-3}}{T}(h-h_{\max})} \tag{6.82}$$

where ν' is the collision frequency at altitude h and T ($= T_n$ for nighttime, and T_e for daytime). Other parameters are as defined previously.

It is worth noting that Brace and Theis (1978) gave an empirical model for daytime electron temperature, T_e, as a function of electron density, in the altitude range 130 to 400 km as

$$T_e(N_e, h) = 1051 + 17.01[h - 161.43]e^{\left(6.094\times10^{-12}N_e[1-0.005497h]-0.0005122h\right)} \tag{6.83}$$

Matsushita (1967) gave approximate collision frequencies of electrons with neutral particles ν'_n, electrons with ions $\nu'_\pm$, ions with neutral particles $\nu'_{\pm n}$, and ions with ions $\nu'_{\pm\mp}$ as follows:

$$\nu'_n = 0.5\, N_n\sqrt{T_n} \tag{6.84a}$$

$$\nu'_\pm = \left[34 + 8.36\log\left(\frac{T_e^{\frac{3}{2}}}{\sqrt{N_e}}\right)\right]\frac{N_\pm}{T_e^{\frac{3}{2}}} \tag{6.84b}$$

$$\nu'_{\pm n} = 3.35\times10^{-21}\frac{N_n}{\sqrt{m_m}} \tag{6.84c}$$

$$\nu'_{\pm\mp} = 3.06\times10^{-14}\frac{\nu'_\pm}{\sqrt{m_m}} \tag{6.84d}$$

where N_n and $N_\pm$ are number of densities of neutral particles, and positive and negative ions respectively. Also, T_n and T_e denote temperature of neutral particle and electron ions respectively in degree-kelvin.

Table 6.3 Calculated Debye length for Example 6.1

	Daytime			Nighttime	
Item	E	F1	F2	E	F
Debye length (mm)	3.65	4.65	2.69	14.25	4.64

A typical temperature profile is shown in Figure 6.2, and expressions (6.1) and (6.2).

The electron collisions render the ionosphere as an absorbing medium having a conductivity, σ, given by

$$\sigma = \frac{N_e v e_e^2}{m_e(\omega^2 + v^2)} \tag{6.85}$$

Without further mathematical derivation, which is somehow tedious, applicable results are given. For the case of electron collision without a magnetic influence from the Earth, the refractive index may be written as

$$n = \sqrt{\left(1 - \frac{e_e^2 N_e}{\varepsilon m_e[\omega^2 + v^2]}\right)} \tag{6.86}$$

6.2.2.5.3.2 With Earth's magnetic field present during electron collision
The theory of propagation of electromagnetic waves through an ionized medium under the influence of an external magnetic field is well founded. In some literature, this theory is called the *magneto-ionic theory*. The next paragraphs attempt to exploit this theory to examine the effect of external magnetic field on refractive index.

Following Millan (1965), the general equation of an electron in an ionized region is defined as

$$m_e\ddot{\mathbf{r}} = -e_e\mathbf{E} - (m_e v)\dot{\mathbf{r}} - \frac{e_e}{c}(\mathbf{r} \times \mathbf{H}) \tag{6.87}$$

where

E = the electric field vector
H = magnetic field vector
r = displacement vector of the electron and the dots on this vector denote differentiation with respect to time
Other symbols are as defined previously.

By assuming that the wave that propagates along the *x*-axis has no component of the Earth's magnetic field along the *y*-axis, then the equations of motion in scalar form may be readily written. The incident electron field is assumed to vary sinusoidally with time. Eventually the solutions to (6.87) are readily obtained leading to the general form of the complex index of refraction, M:

$$M^2 = 1 + \frac{2}{2\gamma - \frac{\gamma_T^2}{1+\gamma} \pm \sqrt{\left[\left(\frac{\gamma_T^2}{1+\gamma}\right)^2 + (2\gamma_L)^2\right]}} \tag{6.88}$$

And the polarization vector of the wave is written as

$$P = \frac{H_z}{H_y} = -\frac{j}{\gamma_L}\left(\frac{1}{M^2 - 1} - \gamma\right) \tag{6.89}$$

where

H_z and H_y correspond to the magnetic field intensity of the wave along z- and y-direction

$\gamma = \alpha + j\beta$ is defined by (6.4a) with reference part, α, and quadrature component, β, expressed by

$$\alpha = -\frac{\omega^2}{\omega_c^2} \qquad \beta = \frac{\omega v}{\omega_c^2} \tag{6.90}$$

$$\gamma_L = \frac{\omega}{\omega_c^2}\left(\frac{He_e}{m_e c}\right)\cos\theta \qquad \gamma_T = \frac{\omega}{\omega_c^2}\left(\frac{He_e}{m_e c}\right)\sin\theta \tag{6.91}$$

$H =$ the magnitude of the magnetic field intensity at any point on the Earth, defined by Chapman and Bartels (1940) as

$$H = 0.31\sqrt{(1 + 3\sin^2\Delta_{glat})} \tag{6.92}$$

where θ and Δ_{glat} correspond to the propagation angle and the geomagnetic latitude, all units in degrees. Other symbols are as defined previously in the text.

The term $(He_e/m_e c)$ is called the *gyromagnetic frequency* of the electron above the Earth's magnetic field. It is obvious in (6.88) that there are two possible values for the complex refractive index, which would indicate two different modes of propagation that travel independently in the deviating medium and each with a polarization vector associated with it. As noted earlier, if the absorption coefficient is zero the quadrature component, β, can be neglected. This situation occurs for higher frequencies, or smaller concentrations of electron densities. For smaller concentrations, the magnitude of α increases, making $\alpha \gg 1$. Consequently, the term $(1 + \gamma)$ in (6.88) approximates to α. With this simplification, the complex refractive index expressed by (6.88) reduces to a real quantity n for two different modes of propagation given by

$$n_{o,x}^2 = 1 + \frac{2}{2\alpha - \frac{\gamma_T^2}{\alpha} \pm \sqrt{\left[\left(\frac{\gamma_T^2}{\alpha}\right)^2 + (2\gamma_L)^2\right]}} \tag{6.93}$$

The subscripts 'o' and 'x' denote ordinary and extraordinary wave respectively. Similarly, the polarization vector P of the wave can be expressed as

$$P_{o,x} = -\frac{j}{\gamma_L}\left(\frac{1}{n_{o,x}^2} - \alpha\right) \tag{6.94}$$

Upon substitution of the terms (6.90) and (6.91) in (6.93), the full expression for the two magneto-ionic components' refractive index can be written as

$$n_{o,x}^2 = 1 - \frac{1}{\frac{\omega^2}{\omega_c^2} - \frac{1}{2}\left(\frac{\omega He_e \sin\theta}{cm_e\omega_c}\right)^2 \pm \frac{\omega He_e \sin\theta}{cm_e\omega_c^2}\sqrt{\left[\left(\frac{He_e \sin^2\theta}{2cm_e\omega}\right)^2 + \cos^2\theta\right]}} \tag{6.95}$$

Alternatively, in terms of the individual magneto-ionic components:

$$n_o^2 = 1 - \frac{1}{\frac{\omega^2}{\omega_c^2} - \frac{1}{2}\left(\frac{\omega He_e \sin\theta}{cm_e\omega_c}\right)^2 + \frac{\omega He_e \sin\theta}{cm_e\omega_c^2}\sqrt{\left[\left(\frac{He_e \sin^2\theta}{2cm_e\omega}\right)^2 + \cos^2\theta\right]}} \tag{6.96}$$

$$n_x^2 = 1 - \frac{1}{\frac{\omega^2}{\omega_c^2} - \frac{1}{2}\left(\frac{\omega He_e \sin\theta}{cm_e\omega_c}\right)^2 - \frac{\omega He_e \sin\theta}{cm_e\omega_c^2}\sqrt{\left[\left(\frac{He_e \sin^2\theta}{2cm_e\omega}\right)^2 + \cos^2\theta\right]}} \tag{6.97}$$

Upon an application of necessary conditions, two cases of quasi-propagation modes can be investigated, namely, *quasi-longitudinal mode and quasi-transverse mode.*

Case I: Quasi-longitudinal propagation

The condition under which quasi-longitudinal propagation mode occurs is when

$$4\frac{\omega^2}{\omega_H^2} \gg \sin^2\theta \tan^2\theta \tag{6.98}$$

where the gyromagnetic frequency (or simply gyro frequency) is

$$\omega_H = \left(\frac{He_e}{m_e c}\right) \tag{6.99}$$

By substituting (6.98) in (6.93), the refractive index reduces to

$$n_{o,x}^2 = 1 - \frac{\omega_c^2}{\omega^2\left(1 \pm \frac{\omega_H}{\omega}\cos\theta\right)} \tag{6.100}$$

At any given frequency, $\omega^2/\omega_c^2 \gg 1$. So, (6.100) can be expanded by its binomial series. Neglecting higher-order terms, the expanded equation becomes

$$n_{o,x} \cong 1 - \frac{\omega_c^2}{2\omega^2}\left(1 \pm \frac{\omega_H}{\omega}\cos\theta\right) \tag{6.101}$$

From this expression, the difference in the refractive index Δn of two magneto-ionic components is expressed as

$$\Delta n = n_o - n_x = \frac{\omega_H \omega_c^2}{\omega^3}\cos\theta \tag{6.102}$$

Or, in view of (6.19a) and (6.99)

$$\Delta n = \frac{N_e He_e^3 \cos\theta}{2\pi^2 c m_e^2 f^3} \tag{6.103}$$

And the polarization vector associated with the refractive index reduces to

$$P_{o,x} = \pm j \tag{6.104}$$

This expression shows that both the ordinary and extraordinary components are circularly polarized.

Case II: Quasi-transverse propagation

The condition under which a quasi-transverse mode of propagation occurs is when

$$\sin^2\theta\tan^2\theta \gg 4\frac{\omega^2}{\omega_H^2} \tag{6.105}$$

which is the reverse of case I. As the frequency is increased, θ rapidly approaches 90°. By substituting (6.105) in (6.93), and expanding the resulting expression by the binomial expansion, while neglecting the higher-order terms, the individual refractive index reduces to

$$n_o \cong 1 - \frac{\omega_c^2}{2\omega^2} \tag{6.106}$$

$$n_x \cong 1 - \frac{\omega_c^2}{2\omega^2}\left[1 + \frac{\omega_H^2}{\omega^2}\sin^2\theta\right] \tag{6.107}$$

As seen in (6.106), the ordinary refractive index is independent of the Earth's magnetic field, H, which is present in (6.107), the extraordinary refractive index. Therefore, it could be said that the term in brackets [.] of (6.107) represents the correction due to the presence of the Earth's magnetic field.

From (6.106) and (6.107), the difference in the refractive index Δn of two magneto-ionic components is expressed as

$$\Delta n = \frac{1}{2}\left(\frac{\omega_H \omega_c}{\omega^2}\sin\theta\right)^2 \tag{6.108}$$

Or, in view of (6.17a) and (6.99)

$$\Delta n = \frac{N_e}{(2\pi m_e)^3}\left[\frac{He_e^2 \sin\theta}{cf^2}\right]^2 \tag{6.109}$$

The polarization vector of the wave in the quasi-transverse mode simply becomes

$$\begin{aligned} P_O &= 0 \\ P_x &= -j\infty \end{aligned} \tag{6.110}$$

This expression shows that both the ordinary and extraordinary components are linearly polarized but the ordinary component is polarized in the direction parallel to the Earth's magnetic field while the extraordinary is polarized in the direction perpendicular to the Earth's magnetic field.

At 6° geomagnetic latitude, plots of the comparison between the two quasi-cases, represented by (6.103) and (6.109), are shown in Figures 6.9 and 6.10 for a skywave radar propagating at frequencies 3 and 30 MHz at $\theta \leq 10°$. By comparison between Figures 6.9 and 6.10, it can be seen that the

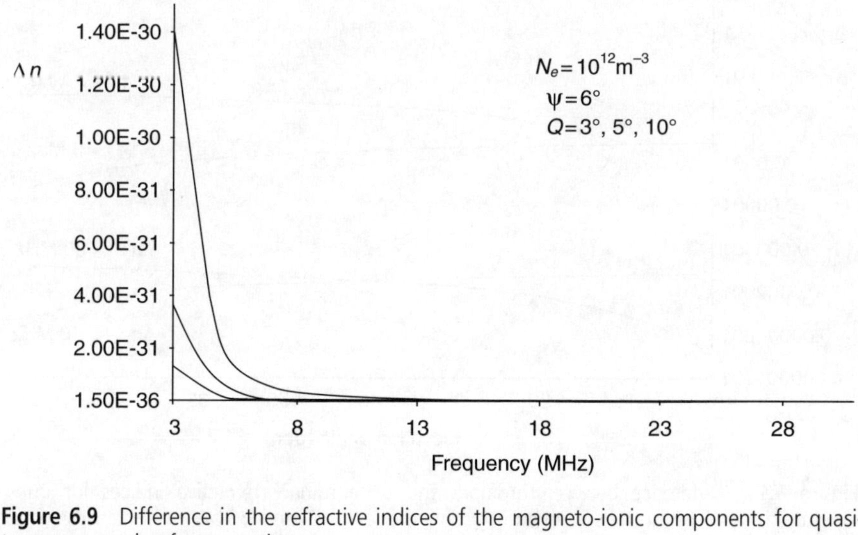

Figure 6.9 Difference in the refractive indices of the magneto-ionic components for quasi-transverse mode of propagation

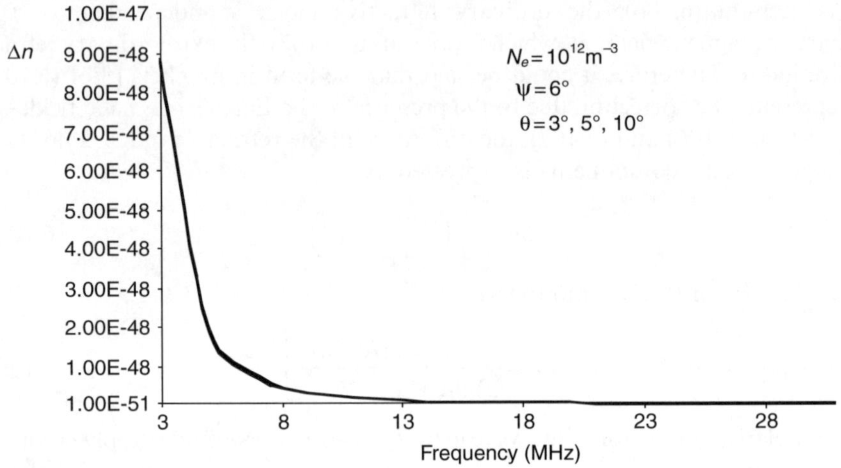

Figure 6.10 Difference in the refractive indices of the two magneto-ionic components for quasi-longitudinal mode of propagation

higher the propagation frequency the smaller the value of the difference between cases. However, the difference is much more discerning when propagating in the lower frequencies ($\leq$10 MHz) for the quasi-transverse mode. As seen in Figure 6.11, case I has a wider band higher with increasing elevation angle than case II. Case I (the quasi-longitudinal propagation mode) holds for nearly all cases of interest in skywave radar propagation particularly the over-the-horizon radar.

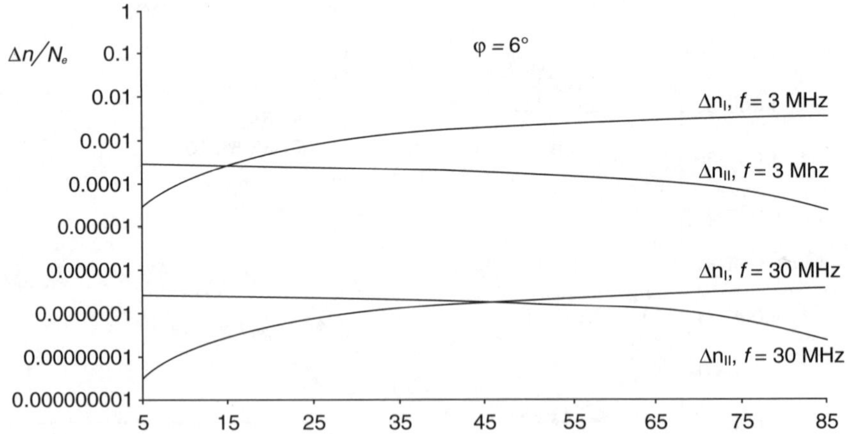

Figure 6.11 Difference between ordinary and extraordinary refractive indices for quasi-longitudinal and quasi-transverse propagation modes at 3 and 30 MHz frequencies, where subscripts I and II denote case I and case II respectively

6.2.2.6 Polarization error

The behaviour of polarization of radar signals in the space–time-frequency domain has an important bearing on radar signal, as well as influencing radar techniques and signal interpretation. For example, the choice of the antenna elements to be used for transmission and reception will determine whether a polarimetric capability is available or not.

As discussed earlier in this chapter, when a linearly polarized wave enters the ionosphere, it splits into two characteristic waves – called *ordinary* '*o*' and *extraordinary* '*x*' waves, which have different phase velocities so that a difference accumulates as they propagate. When they are summed at any point, the polarization of the resultant wave depends on the phase difference. If the two characteristics' waves suffer similar attenuations, their net effect is simply a rotation of the axis of linear polarization called the *Faraday rotation effect*. This description can be formalized as follows.

Suppose that the electric field intensities of two linearly polarized progressive waves can be expressed by

$$E_{o,x} = Ae^{j(\omega t - \gamma_{o,x} s)} \tag{6.111}$$

where A is the field constant amplitude, and for brevity is put as unity. The subscripts 'o' and 'x' denote the two magneto-ionic components of the fields traversing path 's'. Also, $\gamma_{o,x}$ represents the phase propagation coefficient, or proportionality constant, of the two magneto-ionic components defined by

$$\gamma_{o,x} = \frac{\omega}{V_{po,x}} \tag{6.112}$$

where $V_{po,x}$ is the phase velocity of each of the two magneto-ionic components, and in view of (6.17),

$$\gamma_{o,x} = \frac{\omega}{c} n_{o,x} \tag{6.113}$$

From this expression, the difference in phase $d\phi$ between the two waves traversing a distance, ds, may be expressed as

$$d\phi = \Delta\gamma ds = (\gamma_o - \gamma_x) ds \tag{6.114a}$$

$$d\phi = \frac{\omega}{c}(n_o - n_x) ds = \frac{\omega}{c}(\Delta n) ds \tag{6.114b}$$

This difference defines the phase shift for a one-way propagation path, which is also the differential phase shift between two magneto-ionic components. The total polarization shift for a two-way path that is within defined distance limits s_1 and s_2 can be defined by

$$\phi(s) = \frac{\omega}{c} \int_{s_1}^{s_2} \Delta n ds \tag{6.115}$$

It is convenient to define the phase difference $\phi(s)$ in terms of layer thickness, say between limits h_1 and h_2. So

$$\phi(h) = \frac{\omega}{c} \int_{s_1}^{h_2} \Delta n r(h, \theta_{elev}) dh \tag{6.116}$$

where

$$r(h, \theta_{elev}) = \frac{r_0 + h}{\sqrt{(r_0 + h)^2 + (r_0 \cos \theta_{elev})^2}} \tag{6.117}$$

$$r_0 = r' + h_e \tag{6.118}$$

θ_{elev} = elevation angle of the antenna beam (deg).
The variable r' is the distance off the equator to the edge of the Earth's surface, defined by (6.54b), and h_e is the altitude from the Earth's surface to the low edge of the ionosphere.

The difference Δn was defined in (6.103) for quasi-longitudinal propagation mode and (6.109) for quasi-transverse propagation mode. By substituting the difference Δn represented by (6.103) and (6.109) for each mode in (6.116), the two-way polarization rotation at any frequency, for each mode and within the validation range of the propagation angle θ, can be estimated. Having demonstrated the effect of refractive indices differences for the two modes of propagation in Figures 6.9 and 6.10, and by assuming identical propagation conditions, one can infer that for at any given range, the polarization rotation for the longitudinal case (case I) will be far greater than the transverse (case II) condition.

Faraday rotation impinges on HF skywave radar performance. For example, if a differential polarization rotation occurs across the propagation-signal bandwidth and if the scattering behaviour is appreciably polarization dependent, the resulting modulation of echo strength may-spread across the echo in the range domain. This may limit range resolution on some targets, defeating the very improvement sought earlier in Chapter 3.

Spatially, it is possible to have a polarization fringe pattern across a given radar footprint. Polarization fringe pattern is particularly recognizable over the ocean because of the strong polarization dependence of the radar cross-section of the sea surface. Spatial fringe-pattern distribution has obvious implications particularly for ship detection because when Doppler spectra are nested in several beams, where polarization is horizontal for a vertically polarized receiver, nulls would be registered in the beams where polarization fringes are noticed. Of course, a large-scale modelling of polarization fringe patterns can be carried out. Examples include those carried out by Barnum (1969) and Croft (1972).

6.2.3 Observing the ionosphere

The problem of inaccurately determining the propagation height, h, necessitates the need to observe the ionosphere. In addition, observing or probing the ionosphere helps to provide:

(a) A real-time propagation advice required by skywave radars. The frequency required to optimally illuminate a given area varies with changes in electron density in the ionosphere and cannot be predicted precisely. Operating an OTHR requires a real-time evaluation of the ionospheric path for frequency selection. A vertical sounder, an oblique sounder, and the radar itself carry out the real-time evaluation. Separate receiver monitors channel occupancy in the HF band to see which channels are available. One of the unoccupied channels that falls in the optimal band can then be selected for operation.
(b) Measurements that support structuring the ionosphere where propagation delays can be converted accurately into target coordinates; that is, from target slant coordinates into ground coordinates after factoring in the propagation errors.

Sounders are some of the tools used to probe the ionosphere. *Sounders* (also called *ionosondes*) are essentially radars. The signal generated, usually a chirp (swept frequency), by its transmitter system is delivered to the antenna array. It is then transmitted in an upward direction at an altitude between 100 and 350 km, depending on operating frequency, in a small volume of a few hundred metres thick and a few tens of kilometres in diameter over the site. The signal is partially absorbed. The intensity of the signal in the ionosphere, for example, is less than $3\,\mu W/cm^2$, which is far less than the Sun's natural electromagnetic radiation reaching the Earth. The small effects that are produced provide information about the dynamics of the plasma and other processes of solar–terrestrial interactions. The receiver measures the *group delay*, or travel time, of the return signals as they bounce back from the ionosphere. There exist unique relationships between the sounding frequency and the ionization densities that reflect it. As the sounder sweeps from the lower to the higher frequencies, the signal rises above the background noise, including commercial radio sources, and records the return signals reflected from the different layers of the ionosphere. The records are ionograms, which are collected at regular time intervals. An example of an ionogram is shown in Figure 6.12.

Radio waves or pulses travel more slowly within the ionosphere than free space therefore the *apparent*, or *virtual*, height is recorded instead of a *true* height. It should be understood that the group delay is not simply related to the actual distance travelled, or the height of reflection.

For instance, consider the reflection process as single reflections from a mirror at the appropriate height, with the pulses travelling to and from the

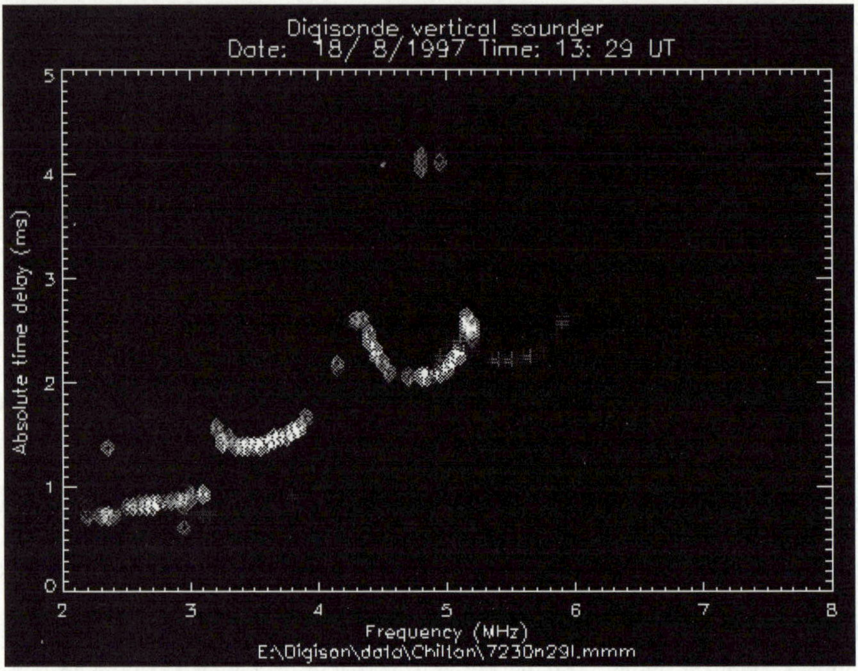

Figure 6.12 Points on vertical ionogram (VI) showing group delays of ionospheric signal at different frequencies. The upper right-hand curving section is the extraordinary, '*x*' component while the rest represents the ordinary '*o*' component. (Crown Copyright Radiocommunications Agency 2002)

mirror. The group delays of the pulses at different frequencies can be converted into the virtual height, h_v, of the mirror using

$$h_v = \frac{ct}{2} \tag{6.119}$$

where c and t correspond to the speed of light and time taken by the pulse to travel to and from the mirror. Virtual height of each layer is denoted by h'E, h'F1 and h'F2 corresponding to that of E, F1 and F2 layer. Points on the ionogram in Figure 6.12 show group delays of ionospheric signals at different frequencies. A group delay of 1.25 ms means that the ionospheric layer the signal is reflected from is at height of 187.5 km. The group delay's axis is directly converted to virtual height, h_v, against the operating frequency, f, for that particular time and location.

6.2.3.1 Interpreting an ionogram

Figure 6.12 is redrawn as Figure 6.13 to make the description clearer. In Figure 6.13, each ionospheric layer shows up as an approximately smooth curve, separated from each other by an asymptote at the layer's critical

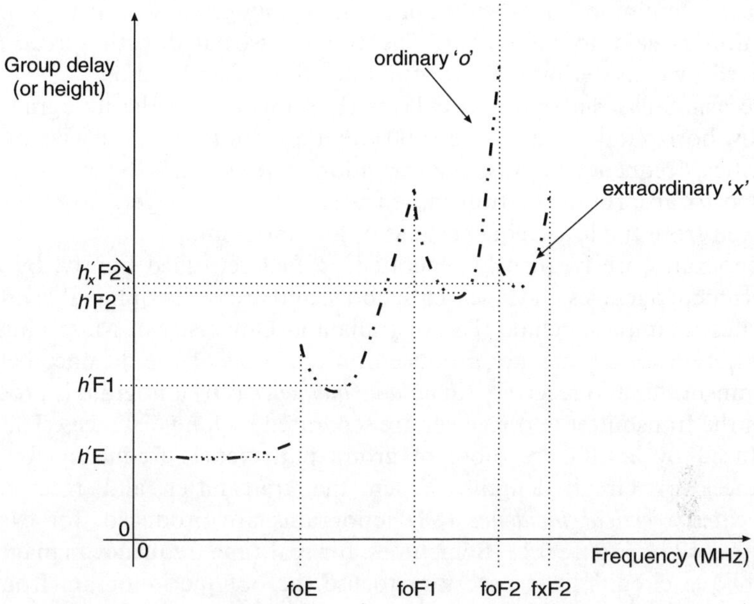

Figure 6.13 A reconstructed ionogram

frequency. The upwardly curving sections at the start of each layer are due to the transmitted wave being slowed, but not reflected, from underlying ionization that has a critical (plasma) frequency close to, but not equalling, the transmitted wave.

The critical frequency of each layer (foE, foF1 and foF2) is scaled from the asymptote, while the virtual height of each layer (h'E, h'F1 and h'F2) is scaled from the lowest point of each curve. The two magneto-ionic components '*o*' and '*x*' of the characteristic wave are also shown in the figure. In this case, the extraordinary component of the F2 layer (fxF2) is shown. Its virtual height and critical frequency are denoted by h'_xF2 and fxF2 respectively.

The extraordinary mode critical frequency, f_{cx}, also has a simple relation to the electron density, which is the sum of the 'ordinary mode' critical frequency (f_c, from (6.19b)) and the magnetic component. Specifically

$$f_{cx} = \frac{1}{2\pi}\sqrt{\frac{e_e^2 N_e}{\varepsilon_0 m_e}} + \frac{He_e}{2m_e} = f_c + \frac{He_e}{2m_e} \tag{6.120}$$

All symbols are as previously defined.

When echoes from other regions of the sky are received with that from the F layer or overhead, the electron concentration in these regions differs from the ionosphere overhead, two traces are observed. Of course, if the geometry is right for echoes to be received from a whole range of locations and the ionospheric conditions vary over the range (such as when a trough is overhead) multiple traces will appear on an ionogram. The F trace in this situation is said to be *spread*. The traces associated with *spread-F* are resolved by considering the horizontal position of each echo.

Occasionally, the sporadic E layer (Es) appears on the ionogram as a narrow horizontal line at around 100 km; it does not exhibit an asymptote at its critical frequency because the transition is too swift.

Due to absorption of transmitted wave by the D layer, no echoes are received from the low-frequency end of an ionogram.

Ionograms are frequently generated, in fact refreshed hourly, by many government agencies and research schools, and are easily obtainable on their websites, examples include IPS Australia and University of Massachusetts.

Different ionograms are produced on the basis of the distance between the transmitter and receiver. An *oblique incidence* (OI) ionogram is produced when the transmitter and receiver are separated by long distances. The plots produced by an IO are those of group path versus frequency for fixed distances or circuit lengths. When the transmitter and receiver are co-located, *vertical incidence* (VI) ionograms are produced, for example Figure 6.12 or Figure 6.13. Sometimes, for real-time frequency management of oblique circuits, it is necessary to use the oblique ionogram from one circuit to manage another circuit allowing for different path lengths. Using transformations based on the path length and the time delay measured from the oblique ionogram easily performs this. By applying similar transformations, vertical incidence ionograms can be converted to equivalent oblique ionograms for any path length providing the circuit control points are reasonably similar to the vertical incident ionosonde location. The advantage of the transformation is that it takes into account all the reflecting layers in the ionosphere.

When the transmitter and receiver are close to each other and the signals being received have been scattered back towards the transmitter by ground backscatter, a *backscatter* (BS) ionogram is obtained. In the case of BS, the circuit length is not specified. The most interesting and useful part of the BS ionograms is the leading edge, which corresponds to the minimum group path at a given frequency. In addition, calculations arising from BS ionograms include the determination of the ground range at a particular elevation angle to define the relationship between the group path and ground range: a relationship that is crucial for *coordinate registration* (CR) in the over-the-horizon radar (OTHR) system. More is said of CR in section 6.2.5.

6.2.4 Skip zone

At higher-elevation radiation angles the rays escape (i.e. the rays are insufficiently refracted and pass through the ionosphere rather than returning to Earth), causing a skip zone of range coverage, see Figure 6.14(a). This shows that the ionospheric refraction process results in a skip zone (distance) from

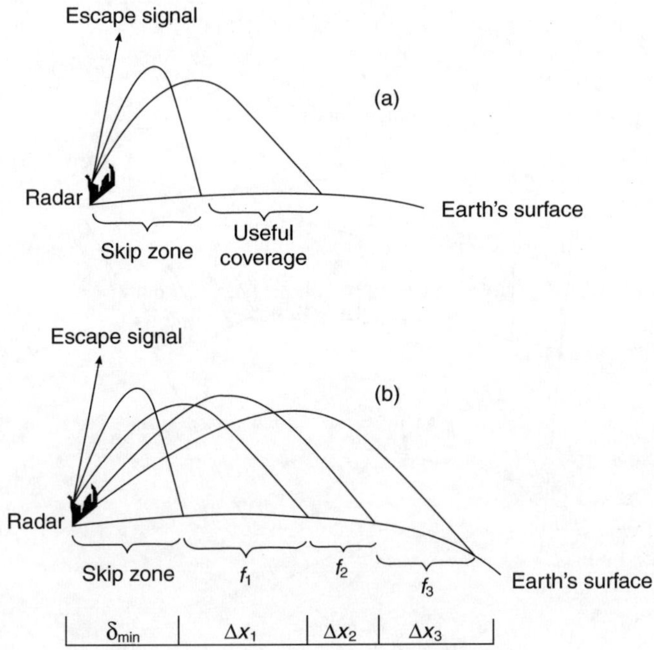

Figure 6.14 Ray paths showing different range extents, Δx_i, illuminated by different operating frequencies f_i

the transmitter to the closest point on ground illumination indicating that the radar site must stand back by at least δ_{min} distance from the closest obligatory surveillance zone. Just beyond the skip zone, energy is returned to the Earth after the reflection height horizon is reached. The useful range coverage, lying between the escape and refraction limits, is where illumination is strongest. It is possible that multiple hops exist, although only one hop is shown in Figure 6.14(a), and energy could circle the Earth.

The skip zone is usually a problem when it exists but it can sometimes be put to good use if secure communications are required. For instance, if we do not want someone to hear our transmissions, we are sometimes able to ensure that the eavesdropper is within the skip zone (McNamora 1991).

As seen in Figure 6.14(b), different range extents (Δx_i) are illuminated by different operating frequencies; implying that longer ranges require higher frequencies. It must be recognized that the trailing edge of an extent may vary as a function of radar parameters and target size, but the start is set by frequency selection and immediately follows the skip zone.

Figure 6.15 shows a plan view of an azimuthal scan (or coverage area) of angle θ (deg). Within the scan are different segmented areas that may be

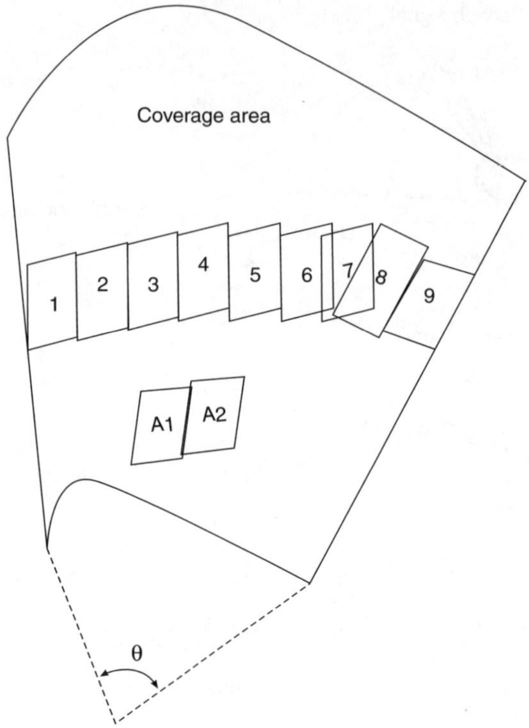

Figure 6.15 An azimuthal sectorial scan. Each segment 1 through to 9, A1 and A2 is illuminated by separate transmit beams each $\Delta\theta$ wide

illuminated by a separate transmitter beam of width $\Delta\theta$ (deg). By simple geometry, the transmitter beamwidth is

$$\Delta\theta = \frac{180\Delta x_{if}}{\pi r_e} \quad \text{(deg)} \tag{6.121}$$

where

Δx_{if} = the range extent at a particular operating frequency (km)
r_e = radius of the Earth (km). If the observed point is not at the equator, then use r' the elliptical distance equation of (6.54b) instead of r_e in (6.121).

It is conceivable that the range of the transmitter footprint could change with azimuth due to ionospheric effect. Each transmitter footprint is then filled with N_x number of contiguous receiver beams, each $(\Delta\theta/N_x)$ wide. The transmitter footprints 1 to 9, A1 and A2 can be interlaced, abutted or overlapped, depending on the interest attached to target(s) within the coverage area. The radar footprint can be moved in range by varying the frequency and moved in azimuth angle by electronic beam steering – a process already discussed in Chapter 4. This footprint is sometimes called instantaneous if only maintained by the radar for a few seconds.

6.2.5 Ray tracing and coordinate registration

The reader might be wondering how accurate are the measurements taken via the skywave radar, in particular OTHR, in the face of these multifaceted reflections? Any radio and radar systems require knowledge of the exact ray path of the radiowave as it travels through the ionosphere. The accuracy of these systems depends on both the accuracy of the ionospheric electron density model and the accuracy of the solutions to the equations used to trace through this model. Numerical and analytical methods have been used to trace through the ray path model. Numerical methods allow taking a real-time snapshot of the ionosphere and ray tracing the leading edges.

The computational demand of the numerical methods may be seen as a drawback that reduces the effectiveness of real-time operational systems. The analytical method on the other hand is fast and could provide more accurate temporal and spatial predictions of the solar–terrestrial environment. An example is the Segmented Method for Analytical Ray-Tracing (SMART), which is claimed to meet operational and computer constraints.

Ideally, if the ionosphere were precisely known along the possible paths between the radar and the target, a simple slant-to-ground transformation technique, or a ray trace electromagnetic propagation model, would have been adequate to generate a look-up table – called CR (coordinate registration) table – of ground coordinates versus slant coordinates. The difficulty is that the trans-ionospheric paths are of variable length and they add variable biases – due to mode ambiguity – in range, azimuth and velocity: examples of these biases have been demonstrated in section 6.2.2.5. Unless these biases are removed to provide a reasonably well-calibrated ground-truth CR system, the formation of tracks is of little value since radar measurements are in slant range, slant azimuth, and radial velocity. With improving knowledge of ray tracings with credible interpolation scheme(s) improved target ground coordinates can be formulated.

Certain techniques have been proposed for improving CR accuracy. These include the use of the following:

- *Use of beacons* – Beacon assisted CR consists of correcting the ground coordinates of each raymode by an amount determined from the radar signal transponded by a beacon of known location. Two limitations of this technique are that (a) it presupposes correct identification of raymode types for both the beacon and target (Krolik and Anderson 1997), and (b) its improved accuracy is likely to be localized; that is, improvement will be around the beacon rather than the global.
- *Use of terrain features* – The use of terrain features is similar in concept to employing beacons except that prominent backscatter from geographical features of known location is used to estimate the correction factors (Zollo and Anderson 1992). The number of terrain features that can be unambiguously identified may limit this technique (Krolik and Anderson 1997).

- *Developing dynamic optimization model* – A real-time dynamic ionospheric model that allows on-the-spot profiling of the ionosphere based on data input from sounders (OI, IV and BS), global positioning system (GPS), transponders and perhaps satellites would be an ideal. An example of a dynamic optimization model is the CREDO, which stands for Coordinate Registration Enhancement Dynamic Optimization (Nickisch and Hauuman 1996). The first generation of CREDO attempted to adjust the ionospheric parameters in real time to minimize the ground range variance for multimode track data. The current CREDO strives to 'fit' ionospheric parameters to ionospheric sounder data (e.g. *vertical incidence* (VI) ionogram and *backscatter* (BS) ionogram).
- *Use of maximum likelihood technique* – Target localization consists of determining the most likely target ground coordinates over an ensemble of ionospheric conditions consistent with the ionospheric sounder data (Krolik and Anderson 1997). While this method attempts to enhance localization accuracy by employing a statistical model for uncertainties in the ionospheric propagation conditions, it may be difficult to extrapolate the solution to a more general case.
- *Multiple location of sounders* – The ionosphere is dynamic. Periodic sampling of the ionosphere by several equidistantly positioned sounders would provide instant situation status. The data then can be used to develop a good fit approximation of the region's ionospheric profile within the sounders' grids. This process would greatly enhance our knowledge of the dynamics of the ionosphere as well as resolving accurately CR measurements.

In essence, the accuracy of coordinate registration (CR) measurements is within the CR operating window. This presupposes selection of appropriate frequency or frequencies that allow optimal illumination of the area to be observed.

6.2.5.1 Comments

Advances made in the understanding of ionospheric behaviour have been encouraging. The key assumption to these improvements has been that the down-range ionosphere is precisely known. This assumption may not be completely true because measurements taken by the ionospheric sounders (BS, VI and OI) are only estimates. A case can therefore be established to increase the number of sounders in designated geographic locations to ensure a better understanding of the ionospheric behaviour, and thus enhance the accuracy of the down-range ionospheric data. Errors in the estimates of down-range ionospheric parameters can seriously degrade the accuracy of the estimated target ground coordinates. It is obvious that we have a challenging research programme ahead.

6.3 Summary

This chapter has discussed the upper part of the atmosphere – the ionosphere – where free electrons occur in sufficient density to have an appreciable influence on the propagation of radio waves. This ionization depends primarily on the Sun and its activities. Since the ionosphere is a dynamic system, better understanding of this part of the atmosphere is required if improvements on coordinate registration at resolving propagation errors, and most importantly those arising from multiple paths, are to be achieved.

Problems

1. There is a need to probe the ionosphere at your home town on 25 March at

   (a) 3.02 pm local time at 115 km, 156 km and 335 km,
   (b) 8:42 pm local time at 132 km and 276 km.

   Estimate the critical frequency and refractive index of the ionospheric layers when propagating at 25 MHz.
2. If there is a facility in your home town or nearby that generates hourly ionograms, compare your critical frequency results in question (1) with that obtained as foE, foF1, foF2, or foF. Can you spot any differences? If yes, why?
3. Explain how the maximum reusable frequency can be determined. Also describe the factors that influence the reusable frequency for a given link at any given time.
4. What is the electron gyrofrequency? Compute a typical value.
5. Is the Earth's magnetic field important in the consideration of high-frequency propagation? Why?
6. How is an ionosphere formed?
7. Do you think it is possible to use the thinning layer of H ions on top of the F layer as a reflecting medium for skywave radio wave propagation? Why?
8. What is the solar zenith angle seen at noon by an observer in your home town on 10 June?
9. You are tasked to measure the virtual heights of the ionosphere at different frequencies. Transmitted waves sometimes travel round the world before being received. How will you know when this gerrymandering has occurred?

7

Skywave radar

Skywave radar is capable of sensing beyond the horizon because it makes use of the ionosphere to refract the radar wave propagated back to earth. A typical example is the over-the-horizon radar (OTHR). Skywave radar utilizes the high-frequency (HF) band, specifically 3 to 30 MHz, because this band enables surface-to-surface radar to target distances well beyond the horizon. Radar to target ranges of 1000 nautical miles and more are typical. Skywave radar achieves its long ranges, in effect, by using the ionosphere as a gigantic mirror.

The conventional microwave radar operates on the line-of-sight principle and propagates through the ionosphere at frequencies of 0.2–40 GHz, whereas the HF band utilized by the OTHR, which is lower than that operated by the microwave radar, interacts with the ionosphere in a way that can be exploited to provide radar coverage at variable distances. Another major difference between skywave and microwave radars is the need to adapt the signal waveform and frequency of the skywave radar to the environment.

Chapters 3 to 5 have provided the fundamental principles governing the design, operation and understanding of the limitations of a radar system. These principles are also fundamental to skywave radar with the added burden of interference due to the environment, which could be harsh. This is due to the OTHR looking down on its targets from the ionosphere. As a result, there are associated constraints:

- the antenna must be very large; one kilometre or more;
- spatial resolution is relatively coarse; typically in tens of kilometres;
- a large backscatter echo from the Earth's surface clutter is produced at the same range as that of the desired targets; and finally
- due to ionospheric electron density distribution, the radar operating frequency and waveform need to be continually assessed.

These constraints might be viewed as an operational nightmare, yet provide advantages. For example, they provide the capability to detect ocean backscatter from water gravity waves with dimensions comparable to those of the radar waves, which in turn provide an opportunity to study and map the sea and surface wave behaviour.

Ships and large aircraft have dimensions that are in the resonant scattering region. When targets are moving, their echoes may be detected by measuring the frequency deviation, or Doppler shift, they cause in the reflected wave. It is believed that aircraft and surface craft with or without high manoeuvrability and speed, and with small radar cross-sections, may also be detected by the skywave radar, including stealth aircraft. This is possible due to aircraft radar cross-section being much more dependent on gross target dimensions than on detail in shape (Headrick 1990).

The main emphasis in this chapter is the skywave radar. Propagating radio waves through the regions comprising the atmosphere result in the degradation of signal-target information. Signal deterioration is due to spatial inhomogeneities that exist in the atmosphere, which vary continuously with time. The spatial variations produce statistical bias errors, which have been discussed in detail in Chapter 6. The influence of the spatial variations is an important consideration, which must be taken into consideration in the formulation and design of skywave radar system.

7.1 Skywave geometry

A skywave geometry is described by Figure 7.1. The regions used by skywave radars are in the ionosphere. When radio waves are beamed from a transmitter and then refracted down from the ionosphere to illuminate a target, the echo from the target may travel by a similar path back to the

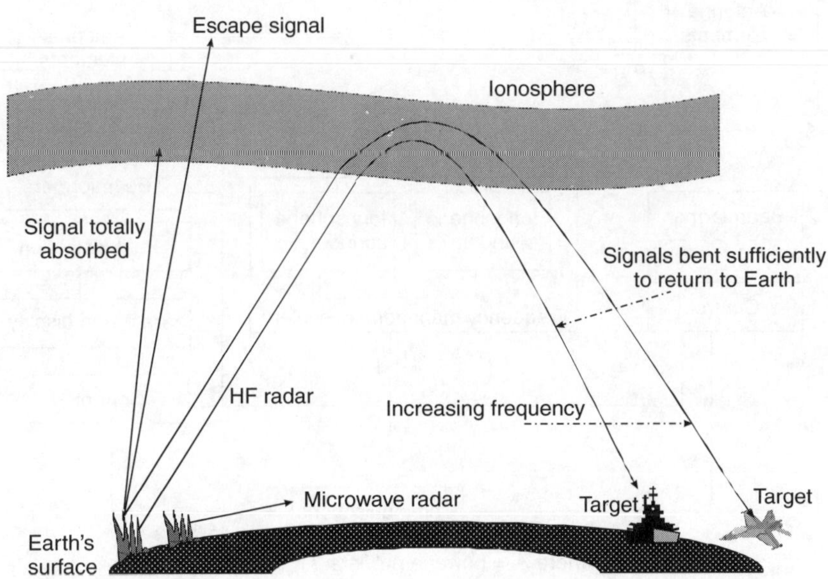

Figure 7.1 Operation of a skywave radar

receiver. Strictly speaking, objects in the target area scatter the incident radar illumination in all directions. Nearly all of the energy will be forward scattered from the ground surface, or sea surface. A small percentage will return via an ionospheric reflection path to a suitable receiver antenna. Propagation effects are prevalent when radio waves traversing the atmosphere manifest themselves as refractive bending, time delays, Doppler errors, rotation of the phase of polarization (called Faraday effect) as well as attenuation. These effects have been discussed in Chapter 6.

7.2 Basic system architecture

By design, a skywave radar system, in particular an over-the-horizon radar (OTHR), generally uses continuous transmission in order to maximize energy. This calls for separate transmit and receive facilities, with the separation being sufficient to avoid direct ground wave coupling between the receivers and transmitters, as well as maintaining the far-field criterion. A basic schematic of skywave radar is shown in Figure 7.2.

Figure 7.2 is similar to the basic structure of a radar system shown in Chapter 2, Figure 2.1, except that there is an additional need for a frequency management system including ionospheric sounders or ionosondes.

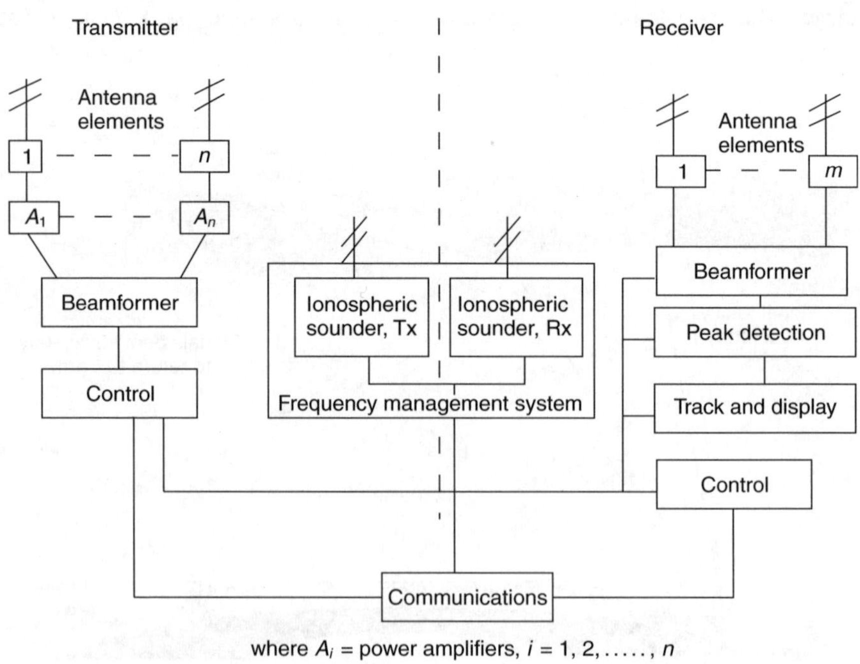

Figure 7.2 A schematic diagram of a skywave radar

As discussed in Chapter 6, these sounders allow a sophisticated understanding of the ionosphere's complexities and aid in the selection of optimal frequency suitable for propagation.

The separation bracket – the mandatory distance between the transmitter and receiver facilities – is necessary to achieve:

- receiver isolation from transmission interference or combination;
- avoidance of high transmitted peak power while assuming and maintaining high energy on the target;
- a convenient way of maintaining radiation in the far field; and
- the use of sophisticated modulated waveforms that may allow clever detection of the *electronic counter countermeasure* (ECCM) process. The reader is referred to Chrzanowski (1990) for a good exposition on radar electronic countermeasures.

Aside from this separation bracket, land buffer zones are required around the highly energized transmitter antennas to ensure that radar emissions do not interfere with electrical equipment. Similar precautions are necessary for the receivers. The receiver antennas need to be protected from extraneous electrical interference by a series of land buffer zones. In addition, the receiver site must be isolated from noise generated by power lines. As such, a continuous operation of internal combustion engines as the power source to the receivers may be necessary. Alternatively, well-shielded underground power lines could be an option.

Although much of the OTHR signal and data processing is common with conventional microwave radar, OTHR antenna considerations are quite different from those arising in general radar. Because of stringent operational requirements and the ionosphere being birefringent and time varying, the antenna system design is dominated by consideration for large physical size for transmit and receive arrays. Table 7.1 shows the array sizes of the Australian (Jindalee) and United States of America (USA) OTHRs. The systems in Table 7.1 use linear arrays. Other array geometry can be used for the OTHRs such as circular arrays. Operational complexity and cost effectiveness associated with any arrangements are essential elements of any choice taken.

Example 7.1 Two frequencies, 3 and 30 MHz, are to be used for propagation. Using the transmitter and receiver array sizes in Table 7.1, calculate the

Table 7.1 Array sizes of three OTHR systems (Sinnott 1989)

OTHR system	Transmitter array	Receiver array
Australia (Jindalee)	127 m	2.8 km
AN/FPS-118 (USAF)	67–345 m (frequency dependent)	1.5 km
AN/TPS-71 (USN)	335 m (total length covering separate bands)	2.5 km

Table 7.2 Estimates of far-field ranges for frequencies at 3 and 30 MHz

Array aperture, D (m)	Range (km) @ 3 MHz	Range (km) @ 30 MHz
127	40.97	409.7
67	6.01	60.1
335	751.91	7519.1

range where the receiver aperture should be located to be in the radiating far field.

Solution

For an antenna to be considered in the radiating far field, the range R from the source to the receptor of aperture D may be represented by equation (3.39); that is, $\Re \geq 2D^3/\lambda$. Putting in the transmitter values from Table 7.1, the receptor array must be located at least at the distance tabulated in Table 7.2.

Table 7.2, column 3 gives values that are somehow unrealistic for $f = 30$ MHz. This suggests that the positioning of the receiver for all frequency settings may not be in the 'strict sense' in the far field.

7.2.1 Transmitter

The typical power requirement of an OTHR transmitter averages between 10 kW and 1 MW. This requirement is necessary to launch high levels of power efficiently. The USA and Australian OTHRs use a frequency modulated continuous waveform (FMCW), consisting of multiple sweeps with linear sawtooth frequency modulation (Lees 1987). If propagation by FMCW is properly constrained, it is intrinsically clean and enables sidelobe reduction to be efficiently controlled.

Transmit antennas are generally arrays of radiating elements, with each element driven by a separate power amplifier. The approach of individual powering of the radiators permits beam steering at low-power amplifier stages. With technological advances in radar technology, digitally switched power supplies can deliver the desired current directly into each antenna element thereby increasing the transmitter capability as well as allowing subdivision of the array to achieve required performance and operations.

The transmitting antenna is a log-periodic curtain in a uniform line array on a wire ground mat to provide ground shield. The log-periodic antenna is a broadband array of closely spaced elements, each 180° out of phase with the next and the spacing changing proportional to its distance from an apex. Basically a log-periodic antenna pattern is by subtraction unlike the planar array in Chapter 4 whose pattern is rather by addition. Log-periodic elements have a distinct structure. This structure is discussed in the next few paragraphs.

Figure 7.3 shows typical log-periodic elements that can transmit over the frequency range 6–30 MHz installed at Alice Springs, Australia. Only a few of the elements, which are near resonance, radiate at a given frequency.

Figure 7.3 A section of the Australian OTHR transmit array

The alternate phasing of the elements allows the array to radiate electrically through shorter elements, which in turn perturbs the pattern less than the longer ones would. The electrical feed serves as the antenna boom, which is usually a parallel transmission line. The operational OTHRs in Australia and the USA use log-periodic antennas capable of covering the entire HF frequency band. The transmitter coverage is potentially enormous: a million square kilometres could be surveyed by the installation.

The log-periodic antenna can be arranged in parallel (as in Figure 7.4a), or radially (as in Figure 7.4b). Figure 7.4 is similar to that given in Sinnott (1987). If log-periodic antennas are arrayed in parallel, the frequency independence is lost, as there is frequency-dependent electric spacing between array elements. Whereas, if log-periodic antennas are arranged radially a frequency-independent array geometry is possible, with the active regions on an arc and separated by a frequency-independent electrical length. It should be noted that there is a limit to the number of such antennas, which can be so arrayed before the edge elements are 'firing' a long way from the boresight of the array.

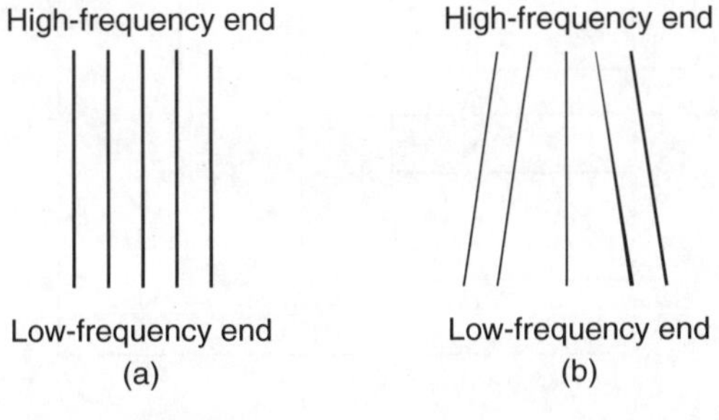

Figure 7.4 A plan view of array geometries: (a) parallel array; (b) radial array

Log-periodic antennas have self-scaling properties. Figure 7.5 shows the structure of a log-periodic antenna with radiators of length l_n and distances d_n. If the antenna's dimensions are scaled by some ratio τ, the antenna will have similar properties at frequencies $f, \tau f, \tau^2 f, \tau^3 f, \ldots, \tau^{n-1} f$. Ratio τ is called the geometric ratio.

The lengths of the radiators l_n and distances d_n increase in a defined manner. For instance

$$\frac{l_1}{l_2} = \frac{l_2}{l_3} = \frac{l_3}{l_4} = \cdots = \frac{l_n}{l_{n+1}} = \tau \tag{7.1}$$

$$\frac{d_1}{d_2} = \frac{d_2}{d_3} = \frac{d_3}{d_4} = \cdots = \frac{d_n}{d_{n+1}} = \tau \tag{7.2}$$

The spacing between adjacent elements is geometrically related by a factor

$$\beta_\tau = \frac{s_1}{2l_2} = \frac{s_2}{2l_3} = \frac{s_3}{2l_4} = \cdots \frac{s_n}{2l_{n+1}} \tag{7.3}$$

The edges of the dipoles lie along two straight lines, which converge at one end, with an apex angle η. The apex angle may be expressed in the form

$$\eta = 2\tan^{-1}\left(\frac{l_2 - l_1}{2s_1}\right) = \frac{1-\tau}{4\beta_\tau} \tag{7.4}$$

where, in practice, $0.7 \leq \tau \leq 0.95$ and $10° \leq \eta \leq 45°$.

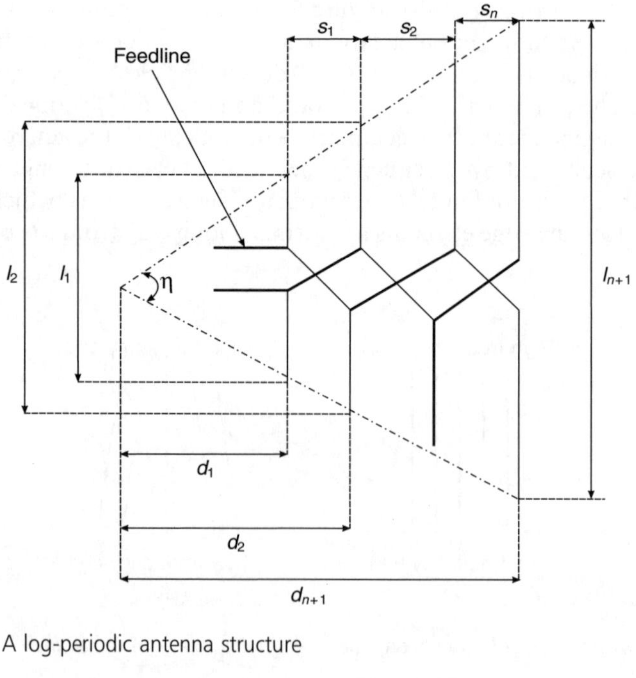

Figure 7.5 A log-periodic antenna structure

The log-periodic antenna characteristic (e.g. impedance and directivity) varies periodically with the logarithm of the frequency. This accounts for its name. It is also found that if the periodic variations are small over a broad band of frequencies, the antenna behaviour is effectively frequency independent. The array structure is fed at the apex end by a balanced line, with connections crisscrossed to adjacent elements, to give the correct phasing of the elements. The problem of aeolian noise is frequent in a log-periodic antenna structure. The aeolian noise attempts to induce harmonics of several orders of magnitude, thereby complicating the elements' phase resolution.

To ensure total absence of grating lobes at the highest frequency, the array spacing must meet the condition stipulated by equation (4.37) in Chapter 4 with a little modification. Specifically,

$$d \leq \lambda\left(\frac{1}{1+|\sin\theta_0|} - \frac{1.5}{N}\right) \tag{7.5}$$

where

d = array maximum allowable spacing (m)
θ_0 = scan angle steered off boresight (deg)
λ = wavelength (m)
N = number of array elements.

The design of log-periodic arrays is often a compromise between the periodic endfires (or curtains), their spacing and the coupling of the transmitters to the radiators (or elements). The arrays must be capable of being electronically beam-steered instantaneously to any part of any preferred sectors.

The elevation beamwidth varies with frequency. If D_r is the length of the array (either transmitting or receiving), then, from (4.22), the array's beamwidth is

$$\theta_{BW} = \frac{0.8858\lambda}{D_r} \quad \text{(radians)} \tag{7.6}$$

Note that $D_r = Nd$, where N and d represent the number of array elements and the separation distance between the elements. However, if the array is steered off boresight or broadside by θ_0, the beamwidth (as well as the array gain) will degrade according to (4.32):

$$\theta_{BW} = \frac{0.8858\lambda}{D_r \cos\theta_0} \tag{7.7}$$

7.2.2 Receiver

The power requirement of transmitters is in the kilo- to megawatt range: the receivers operate at microwatt levels. In view of the high external noise

Figure 7.6 A section of Australian OTHR receiving antenna

environment, the efficiency of the receive system will be relatively low. A typical OTHR receiver is shown in Figure 7.6.

Two steerable receiving antennas are arrayed over distances of 2.3 and 2.5 km to receive the HF illuminating signals transmitted and returned via the ionosphere. The receiver contains a uniform linear array of phased monopole pairs on a wire ground mat. Each monopole pair has a receiver and analogue-to-digital converter attached to it. Each monopole pair feeds its own nearby receiver front-end which, with two frequency conversions, transmits signals via described propagation paths (e.g. satellite, optical fibres, etc.) to the receive back-ends. The basic configuration of the receivers is similar to that described in Chapter 2, section 2.1.2. The digitai beamformer forms the required number of beams, which are then Doppler processed to separate the moving targets from the ground clutter. More is said of beamforming in section 7.3. As a result of technological advances, the receiver backends are digital ensuring high-speed signal and data processing, for instance, digital bandwidth compression, digital filtering and down-sampling.

A myriad of data is often acquired during any radar scans or sweeps. An example of this is an OTHR, which is particularly noted for its wide-area scanning or sweeping. The unprocessed data acquired often occupy a large facility. Pre- and post-processed data could also be large and might require a large transfer and processing time. In a real-time operational situation, in particular during tracking, time is a critical element if the true-target profile under investigation is to be quickly ascertained in real time. To ensure fast transportation and delivery of data to its intended destination, a compression process is used.

Data compression is the process of encoding a body of data (say D_M) into a smaller body of data (say $\partial(D_M)$). It must be possible for the compressed data $\partial(D_M)$ to be decoded (reconstructed) back to the original body of data D_M or some acceptable approximation. The data compression method has been discussed in detail in Chapter 2.

Woodman and Chau (2001) proposed another data compression method, which works in a similar fashion to complementary phase coding used in pulse compression – already discussed in Chapter 2 – for coherent radars. Their method involves transmitting a large array of phase-coded antennas at full power and later synthesizes it by linear superposition and proper phasing. Full decoding is done by appropriate algorithms, which add the power and cross-power estimates of the signals of each code, so that no extra burden is added other than the summations.

7.2.3 Frequency management system

As demonstrated in Chapter 6, the fundamental mechanism enabling skywave HF radar to detect targets at long ranges, or to be used for remote sea-state sensing, is the ability of the ionosphere to refract electromagnetic energy. The variability of the skywave transmission medium requires different operating frequencies at different times (Earl and Ward 1986). So, a successful operation of a skywave radar, in particular the over-the-horizon radar, is dependent upon the application of a real-time frequency management system (FMS).

As seen in Figure 7.2, the FMS comprises ionospheric sounders and a spectral surveillance subsystem. The ionospheric sounders are backscatter sounder (BSS), oblique incidence (OI) sounder and vertical incidence (VI) sounder, which have already been discussed in Chapter 6. The sounders' prime requirements are to provide real-time propagation advice and ionospheric structure measurements sufficient for coordinate registration (that is, enabling conversion from radar measurement space to target ground coordinates). Often, between OI and VI, and in conjunction with BSS, the OTHR may be designed to self-calibrate and to achieve acceptable target location accuracy.

The spectral surveillance subsystem comprises the *background noise analyser* (BNA) and *clear channel occupancy analyser* (COA).

The BNA is used to determine the level of noise gathered by both omnidirectional and directional antennas looking towards the surveillance area. Apart from man-made noise, the source of BNA at low frequencies is from lightning discharges, while at high frequencies the source is from the galaxy. The BNA evaluation process starts by collecting spectrally processed data in N frequency bins from m contiguous measurements from x quietest channels for, say, p discrete scans. The m measurements are then summed to reduce data variance and normalized for system gain. Uncorrupted data are then averaged and interpolated into IF values. The interpolated values are then converted to power and formatted for transmission.

The use of BSS with BNA becomes a potent tool in clutter-to-noise ratio and maximum observable frequency. The BSS evaluation process involves collecting range correlated power measurements from Y range cells from B bandwidth bands, each band containing P spectra. Each of the B bands is analysed for noise contamination, and those bands that are uncontaminated are averaged to give, say, C integrated spectra. If all the spectra over a particular frequency band is corrupted, the data would be interpolated from adjacent cells. The data is further analysed to estimate the maximum observable frequency and the BSS ionogram is then converted to power and formatted for transmission. When BNA is combined with the BSS, data yields the ability to predict the achievable clutter-to-noise ratio, which is a direct indicator of the achievable signal-to-noise ratio. Technically, BNA is fundamentally involved in the selection of the optimum frequency band for radar operation.

The frequency channels distributed over a larger part of the HF band are often congested. This is due to broadcast stations, fixed-service point-to-point transmitters, essential services operators (ambulance, police, defence, etc.), and many other spectrum users having regular schedules. The channel occupancy analyser (COA) provides a real-time description of spectrum availability by scanning the HF spectrum every couple of minutes and allocates specific channels for radar use that are guaranteed to be noise-free – that is, free of radio frequency interference from other transmissions – and unoccupied. The spectral surveillance subsystem alternates between measurements of channel occupancy and background noise. The radar assigned for measuring the background noise data has directional antenna, and noise is measured on each of the designated number of beams comprising the directional antenna. Whereas, the channel occupancy data is measured on an omnidirectional antenna in order that transmissions are not masked in the nulls of the receiving antenna.

A method for evaluating channel occupancy is as follows. Measure the power level in all the $N - 1$ kHz channels in the HF range used for propagation. Obtain estimates based on the average of m passes from contiguous samples over the spectrum. Develop a convenient algorithm to classify channels as either clear or occupied by a cumulative-weighted index. The quietest x kHz bandwidth channel in each of the x bands is detected for use in the background noise analyser.

If Q is allowed to denote the channels occupied by signals greater than some threshold, and there are N channels independently occupied, then the probability of finding N adjacent channels available can be represented by

$$P_N = (1 - Q)^N \tag{7.8}$$

Stehel and Hagn (1991) described a method of linking Q to the threshold for a European situation. This may not be easily extrapolated to other locations.

Example 7.4 Consider eight independently occupied 1 kHz channels. If there is a 95 per cent chance of finding a clear channel, find the probability of finding 8 kHz adjacent channels for a signal sweep.

Solution
$N = 8$
$\bar{Q} = 95$ per cent $= 0.95$
$Q = 1 - \bar{Q} = 0.05$
From (7.8), the probability of finding 8 kHz adjacent channels is
$P_8 = (1 - 0.05)^8 = 0.6634$ (66.34 per cent)

7.2.4 Communications

The need for a good communication system cannot be overstated. It is clear that the radar facilities depend on reliable high-bandwidth communications for delivering and transferring information between transmit and receive chains for effective management of the systems. Between the transmitter and receiver facilities, primary communications may be delivered via radio link. Within each chain or facility, internal data communications could be in the form of a local area network (LAN). The design of the overall communication network required depends on the functions for which the radar system is designed, the capital outlay and reliability of the service expected.

Strict time synchronization between transmitter and receiver facilities is necessary. Use of atomic standards with, perhaps, frequent referencing back to global positioning system (GPS) data may be necessary.

7.2.5 Signal processing and peak detection

Signal processing takes several forms of data acquired by skywave radar such as range processing and range sidelobe suppression, Doppler processing, beam processing, and data conditioning.

Range processing and range sidelobe suppression are performed on each of the repetitive frequency sweeps, or preferred transmitted waveforms, by means of a weighted correlation against a reference waveform generated by the local oscillator. For a given bandwidth, the matched filtering approach is sufficient despite possible distortions to the point of response function by the transmission medium. In principle, compensations for some of those distortions could be included in the correlation process (Jarrott and Soame 1994). A sidelobe suppression technique has already been discussed in Chapter 3, section 3.2.4.

Fourier analysing and weighting across a set of repetitive frequency sweeps or preferred transmitted waveforms that constitute a coherently processed dwell may carry out Doppler processing. In fact, the transmission medium's distorting effects do not invalidate this approach. As observed in Chapter 6, ionospheric distortions creep into received signals. But when lengthy dwells (for example, during sea or ocean surveillance) are employed that extend beyond the prevailing ionospheric coherence time, unacceptable degradation of the 'coherent processing' occurs. Often, strictly add-on techniques such as the identification and subsequent removal of phase modulations are sufficient to restore the integrity of the signal clutter-to-noise ratio (Netherway *et al.* 1989; Southcott *et al.* 1998).

Fourier analysing each range cell, formed by the uniform-linear array, together with aperture weighting are adequate in performing beam processing. In real life, the assumptions that noise fields are spatially isotropic and their presence is at most on one target are far from true. Adaptive data beamforming could be effective in maximizing signal-to-noise ratios (SNR), only if formulated for robustness (Kassam and Poor 1985), otherwise any expected gains from adaptive beamforming would be lost to undue system sensitivity. More is said about beamforming in section 7.3.

Data conditioning, or a simple clean-up process, supplements those techniques previously addressed. Data conditioning is needed because of the presence of impulsive phenomena and radio frequency interference (RFI) in the data. An effective example of a clean-up process is the data whitening technique. The whitening technique does not restore the SNR but prevents local false detections. More is said about whitening in Chapter 10. Impulsive phenomena and other non-white or non-stationary phenomena can be located by their signatures, which are often localized in one or more domains, namely, the time domain, the range-azimuth domain, or the Doppler domain. Recognition always implies the temporary suppression of the otherwise dominating surface clutter, and when suppressed, frees up useful dwells. The data-conditioning process helps in estimating and removing ionospheric biases in any Doppler estimates. Without this correction, ship tracking may be impeded for a considerable length of time when ionospheric Doppler is notable.

Due to the clutter nature of the skywave data, *constant false-alarm rate* (CFAR) processing is critical. This is achieved by estimating the background energy of each data point, using suitably small local neighbourhoods so that data whitening can be done prior to peak selection (Jarrott and Soame 1994). More is said about CFAR in Chapter 10. CFAR processing is an important design aim, confining the radar to output a limited and predetermined number of false detections in a given period of time. If the number of false detections is low then the threshold may have been set high and consequently the probability of detecting a real target is reduced. Conversely, if the false alarms are too frequent, then the detection probability is improved but subsequent parts of the radar system may be overloaded.

Essential signal-peak detection processes are fast Fourier transform (FFT) and thresholding. The basic concept of FFT has been discussed in Chapter 1 and that of thresholding is discussed in Chapter 9 with further

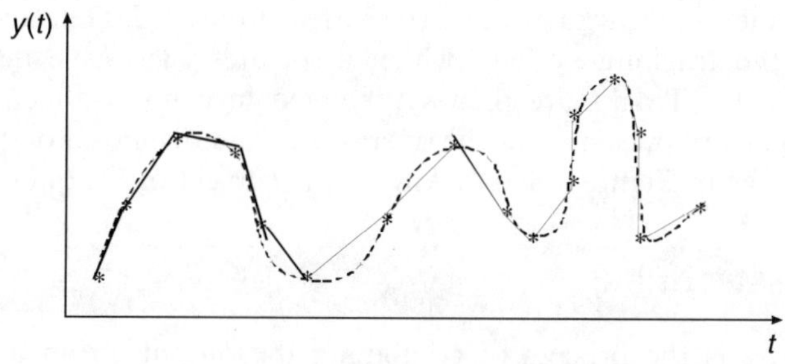

Figure 7.7 An example of linear interpolation between sample points. The solid line curve is the linear interpolation of the original signal represented by the dashed curve

application in Chapters 11 and 12. Peak detections are declared whenever the signal estimates exceed a preassigned value, or threshold. This presupposes good system's peak detection and beamforming capabilities. Individual points that exceed the threshold are resolved into peaks and the position of each peak is refined by *interpolation* across the adjacent range, Doppler and azimuth bins. Interpolation is a commonly used procedure for reconstructing a function either approximately or exactly from samples. One simple interpolation procedure is linear interpolation, whereby adjacent sample points are connected by a straight line as shown in Figure 7.7. In more complex interpolation formulas, higher-order polynomials or other applicable mathematical functions may be used to connect sample points. More is said about interpolation in Chapter 10.

7.2.6 Track and display

After selecting and interpolating a large number of candidate signal peaks per dwell in the CFAR selection, the detected peaks are then made suitable for transmission to the next stage of the radar chain (track and display)

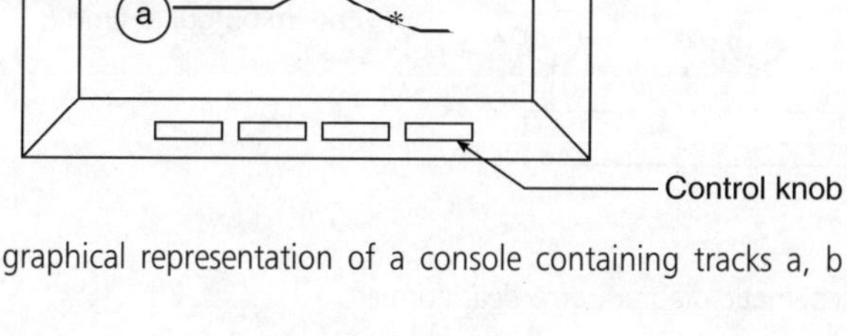

Figure 7.8 A graphical representation of a console containing tracks a, b and c, and the menu icon

where the emerging tracks of the targets are synthesized. Detected peaks are passed to the tracking system, which associates successive detections to establish tracks. Tracks are then synthesized in a multi-dimensional data format, which are presented to the operator(s) and displayed on the console, for example as in Figure 7.8. Tracking is the subject of Chapter 12.

7.3 Beamforming

Beamforming is the process of combining the outputs from a number of antenna elements arranged in an array of arbitrary geometry, so as to enhance signals from some defined spatial regions while suppressing those from other regions. This process can be implemented in a variety of ways. This could take the form of a digital processing technique or a hardware cabling method.

In its simplest form, the cabling method depends on a well-defined geometric structure for the array and location of possible sources. The method involves accurate switching and matching of cables of known lengths with antenna feedlines.

Digital beamforming is based on capturing the base signal at each of the antenna elements (i.e. at the array aperture) and converting to discrete signals thereby permitting formation of multiple, simultaneous antenna beams. 'Base signal' is used in this context because beamforming can be performed in any of the signal bands: baseband, narrowband, or broadband. A baseband signal is one in which the spectrum is primarily concentrated at the designated frequencies and in which no translation of the spectrum has been performed. Baseband signals contain amplitudes and phases of the signals received at each element of the antenna array. Digital beamforming is a process that opens the door to a large domain of signal processing techniques. The process also provides flexibility in the type of beam pattern that can be produced.

A schematic diagram of a beamformer is shown in Figure 7.9. Beamforming process involves weighting the digitized signals, and adjusting their phases and amplitudes such that when added together they form the desired beam. For instance, in Figure 7.9, the output y_k, at time k, is a linear

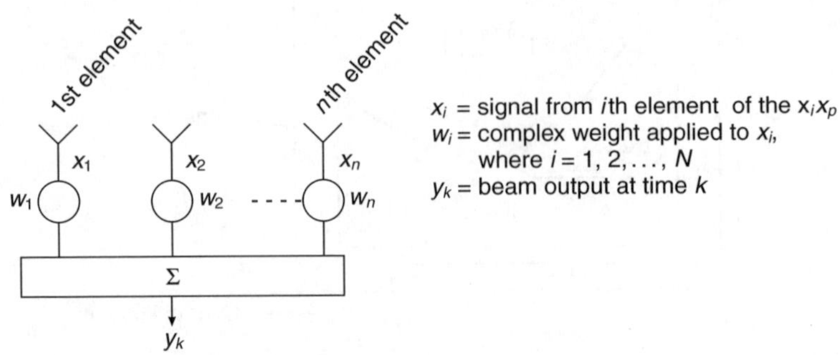

Figure 7.9 A schematic diagram of a beamformer

combination of the data at the N sensors at the same time k. It should be noted that the beamformer's response is a function of frequency, ω and the direction of arrival. Throughout the beamforming discussions, the functional variables (azimuth or elevation, ϕ, and angle of incidence, θ) are assumed to be available. In practice, for multi-dimensional data streams, beamforming is done in pairs: $y_k(\phi, \omega)$ and $y_k(\theta, \omega)$, primarily to simplify control, data processing and estimation.

For simplicity and following the established convention, the sampled data is multiplied by conjugates of the weights, written in a mathematical form as

$$y_k = \sum_{p=1}^{N} w_p^* x_p(k) \tag{7.9}$$

where * denotes complex conjugate,

x_p = the received data from pth element of the array $x_i x_p$ and
w_p = the complex weight applied to x_p.

In line with many engineering applications, the data and weights are assumed to be complex since often a quadrature receiver is used at each sensor to generate in-phase (I) and quadrature (Q) data. The expression in (7.9) can be written in the vector form as

$$y_k = \mathbf{w}^H \mathbf{x}(k) \tag{7.10}$$

where superscript H denotes Hermitian[1] transpose and the subscript k indicates sampling time or index. The process represented by (7.9) is often described as 'element-space beamforming' because the data x_p from the array are directly multiplied by a set of weights w_p to form a beam at a designated steering angle.

Planar sensor arrays can be considered to be sampled apertures. As such they can be viewed as multi-dimensional spatial filters and require a multi-dimensional beamforming technique. For multi-dimensional antenna arrays, the beamforming concept described by (7.9) can be easily extended. As an illustration, consider three-dimensional ($N \times L \times M$) or volumetric arrays, the beamformer output at time k is given as

$$y_k = \sum_{p=1}^{N} \sum_{j=1}^{L} \sum_{i=1}^{M} w_{pji}^* x_{pji}(k) \tag{7.11}$$

assuming that there are no delays in each of the N and L sensors. Since a beamformer represents a linear combination of the sensor data, its

[1] If A is a matrix of order $(m \times n)$ with complex elements, a_{ik}, then the complex conjugate A^* of A is found by taking the complex conjugates of all the elements. A Hermitian matrix is a square matrix which is unchanged by the transpose of its complex conjugate, i.e. A is Hermitian if $(A^*)^H = A$.

implementation can be represented by decomposing **w** into a product of matrices and a vector, such as

$$\mathbf{w} = \left[\prod_{i=1}^{v_o} \mathbf{v}_i\right] \mathbf{w}_v \tag{7.12}$$

where $\mathbf{v}_i$ is a series of matrix transformations of comfortable dimensions and $\mathbf{w}_v$ is the vector. As a general rule, the matrix transformations are selected to enhance performance and/or reduce computational complexity (Van Veen and Buckley 1988). The FFT implementation of the DFT is analogous to (7.12) since the DFT matrix can be expressed as a series of simple computations (see Chapter 1 for details).

Instead of direct weighting of sampled data from each element, the data signals from the elements can first be processed by a multiple-beam beamformer to form a set of orthogonal beams (Litva and Lo 1996). The output of each beam can then be weighted and the result combined to produce the desired output. A process that performs this function is called beam-space beamforming. The required multiple beamformer usually produces orthogonal beams. Beams are mutually orthogonal when the average value over all angles of the product of one beam response with the conjugate of the other is zero. The beam-space beamforming technique is implemented by feeding the baseband signals from the antenna elements into the FFT processor, which generates N simultaneous orthogonal beams. A subset of the orthogonal beams is then weighted to form the desired output. This process is explained as follows.

Following the linear array concept of Chapter 4, assume that the antenna elements are equally spaced at distance, d. Also assume that a plane wave incidents on the receiver elements at beam angle from broadside. For simplicity, the individual elements are assumed to have an isotropic response in azimuth. A broadside azimuth beam is formed when the signals at all of the elements are in phase. Other beams, of approximately equal amplitude to the first, are formed at all angles for which the phase difference between elements is an integral number of wavelengths. As also discussed in Chapter 4, nulls are created in the direction of the interfering sources (in the case ϕ_b). For an N-element linear array, N overlapped orthogonal beams $v(\phi_b)$ can be formed using

$$v(\phi_b) = \sum_{p=1}^{N} x_p e^{-j\frac{2\pi}{N}p\phi_b} \tag{7.13}$$

If the pth desired output is assumed to occur from a combination of the weighted $(b-1)$th and $(b+1)$th beams, then the output may be written as

$$y_p = w_{b-1}^p v(\phi_{b-1}) + w_{b+1}^p v(\phi_{b+1}) \tag{7.14}$$

The beamformer described above deals with fixed beam patterns (i.e. with fixed weights that are time invariant) for a given specification. This is

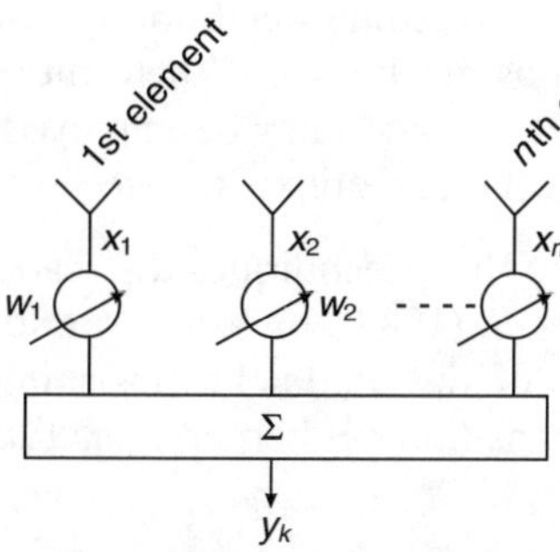

Figure 7.10 A schematic diagram of an adaptive beamformer

cognisant of the conventional beamforming technique. A conventional beamformer can be optimized if a specified optimization criterion is given. The process of optimization can be compared to optimum filtering, detection, and estimation. In a real-world application, fixed weights are impracticable for optimum performance to be achieved. Thus, the weights have to be adaptively selected. The term adaptive means that the weights are changing with time. An adaptive beamformer – comparable to adaptive filters – will sense its operating environment and automatically optimize a prescribed objective function of the array pattern by adjusting the elemental control weights. Figure 7.9 can now be modified to indicate the adaptive nature of the beamformer as shown in Figure 7.10.

The arrows in Figure 7.10 indicate that the elemental weights are time variant (i.e. changing with time). The choice of the weight vector **w** is based on the statistics of the signal vector **x** received at the array. Basically, the aim is to optimize the beamformer response with respect to a prescribed criterion so that its output y_k contains the least contribution from noise and other interference. An algorithm designed for that purpose would specify the means by which the optimization is to be achieved.

Beamforming optimization implies selecting the weights based on the statistics of the data received at the antenna array so that the beamformer output response contains little or no contributions from noise and signals arriving from directions other than the desired signal direction. A number of criteria for selecting the optimum weight include:

- *Minimum mean-square error* – A technique that minimizes the error between a beamformer output and the desired signal on the notion that a reference signal is known.
- *Linearly constrained minimum variance* – A method used where the reference signal is unknown. It involves constraining the response of the beamformer so that those signals from the desired direction are passed with specific gain and phase. The weights are chosen to minimize output variance, or power, subject to response constraint.

- *Least mean squares* – The preceding weight optimization techniques assume that their optimal solutions are known. Where these optimal solutions are unknown, adaptive weight estimates may be employed that use the least mean squares technique. This technique utilizes the steepest descent technique.

For a general description of these techniques the reader is advised to consult Ahmed (1987), Treichler *et al.* (1987), or Widrow and Stearns (1985).

The dynamic range D_y of the digital beamforming due to thermal and quantization noise is given as (Schoenberger *et al.* 1982)

$$D_y = \frac{32^{2b}}{26} N \tag{7.15}$$

where N and b correspond to the number of array elements and digitization bits. Another study (Stehel and Hagn 1991) gave the dynamic range as

$$D_y = 2^{2(b-1)} N_c \tag{7.16}$$

where N_c and b correspond to the number of parallel channels and number of analogue-to-digital bits.

7.3.1 Beam control and calibration

The beam pattern may be controlled by effective reduction in the pattern sidelobes and optimally changing the aperture weights so as to steer the beam in any preferred direction negating the sidelobes' influence. A variety of techniques for performing weight optimization and sidelobes' suppression or cancellation have been discussed in the previous section and Chapter 3 respectively. Our attention now focuses on calibration techniques.

Calibration is the removal of equipmental, or propagation, deviations from reality, which, if necessary, are estimated or measured in real time. Calibration may also imply transformation of radar output data into international measurement units (Jarrott and Soame 1994).

Calibration can be performed by injecting a test signal at a particular time into the inputs of the multi-channel receiver associated with the digital beam former. The test signal can be supplied by an auxiliary antenna in the near field of the main antenna or by precise coupling lines across the antenna face. Both techniques ensure that antenna element and feed errors are contained within the calibration loop, thereby offsetting channel-matching errors. The use of an auxiliary antenna requires that the antenna has a defined directivity so that it can illuminate the face of the main antenna without also illuminating other structures within the vicinity. This safeguards re-radiation towards the antenna face, which could destroy the prescribed distribution of the calibrating field. The auxiliary antenna must also be physically offset sufficiently far from the field of view of the main antenna. This is necessary to prevent undue influence on the distribution across the main antenna arising from target echoes.

Let us define the test signal distribution across the receiver inputs as (Barton 1980)

$$X_r'^T = [x_1', x_2', x_3', \ldots, x_n'] \tag{7.17}$$

Let the response within the processor be written as

$$X_r''^T = [x_1'', x_2'', x_3'', \ldots, x_n''] \tag{7.18}$$

A diagonal matrix operator **C** will be formed in which

$$c_{rr} = \frac{x_r'}{x_r''} \tag{7.19}$$

Signals received during normal operation are weighted by modified weight values. As such, a nominal weight vector **w**′ required for a particular beam shape and pointing angle is corrected such that the weight vector actually applied is

$$\mathbf{w} = \mathbf{C}\mathbf{w}' \tag{7.20}$$

It should be noted that time of arrival changes across the antenna aperture, and cable delays between the antenna elements and the receivers are likely to cause frequency, amplitude, and phase variations at the beamformer.

7.3.2 Conclusion

The beamforming process attempts to preserve the total information available at the antenna aperture. With the digital beamforming technique, the weight w_i can easily be exploited by changing its value to steer the beam in any preferred direction and manipulate its shape to optimize the system performance. By carefully selecting the aperture weights, which may be complex, beamforming can be equated to a simple discrete Fourier transforms (DFT): this presupposes effective beam-pattern control and calibration.

7.4 Radar equation: a discussion

A form of the radar equation developed for line-of-sight radars, in Chapter 5 equation (5.56), is applicable to skywave radars but with a different emphasis on certain notations and definitions. As earlier discussed in Chapters 3 through to 5, the ability of the radar to detect target power depends on the background noise power that competes with the target power, S_T. The target power is approximately equivalent to the received power, S. It should be noted that the received power S is emphasized because not all signals are targets and not all backgrounds are noise. These background interferences are denoted as N_B. Like equation (5.56), the received power to background interference ratio depends on:

$$\frac{S}{N_B} = \underbrace{\left\langle \frac{1}{(4\pi)^3} \right\rangle}_{cons\ \tan\ t} \underbrace{\left(\frac{1}{N_0 F_a L_i^2} \right)}_{environment} \underbrace{\left\{ \frac{T_i P_t G_t D_r}{L_s} \right\}}_{\substack{radar \\ capability}} \underbrace{\left[\frac{\sigma |F^4|}{R^4} \right]}_{\substack{t\ \arg\ et \\ characteristic}} \lambda^2 \tag{7.21}$$

where

T_i = total integration time
L_i = ionospheric losses
L_s = total system losses
N_0 = external noise, nominally derived from thermal noise (kB_nT_0) but multiplied by the noise density factor
F_a – analogous to antenna noise factor
F_N – more is said later in the text.

Reading (7.21) after the equality sign from left to right, the following is described.

The first term is just the proportionality constant.

The second term is the environmental factor comprising external noise and ionospheric losses including two-way propagation path, polarization mismatch, and ionospheric anomalies discussed in Chapter 6.

In the HF spectrum, noise levels are generally expressed by a factor, denoted by F_a in (7.21), which is similar to antenna noise factor F_N included in (5.59). The noise density factor, F_a, describes the antenna referred external noise density in excess of thermal noise (kB_nT_0). The noise density factor is a strong function of frequency and varies with time of day, season, sunspot number, location, etc. Kingsley and Quegan (1992) gave a rough mean value for F_a as

$$F_a = \begin{cases} 60 - 2f_{\text{MHz}} & \\ 45 - f_{\text{MHz}} & \end{cases} (\text{dB}) \quad \begin{matrix} 5 < f \leq 15\,\text{MHz} \\ 15 < f < 28\,\text{MHz} \end{matrix} \tag{7.22}$$

The nighttime F_a values are about 10 dB below the daytime given approximately by (7.22). A more sophisticated approach to obtaining F_a is given in (Weiner 1991).

The third term is the radar capability, or *radar figure of merit* (FOM), comprising the transmitter power P_t and gain G_t, receive beam directivity D_r, effective processing (or integration) time T_i, and system loss L_s. The directivity of the receive array is used, in some cases, instead of the receive array gain, which can be found from (5.12). An inquiring mind might ask: what of the influence of the currents flowing in the ground mat, ohmic heating, cable losses, etc.? Of course, these losses are present but are included in the system loss. The system loss is separated from the total losses L_{tot} in (5.59) in order to distinguish it from environmentally induced losses. The system loss is localized and can be reduced by the system designer.

The fourth term is the target characteristic comprising radar cross-section, σ, range, R, and ground reflection effect, $|F^4|$. The ground reflection effect increases in apparent σ due to illumination via ground reflection as well as by direct path – already discussed in Chapter 5, section 5.1.7.1.6. In some literature, the ground reflection effect is denoted by *Mu*. This ground reflection effect increases, and gets worse, with increasing frequency because higher frequencies are associated with longer ranges.

The last term, λ^2, is the radar wavelength. Though not particularly associated with any of the identified terms in (7.21), it, however, influences the environment, target characteristics, and the radar capability terms; they change with λ. Of course, the radar capability term, or FOM, increases with frequency because receive beam directivity increases, which about cancels the explicit λ^2 term.

The waveform of the transmitted signal does not enter into the radar equation. This implies that the signal can be selected for other considerations such as range and Doppler resolution. The choice of signal, however, does play a major part of radar (or sonar) signal processing and good discussions can be found in Cook and Bernfeld (1967) and Vakman (1968).

Before closing the discussion on radar equations, it is beneficial to briefly talk about the background interference, N_B. In practice, 'background' implies the content of cells around the target cell in three dimensions (range, azimuth and Doppler). The detection process, to be discussed in Chapter 10, must make an estimate of the power in the cells in the neighbourhood of the potential target cell. The simplest being the mean power in the defined neighbourhood cells. Often, as it becomes obvious in Chapters 10 and 12, the signal-to-background interference ratio (S/N_B) value is nominated (like a threshold) to give some satisfactory standard of detection probability with acceptable probability of false alarm.

7.5 Applications of skywave radar

There are many ways to exploit the largely untapped potential of existing military skywave radar, in particular OTHR. If properly harnessed, OTHRs can be used for the following applications.

7.5.1 Shortwave radio forecasting and ionospheric models

Resurgence in military and commercial use of the crowded shortwave radio spectrum has prompted a renewed interest in HF channel identification. The resurgence also helps having a better understanding of solar and geomagnetic influences on climatic changes and ionospheric models. With increased knowledge of climatological models, we would be in a good position to:

- develop better ionospheric models for HF radio frequency management over large parts of the Earth that are inaccessible to conventional ionospheric sounders;

- develop shortwave prediction models and warnings that are useful for optimizing point-to-point radio communication;
- map the occurrence of transient phenomena in the ionosphere, such as polar and equatorial disturbances, and sporadic-E ionization;
- measure the spectral broadening of ionospherically propagated ground clutter, which can be used as an indicator of the information capacity of an ionospheric path; and perhaps
- probe the structure and dynamics of the solar corona thereby enabling a good prediction of space weather. The OTHR will act as a test bed for developing HF solar radar technology.

7.5.2 Climatic monitoring and forecasting (Georges *et al.* 1993; Sinnott 1987)

An OTHR offers a unique capability for continuously mapping sea-surface winds and waves over very large ocean surface areas. In particular, OTHR's ability to map sea-surface wind direction permits an accurate assessment of the location, shape and growth of the tropical waves, which often intensify and develop into tropical storms and hurricanes. An OTHR could also map synoptic and large-scale meteorological features that determine whether tropical or subtropical waves will grow or die. To this author's knowledge, no present or planned observing system offers this capability. Both the Australian Jindalee and USA Navy's OTHR systems have demonstrated the radar usefulness for weather services and fleet numerical predictions, respectively.

7.5.3 Air traffic control, and search and rescue

Since the primary objective of OTHR is for surveillance purposes, i.e. tracking of aircraft and surface craft, this capability could be adapted directly to air traffic control. The ability of the OTHR to see beyond ocean regions inaccessible to conventional radar and the capability to map surface currents with high resolution over very large ocean areas could support search and rescue missions.

7.5.4 Monitoring climate change (Anderson 1986; Croft 1972; Georges *et al.* 1998)

The influence of ocean currents has been known to affect our weather and climate because:

(a) sea state affects ocean albedo, which in turn affects the absorption of solar radiation;
(b) sea surface roughness affects the uptake of greenhouse gases; and
(c) Surface wind stress affects ocean circulation and the global heat fluxes and budget.

OTHRs can be exploited to:

- monitor surface winds, waves and currents over large ocean areas, which could assist considerably in understanding their role in global environmental change;
- parameterize sea state, crucial to the estimation of the effects of air–sea interaction on global climate change, and on the El Niño phenomenon;
- validate and initialize numerical weather prediction models; and
- predict the trajectories of surface-borne pollutants, and monitor bursts of strong currents that can cause economic damage, such as damage to offshore oil platforms, which could cause oil spills.

7.6 Summary

The fundamental mechanism that enables the skywave radar to be used for long-range surveillance is the ability of the ionosphere to refract electromagnetic energy. A particular discussion on skywave radar was centred on the over-the-horizon radar (OTHR). The basic structure of an OTHR system, and its component parts, has been explained.

Although the OTHR concept is simple, using the ionosphere as reflectors for the radar requires an understanding of the complexities of the ionosphere. Despite this, each propagation mode has been observed to be well behaved thereby enabling the OTHR to discern between the different propagation paths. Finally, the capability, channel occupancy, and potential of the OTHR have been discussed.

Problems

1. If the site of a skywave radar is not well isolated from the urban area, determine the effect(s) on the spectral surveillance subsystem measurements.
2. Why are OTHR systems much more demanding on the quality of their propagation paths?
3. Ten 1 kHz channels are sampled in the process of establishing the optimum frequency band. Based on previous observations, there is a 90 per cent chance of finding a clear channel. What is the probability of finding 10 kHz adjacent channels for any signal sweep?
4. In the process of discerning between single path and multipath circuits, are the data from the sounders sufficient for determining frequencies that are free from multipath and spectral broadening? If not, what would you suggest?
5. An array antenna of N elements is required to have a beamwidth of 13.2° when spaced equidistantly at 0.3λ. Calculate the elements required when propagation is conducted at 13.5 MHz. For these elements, spacing and

frequency, if the receiving antenna is steered about 12° off the boresight, will the beamwidth be the same?

6. For an aperture of 2.5 km, design a receiver capable of receiving data in the 5 to 15 MHz frequency band and capable of being steered up to 15° off the boresight without an occurrence of grating lobes. Clearly state your assumptions.
7. Example 7.1 demonstrates that the positioning of the receiver for all frequency settings may not in the 'strict sense' be in the far field. Radar equations are developed on the premise of the far-field situation. What effect will the non-conformance have on radar measurements?

Part III

Peak Detection and Background Theories

The issue of what to do with data acquired by radar becomes relevant after the data have been processed, which might have been corrupted prior to being processed and when the data true nature is known. Data processing involves the transformation of a set of coordinated physical measurements into decision statistics for some hypotheses. Those hypotheses, in the case of radar, are whether targets with certain characteristics are present with certain position, speed, and heading attributes. To test the trueness of the hypotheses requires knowledge of probability and statistical theory and decision theory together with those espoused in Chapter 1 – the reader will be in a better position to know the other process involved in signal-peaks detection. Hence, this part is structured into three chapters: 8, 9 and 10.

Chapter 8 reviews some of the important properties and definitions of probability and random processes that bear relevance to the succeeding topics in Part IV. By this approach, the author consciously attempts to reduce complex processes involved in synthesizing radar system signals to their fundamentals so that their basic principles by which they operate can be easily identified. The basic principles are further built on in Chapter 12 to solve more technical tracking problems.

Chapter 9 investigates one type of optimization problem; that is, finding the system that performs the *best*, within its certain class, of all possible systems. The signal-reception problem is decoupled into two distinct domains, namely detection and estimation. The detection problem forms the central theme of Chapter 10 while estimation is discussed in Part IV, Chapter 11. Detection is a process of ascertaining the presence of a particular signal, among other candidate signals, in a noisy or clutter environment.

8

Probability theory and distribution functions

In radar applications, such as tracking, the signals plus interference received are stochastic in nature and can often be described only by statistical means. Indeed it is the stochastic nature of some of these signals that reflects their ability to impart information, although noise, clutter, or other interference may mask the desired signals. The word 'stochastic' is used as a synonym for 'random'. Both are interchangeably used in the literature.

Use of probability measurements arises from the need to extract plausible explanations from events, which may have too much information of the undesirable kind. Thus, this chapter attempts to provide the readers with a sufficient background in probability theory as a precursor to the understanding of the subsequent chapters. It does not, however, attempt to rigorously treat the probability theory, but only attempts to review some of the important properties and definitions of probability and random processes upon which succeeding chapters are built.

This chapter also includes a discussion on distribution functions and their properties that involve more than one random variable. Applications of the distributions, which are often encountered in signal processing, are given.

8.1 A basic concept of random variables

A random variable, or variant as it is sometimes called, is a function defined on a sample space, Ω. A sample space is the combination of all possible outcomes of a random experiment. For example, suppose a coin is tossed thrice. A coin usually has two outcomes: head (H) or tail (T). One possible outcome of the experiment is that all tosses result in tails. The complete possible outcomes are: HHH, HTH, HHT, HTT, THH, TTH, THT, TTT. If, in shorthand form, $\omega_1, \omega_2, \omega_3, \omega_4, \omega_5, \omega_6, \omega_7, \omega_8$, respectively, denote the outcomes, then the sample space containing the outcomes of the experiment can be written as $\Omega = \{\omega_1, \omega_2, \omega_3, \omega_4, \omega_5, \omega_6, \omega_7, \omega_8\}$.

A particular outcome of the experiment is known as a sample point. Suffice to say that within each outcome a sample point can be assigned.

Where a sample space contains a finite number of sample points, the sample space is considered to be *discrete*. For example, in throwing a dice, the sample space comprises six discrete sample points denoted by the numbers 1, 2, 3, 4, 5 and 6. When a discrete sample space contains an infinite number of sample points, then the sample space is considered to be *continuous*. An example of such a sample space is the thermal noise voltage, which thermally excites electrons in a finite conductor. In essence, a random variable can be discrete or continuous in any sample space. An *event* is a combination of possible outcomes, which is a subset of the sample space. For example, obtaining an even or odd number in throwing the dice is an event.

In general, the values of a variant, or a random variable, may be real or complex. Where multiple variables are involved, vectors can be used to represent the variables. Given that a random variable is a function defined on a sample space, it is logical to associate probabilities with the values of the random variable. A method of associating the probabilities is called the *probability function*. For instance, for all possible values associated with a discrete random variable x, the random variable's associated probability function $p(x_i)$, may be defined as

$$p(x_i) = P(x = x_i) \geq 0 \tag{8.1}$$

where $i = 1, 2, \ldots, n$, and $P(x)$ denotes the 'probability of variable x'.

8.2 Summary of applicable probability rules

A brief review of some important rules of probability is discussed in this section.

Rule 1. If $P(x)$ and $P(\bar{x})$ correspond to the probabilities of event x occurring and not occurring, then

$$P(\bar{x}) = 1 - P(x) \tag{8.2}$$

Rule 2. If x and y denote two independent events, then the probability that both events will occur is the product of their respective individual probability:

$$P(xy) = P(x)P(y) \tag{8.3}$$

This type of probability is known as *joint probability*. It follows from (8.3) that if n independent events occur jointly, then the probability of joint occurrence is the product of the events' individual probabilities:

$$P\left(\prod_{i=1}^{n} x_i\right) = P(x_1)P(x_2)\cdots P(x_{n-1})P(x_n) \tag{8.4}$$

N-dimensional variables arise in a number of communications and radar problems, for example in the range-cell averaging techniques for determining noise statistics in *constant false alarm rate* (CFAR) receivers. The basic concept and measurement of CFAR will be discussed in Chapter 10.

Rule 3. If x and y are two events mutually exclusive, written as $m(xy) = 0$, then the events probability is zero; that is, $P(xy) = 0$. Also, the probability that any one of these events will occur is the sum of their individual probabilities; that is,

$$P(x \text{ or } y) = P(x + y) = P(x) + P(y) \tag{8.5}$$

It follows from (8.5) that the probability of occurrence of one of n mutually exclusive events can be expressed as

$$P\left(\sum_{i=1}^{n} x_i\right) = \sum_{i=1}^{n} P(x_i) \tag{8.6}$$

Rule 4. If the events x and y are not necessarily mutually exclusive; that is, $m(xy) \neq 0$, then the probability that at least one of the two events will happen is the sum of their individual probabilities less their joint probability. Concisely written as

$$P(x \text{ and/or } y) = P(x \cup y) = P(x) + P(y) - P(xy) \tag{8.7}$$

Following the above reasoning, the probability that at least one occurrence in more than two events can be deduced as follows:

1. at least one of three events:

$$P(x \cup y \cup z) = P(x) + P(y) + P(z) - P(xy) - P(xz) - P(yz) + P(xyz) \tag{8.8}$$

2. at least one of n events:

$$P(x_1 \text{ and/or } x_2 \text{ and/or} \cdots \text{and/or } x_n) = P\left(\bigcup_{i=1}^{n} x_i\right) \tag{8.9a}$$

$$P\left(\bigcup_{i=1}^{n} x_i\right) = \sum_{i=1}^{n} P(x_i) - \sum_{j>i}^{n} P(x_i x_j) + \sum_{k>j>i}^{n} P(x_i x_j x_k) - \sum_{m>k>j>i}^{n} P(x_i x_j x_k x_m) + \cdots \tag{8.9b}$$

Equation (8.9b) is equivalent to the probability that at least one of the n events will take place is one minus the joint probability that *none* of these event will happen; that is,

$$P(x_1 \text{ and/or } x_2 \cdots \text{and/or } x_n) = 1 - P(\bar{x}_1 \bar{x}_2 \bar{x}_3 \cdots \bar{x}_n) = 1 - \Pr\left(\prod_{i=1}^{n} \bar{x}_i\right) \tag{8.10}$$

If these events are *independent*, in view of (8.2), (8.4) and (8.10), the probability that at least one of the n events will occur is

$$P(x_1 \text{ and/or } x_2 \cdots \text{ and/or } x_n) = \sum_{i=1}^{n} P(x_i) \tag{8.11}$$

If the events are *not* independent, the probability becomes conditional. A *conditional probability* of an event x_1 with respect to another event x_2 (written as $P(x_1 \mid x_2)$) is the probability that x_1 will take place given x_2 has occurred. Consequently,

$$P(x_1 \mid x_2) = \frac{P(x_1 x_2)}{P(x_2)} \tag{8.12}$$

By this expression, the general form of the expression in (8.3) can be written as

$$P(xy) = P(x)P(y \mid x) \tag{8.13}$$

which by extension to three events yields

$$P(xyz) = P(x)P(y \mid x)P(z \mid xy) \tag{8.14}$$

The term $P(z \mid xy)$ can be interpreted as the conditional probability of z given the occurrence of both x and y. The generalized form of (8.14) is quite useful in the optimal estimation problem where limited information of any given set of received random variables is known. This will become obvious to the reader later in the text when the statistical estimates of target state variables are being formulated.

To develop the case of a pair of events where the point x_i ($= x_1, x_2, \ldots, x_{n-1}, x_n$) may take on n discrete values, suppose that the probability of event y depends on knowledge of the previous event occurring in one of the n distinct ways. The probability of y, which is unconditional, can thus be expressed as the sum of conditional probabilities weighted by their respective probabilities, $P(x_i)$. That is,

$$P(y) = \sum_{i=1}^{n} P(y \mid x_i)P(x_i) \tag{8.15}$$

where x_i are mutually exclusive and $\sum_{i=1}^{n} P(x_i) = 1$.

8.2.1 Bayes' theorem

Bayes' theorem follows naturally from the conditional probabilities explained in (8.15). This theorem allows for a method of combining the initial, or *prior*, probability concerning the occurrence of some event with related experimental data to obtain an amended or *posterior* probability. Bayes' theorem can be explained as follows using some of the above rules. Suppose that the probability $P(x_i)$ of values x_i are known, where

$i = 1, 2, \ldots, n$. Suppose also that an event y occurs in conjunction with values x_i occurring. The question becomes: how has the event y actually occurred and what will its impact be on the individual probability of x_i? This translates to finding $P(x_i \mid y)$ for all values of i. In view of (8.13),

$$\begin{aligned} P(x_i y) &= P(y \mid x_i) \\ &= P(y)P(x_i \mid y) \end{aligned} \tag{8.16}$$

Rearranging (8.16), to yield

$$P(x_i \mid y) = \frac{P(x_i y)}{P(y)} = \frac{P(x_i)P(y \mid x_i)}{P(y)} \tag{8.17}$$

Upon substituting (8.15) in (8.17):

$$P(x_i \mid y) = P(x_i) \frac{P(y \mid x_i)}{\sum_{i=1}^{n} P(y \mid x_i)P(x_i)} \tag{8.18}$$

which produces the expression known as the Bayes' theorem. A closer look at this expression reveals that two terms are prevalent. Reading from left to right after the equality sign:

(i) the first term, $P(x_i)$, is the initial or prior probability; and
(ii) the second term, $P(y \mid x_i)/\sum_{i=1}^{n} P(y \mid x_i)P(x_i)$, is the amended or posterior probability. This probability corrects the *prior* probability on the basis of data in hand.

Bayes' theorem is easily applied in real life in discerning which event probability is to be used to assign weights to radar received signal as coming from clutter or from the target. An example can be formalized as follows.

Example 8.1 Suppose a target is observable and its presence (or absence) in a surveillance region can be denoted as B_1 (or B_2). There is always a tendency that one can erroneously classify the target to be in the surveillance region while it is not or vice versa. Let A_1 denote the signal peak being correctly associated with the target and A_2 is when the signal peak is not correctly associated with the target. Because of the potential misplacement in the target-peak association, errors are likely to occur. On the assumption that the target is observable, the probability of associating detected peaks to the target involves setting up *a priori* observability correctly. So, the error probabilities are written as

$$\begin{aligned} P(A_1 \mid B_2) &= q_{01} \\ &= 1 - p_{01} \end{aligned} \tag{8.19a}$$

This is the probability that a peak was detected when there was no target in the region. Also,

$$P(A_2 \mid B_1) = q_{10} = 1 - p_{10} \tag{8.19b}$$

which is the probability that no peak was detected when there was a target in the region.

Let's define p as the probability that the target is observable and $q(=1-p)$ that the target is not observable. By this definition, the *a priori* probability that a target was observed in the region or not is given as

$$P(B_1) = p \tag{8.20a}$$

$$P(B_2) = q = 1 - p \tag{8.20b}$$

What is left to be evaluated is the *a posteriori* probability $P(B_j \mid A_i)$ where $j, i = 1, 2$. From (8.18):

$$P(B_j \mid A_i) = \frac{P(B_j)P(A_i \mid B_j)}{P(B_1)P(A_i \mid B_1) + P(B_2)P(A_i \mid B_2)} \tag{8.21}$$

The four *a priori* probabilities can be written as

$$P(B_1 \mid A_1) = \frac{pp_{10}}{pp_{10} + q(1 - p_{01})} \tag{8.22a}$$

$$P(B_1 \mid A_2) = \frac{p(1 - p_{10})}{qp_{01} + p(1 - p_{10})} \tag{8.22b}$$

$$P(B_2 \mid A_1) = \frac{q(1 - p_{01})}{pp_{10} + q(1 - p_{01})} \tag{8.22c}$$

$$P(B_2 \mid A_2) = \frac{q(1 - p_{01})}{qp_{01} + p(1 - p_{10})} \tag{8.22d}$$

Remembering that

$$\begin{aligned} P(A_1) &= P(B_1)P(A_1 \mid B_1) + P(B_2)P(A_1 \mid B_2) \\ &= pp_{10} + q(1 - p_{10}) \end{aligned} \tag{8.22e}$$

$$\begin{aligned} P(A_2) &= P(B_1)P(A_2 \mid B_1) + P(B_2)P(A_2 \mid B_2) \\ &= p(1 - p_{10}) + (1 - p)p_{01} \\ &= 1 - P(A_1) \end{aligned} \tag{8.22f}$$

A special case occurs when the error probabilities are equal; that is, when $q_{01} = q_{10}$ and $p_{01} = p_{10}$. Also if the *a priori* are equal, that is, $q = p = 0.5$, then:

$$\begin{aligned} P(A_2) &= P(A_2) = 0.5 \\ P(B_1 \mid A_2) &= P(B_2 \mid A_1) = 1 - p_{10} \\ P(A_1 \mid B_1) &= P(A_2 \mid B_2) = p_{10} \end{aligned} \tag{8.22g}$$

Example 8.2 This example is similar to the problem given in Gangolli and Ylvisaker (1967). Suppose three containers numbered 1, 2 and 3 contain, respectively, one red and one black ball, two red and three black balls, and four red and two black balls. Consider an experiment consisting of the selection of a container followed by the draw of a ball from it. Let 'Red' be denoted by R and the 'Black' by B. One could arrange a sample space Ω as follows:

$$\Omega = \{(1,R),(1,B),(2,R),(2,B),(3,R),(3,B)\}$$

For the events dictated by the selection process, one could arrange them as follows:

$$B_1 = \{(1,R),(1,B)\}$$
$$B_2 = \{(2,R),(2,B)\}$$
$$B_3 = \{(3,R),(3,B)\}$$

These events are mutually exclusive and exhaustive.

If $A = \{(1,R), (2,R), (3,R)\}$ and each ball in the container is equally likely to be drawn, then

$$P_{B_1}(A) = \frac{1}{2}, \quad P_{B_2}(A) = \frac{2}{5} \quad \text{and} \quad P_{B_3}(A) = \frac{2}{3}$$

If the container is not observed but a red ball is drawn, what is the probability that it was drawn from container 1, 2, or container 3?

The question is technically: what are the *a posteriori* probabilities $P_A(B_1)$, $P_A(B_2)$, and $P_A(B_3)$?

The solution to this problem depends on the *a priori* probabilities: $P(B_1)$, $P(B_2)$, and $P(B_3)$. For this problem, suppose $P(B_1) = P(B_2) = P(B_3) = 1/3$.

Using the Bayes' theorem, that is

$$P_A(B_i) = \frac{P(B_i)P_{B_i}(A)}{\sum_{j=1}^{m} P(B_j)P_{B_j}(A)} \qquad i = 1,2,\ldots,m$$

the following values are obtained for the *a posteriori* probabilities:

$$P_A(B_1) = \frac{15}{47} \quad P_A(B_2) = \frac{12}{47} \quad P_A(B_3) = \frac{20}{47}$$

Of course, $\sum_{i=1}^{3} P_A(B_i) = 1$.

8.3 Probability density function

The probability function concept described in the previous section applies strictly to discrete random variables but becomes less meaningful for continuous random variables. Instead, the probability density function is used. The probability density function is also called the *probability density, density function*, or simply *density*. If in the sample space an arbitrary large number of experiments are performed, at any given time t, a *probability density function* (pdf), denoted by $f_x(x(t))$, can be formed, which will be a continuous histogram of such event x at time t. Statistics derived from such experiments are called *ensemble* statistics. By definition, the probability[1] that a random variable x lies in an infinitesimal interval between x_i and $x_i + \Delta x$ is given as

$$f_x(x) = \lim_{\Delta x \to 0} \frac{P(x_i \leq x \leq x_i + \Delta x)}{\Delta x} = \frac{d}{dx} F(x) \tag{8.23}$$

where $F(x)$ is the characteristic function of x. Equation (8.23) is equally extendable to a single random variable having n-dimensional space:

$$f_x(x_1, x_2, \ldots, x_n) = \frac{d^n}{dx_1, x_2, \ldots, x_n} F(x_1, x_2, \ldots, x_n) \tag{8.24}$$

where $f_x(x_1, x_2, \ldots, x_n)$ is the joint probability distribution function of $(x_1, x_2, \ldots, x_n)$. The expression given by (8.24) is called the joint pdf. The equation exists whenever its right side exists. The joint pdf must satisfy the following conditions:

(i) $f_x(x_1, x_2, \ldots, x_n) \geq 0$ (8.25a)

(ii) $\int_{-\infty}^{\infty} \int_{-\infty}^{\infty} \cdots \int_{-\infty}^{\infty} f_x(x_1, x_2, \ldots, x_n) dx_1 dx_2 \ldots dx_n = 1$ (8.25b)

If $x_1, x_2, \ldots, x_n$ are continuous random variables, then the marginal density is defined as

$$f_x(x_i) = \int_{-\infty}^{\infty} \int_{-\infty}^{\infty} \cdots \int_{-\infty}^{\infty} f_x(x_1, x_2, \ldots, x_n) dx_1 dx_2 \ldots dx_n \tag{8.26}$$

[1] It should be noted that, although $f(x)$ is not a probability *per se*, the phrase probability density function originates from the fact that the product $f(x)\Delta x$ approximates to $P(x_i \leq x \leq x_i + \Delta x)$ if Δx is small. Therefore, the probability that the random variable x lies in the interval ξ_1 and ξ_2 can be expressed as

$$P(\xi_1 < x \leq \xi_2) = \int_{\xi_1}^{\xi_2} f(x) dx$$

In general, when the random variable depicts points of a random signal or process that is a function of time, the probability density function of various orders may be easily defined. For example, the probability that the random variable x lies between ξ and $\xi + \Delta\xi$ at the time $t = t_1$ can be written as $p(\xi, t_1)$. Conversely, if at t_1 and t_2 the variable x lies, respectively, between $\{\xi_1$ and $\xi_1 + \Delta\xi_1\}$ and $\{\xi_2$ and $\xi_2 + \Delta\xi_2\}$, the corresponding probability can be defined as $p(\xi_1, t_1; \xi_2, t_2)$.

If $x_1, x_2, \ldots, x_n$ are continuous and independent, then

$$f_x(x_1, x_2, \ldots, x_n) = \prod_{i=1}^{n} f_{xi}(x_i) \tag{8.27}$$

It should be noted that when considering the joint pdf of two or more random variables, the differential term in (8.23), or in (8.24), would be replaced by a normalizing factor called the *Jacobian, J*. The definition of the Jacobian J can be explained by an illustration, as follows.

Consider two random variables X and Y with corresponding functions defined by $\mathbf{Z} = f_1(X, Y)$ and $\mathbf{P} = f_2(X, Y)$. The values of $Z = \mathbf{Z}(\varpi)$ and $P = \mathbf{P}(\varpi)$ depend on the outcome of the event ϖ. These values in turn determine the values of $X(\varpi)$ and $Y(\varpi)$. If the distributions of X and Y are given, the joint probability distribution of Z and P can be estimated by:

(i) finding the real solutions to the equations $z = f_1(x_j, y_j)$ and $p = f_2(x_j, y_j)$ for all i;
(ii) evaluating, at each root, the Jacobian J of the transformation from (x, y) to (z, p);

where

$$J = \begin{vmatrix} \dfrac{\partial p}{\partial x_j} & \dfrac{\partial p}{\partial y_j} \\ \dfrac{\partial z}{\partial x_j} & \dfrac{\partial z}{\partial y_j} \end{vmatrix} \tag{8.28}$$

(iii) calculating the joint pdf of Z and P.

$$f_{zp}(x, y) = \sum_{j=1}^{n} \frac{f_{xy}(x_j, y_j)}{|J_j|} \tag{8.29}$$

The J factor plays the same role as the differential term in (8.23), or in (8.24).

Example 8.3 Consider a radar circular display screen of unity radius having the locations of a radar target uniformly distributed over the circle radius. Determine the joint and marginal pdf of the range and azimuth of the target when the density function is described by

$$f_{xy}(x, y) = \begin{cases} \frac{1}{\pi} \sqrt{x^2 + y^2} & \leq 1 \\ 0 & \text{elsewhere} \end{cases} \tag{8.30}$$

Solution

As in the radar tracking problem, the radar variable is the target range R having a pdf of f_R. Being circular, the range can be formalized in terms of the x, y coordinates, the target elevation angle, θ, and the screen radius, r, as

$$R = \sqrt{x^2 + y^2}$$
$$x = r\cos\theta$$
$$y = r\sin\theta \qquad (8.31a)$$
$$\theta = \tan^{-1}\left(\frac{y}{x}\right)$$

The Jacobian factor J is dependent on the target's variables R and θ, and in view of (8.31a) and (8.28),

$$J = \begin{vmatrix} \frac{\partial R}{\partial x} & \frac{\partial R}{\partial y} \\ \frac{\partial \theta}{\partial x} & \frac{\partial \theta}{\partial y} \end{vmatrix} = \begin{vmatrix} \cos\theta & \sin\theta \\ \frac{-\sin\theta}{r} & \frac{\cos\theta}{r} \end{vmatrix} = \frac{1}{r} \qquad (8.31b)$$

Within the limits $0 \le r \le 1$ and $0 \le \theta \le 2\pi$, the joint pdf:

$$f_{R\theta}(r, \theta) = \frac{f_{xy}(x, y)}{|J|} = \frac{r}{\pi} \qquad (8.31c)$$

Within the limit $0 \le r \le 1$, the marginal pdf

$$f_R(r) = \int_0^{2\pi} f_{R\theta}(r, \theta)d\theta = \int_0^{2\pi} \frac{r}{\pi} d\theta = 2r \qquad (8.31d)$$

For a discrete case, consider a set of random variables $x_1, x_2, \ldots, x_n$, having corresponding discrete points $k_1, k_2, \ldots, k_n$. To express a set of probability density functions in discrete format, certain conditions similar to that stipulated in (8.25) must be met:

(i) $P(X_1 = k_1, X_2 = k_2, \ldots, X_n = k_n) \ge 0$ (8.32a)

(ii) $\sum_{k_1}\sum_{k_2}\ldots\sum_{k_n} P(X_1 = k_1, X_2 = k_2, \ldots, X_n = k_n) = 1$ (8.32b)

Also the marginal density can be written by taking one sample at a time:

$$P(X_1 = k_1) = \sum_{k_2}\sum_{k_3}\ldots\sum_{k_n} P(X_1 = k_1, X_2 = k_2, \ldots, X_n = k_n) \qquad (8.33)$$

By taking two samples at a time, the marginal density is written as

$$P(X_1 = k_1, X_2 = k_2) = \sum_{k_3}\sum_{k_4}\ldots\sum_{k_n} P(X_1 = k_1, X_2 = k_2, \ldots, X_n = k_n) \qquad (8.34)$$

Hence, for j samples, the marginal density can be written as

$$P(X_1 = k_1, X_2 = k_2, \ldots, X_j = k_j) \\ = \sum_{k_{j+1}} \sum_{k_{j+2}} \cdots \sum_{k_n} P(X_1 = k_1, X_2 = k_2, \ldots, X_n = k_n) \quad (8.35)$$

providing that $j < n$.

8.4 Moment, average, variance and cumulant

Moments, averages, mean, or expected values (or expectance) are synonymous with random processes. These terms are denoted by many notations, which are incorporated in the definition below. For example, the first moment, m_1, of $x(t)$ is

$$m_1 = E[x(t)] = \int_{-\infty}^{\infty} x(t) f(x, t) dx \quad (8.36)$$

which is just the average value of $x(t)$ and where $f(x,y)$ denotes the probability density function of x at time t. There are different notations used in the literature to represent average including μ, av$[x(t)]$, $\bar{x}(t)$, or $<x(t)>$.

The second moment (also called *covariance*, or the *mean square value*) about the mean is a measure of the dispersion or spread of the random variable $x(t)$ on the sample space, defined as

$$m_2 = E[x(t_1)x(t_2)] = \int_{-\infty}^{\infty} \int_{-\infty}^{\infty} x_1 x_2 f(x_1 x_2 : t_1 t_2) dx_1 dx_2 \quad (8.37)$$

where $f(x_1, x_2 : t_1, t_2)$ is the joint probability density of the pair of random variables $[x(t_1), x(t_2)]$.

Generalizing therefore, the kth moment about the origin of a random variable $x(t)$ is the statistical average of the kth power of $x(t)$ defined as

$$m_k = E[x(t)]^k = \int_{-\infty}^{\infty} x^k(t) f(x, t) dx \quad (8.38)$$

When order $k \geq 3$, the moments are called higher-order moments statistics, which form the basis of higher-order statistical signal processing. If the random variable $x(t)$ is not described about the origin, its moment properties can be described in terms of *cumulant* (Rosenblatt 1985). The cumulant, denoted by c_k, of the kth order is found by successively differentiating the natural logarithm of the characteristic function and evaluating the derivative at the origin. A good overview of this approach can be found in Boashash *et al.* (1995).

The concept of moments can be extended to *bivariant* cases of different orders k, n, involving two random variables $x(t)$ and $y(t)$ having corresponding powers k and n. Assume that the variables $x(t)$ and $y(t)$ lie, respectively, within the intervals x and $x + dx$ at time t_1, and y and $y + dy$ at t_2, then their joint $(k + n)$th moment can consequently be expressed as

$$m_{k+n} = E\left[x^k(t_1)y^n(t_2)\right] = \int_{-\infty}^{\infty}\int_{-\infty}^{\infty} x^k(t)y^n(t)f(x, t_1; y, t_2)dxdy \qquad (8.39)$$

for m_x and m_y having zero means, and where $f(x, t_1; y, t_2)$ is the joint probability density function of $x(t)$ and $y(t)$.

The kth moment (8.38) can be efficiently calculated through the introduction of a function $\Im(u)$, called the *characteristic function* of the random variable $x(t)$, defined as

$$\Im(u) = \int_{-\infty}^{\infty} e^{jux}f(x)dx \qquad (8.40)$$

This equation is similar to the inverse Fourier transform definition in (1.11). Using the kth moment definition of (8.38), a series expansion is obtained:

$$\Im(u) = \sum_{k=0}^{\infty} \frac{[ju]^k}{k!} m_k \qquad (8.41)$$

Since $\Im(u)$ is the inverse Fourier transform of the probability density $f(x, t)$, it can easily be calculated and also provide the higher-order moments of the signal.

Drawing from Rule 2 in section 8.2 that implies the same relationship for the characteristic function:

$$\Im(u_1, u_2) = \int_{-\infty}^{\infty}\int_{-\infty}^{\infty} e^{j(u_1x_1+u_2x_2)}f(x_1, x_2)dx_1dx_2 \qquad (8.42)$$

Using the correlation principles discussed in Chapter 1, section 1.3.4, this expression can be split into two as a product of two characteristic functions:

$$\Im(u_1, u_2) = \Im(u_1)\Im(u_2) \qquad (8.43)$$

This expression implies that the correlation concept is linearly dependent.

In essence, the bivariant functions can be resolved in similar manner as in (8.43). In this case, $\Im(u_1)$ and $\Im(u_2)$ would denote the characteristic functions of variables $x(t)$ and $y(t)$ respectively.

When dealing with multivariable systems, the random variables encountered can be represented by vector quantities. As an illustration, for n observations, let $\mathbf{x}(t)$ denote a column vector

$$\mathbf{x}(t) = \begin{bmatrix} x_1(t) \\ x_2(t) \\ x_3(t) \\ \vdots \\ x_n(t) \end{bmatrix} \qquad (8.44)$$

where its transpose is $[\mathbf{x}(t)]^T = [x_1(t), x_2(t), x_3(t), \ldots, x_n(t)]$. From (8.36) the expectance of the vector can be written as

$$E[\mathbf{x}(t)] = \begin{bmatrix} E[x_1(t)] \\ E[x_2(t)] \\ E[x_3(t)] \\ \vdots \\ E[x_n(t)] \end{bmatrix} = \begin{bmatrix} \int x_1 f(x_1, t)dx_1 \\ \int x_2 f(x_2, t)dx_2 \\ \int x_3 f(x_3, t)dx_3 \\ \vdots \\ \int x_n f(x_n, t)dx_n \end{bmatrix} \tag{8.45}$$

By suppressing the time dependence of vector $\mathbf{x}(t)$, its covariance matrix can also be expressed as

$$\mathrm{cov}[\mathbf{x}] = E\left[(\mathbf{x} - E[\mathbf{x}])\left(\mathbf{x}^{\mathrm{T}} - E[\mathbf{x}^{\mathrm{T}}]\right)\right] \tag{8.46}$$

If $E[\mathbf{x}] = \mathbf{0}$, and in view of equation (8.45), the covariance matrix can be expressed as

$$\mathrm{cov}[\mathbf{x}] = E\left[\mathbf{x}\mathbf{x}^{\mathrm{T}}\right] = \begin{bmatrix} E[x_1x_1] & E[x_1x_2] & E[x_1x_3] & \ldots & E[x_1x_n] \\ E[x_2x_1] & E[x_2x_2] & E[x_2x_3] & \ldots & E[x_2x_n] \\ E[x_3x_1] & E[x_3x_2] & E[x_3x_3] & \ldots & E[x_3x_n] \\ \vdots & \vdots & \vdots & \vdots & \vdots \\ E[x_nx_1] & E[x_nx_2] & E[x_nx_3] & \ldots & E[x_nx_n] \end{bmatrix} \tag{8.47}$$

The covariance matrix is symmetric and positive definite.

Before closing this section, it is worth noting that both the expectance and the covariance matrix could be conditional. Just as the conditional probability concept discussed above, the conditional expectance of a random variable can also be developed for mono-, bi-, and/or multivariant cases. Using previous developments in sections 8.2 and 8.3 together with expectance definitions in section 8.4, both the scalar and vector cases can be formulated.

Another useful property associated with conditional expectance is that, for example, if a two-dimensional random variable (X, Y) has a conditional expectance for X given Y as $E(X \mid Y)$ and the variables X and Y are independent, then

$$\begin{aligned} E[E(X \mid Y)] &= E(X) \\ E[E(Y \mid X)] &= E(Y) \end{aligned} \tag{8.48}$$

A problem of importance in radar tracking and control systems is in determining the parameters of a model given observations of the physical process being modelled. In most cases, the system parameters cannot be determined by *a priori*, or they vary during an operation. In such cases, an application of probability concepts to the system parameter estimation becomes a handy tool indeed. A practical example is determining which radar signals come from targets in a surveillance area of interest, while these signals are noise corrupted. The basic assumption frequently utilized in such multi-target conditions is that the targets are independent of one another. As such, an estimated target's state, ζ, can be developed at any sampling instant, or time t given the returns η up to time j. Likewise, the covariance associated with such an estimate can be obtained. Within the bounds of such returns, in view of (8.12), the conditional probability distribution function of target returns could be written as

$$p(\zeta_t \mid \eta_j) = \frac{p(\zeta_t \mid \eta_1, \eta_2, \eta_3, \ldots, \eta_n)}{p(\eta_1, \eta_2, \eta_3, \ldots, \eta_n)} \quad j = 1, 2, \ldots, n \tag{8.49}$$

where $p(\eta_1, \eta_2, \eta_3, \ldots, \eta_n) = P(\prod_{i=1}^{n} \eta_i)$ is the joint probability distribution function of η_j.

The question of which probability models to use in a particular problem is an important one, and should be answered carefully using all available data and background information. The answer cannot be dictated by mathematics, but must be arrived at by careful examination of the physical situation.

Before proceeding to the topic of distribution functions, it is necessary to explain briefly the notion of stationarity and ergodicity.

8.5 Stationarity and ergodicity

A signal is said to be stationary if its mean, expected, or ensemble average, value at different times is constant. If stationarity exists not for all distribution functions p_n, but only for $n \leq k$, then the process is said to be stationarity to order k. The case $k = 2$ is called, obviously, stationarity to order 2, but more often *weak stationarity* or *stationarity in the wide sense*. If the stationary property of the signal can be limited to its first- and second-order moments, the signal is wide sense stationary when characterized as

$$E[x(t)x(t + \tau)] = R_x(\tau) \tag{8.50}$$

where $R_x(\tau)$ is the autocorrelation function of the signal (already discussed in Chapter 1, section 1.3.4).

In general, in a weak or wide sense stationary, $<x(t)>= \mu = \text{constant}$ since p_1 does not depend on time t and the correlation depends on the time difference only as p_2 does (Adomian 1983).

The statistical parameters are, in general, difficult to estimate, or measure, directly because of the ensemble averages involved (Bellanger 1987). A reasonably accurate measurement of ensemble averages requires that many process realizations are available or that the experiment is repeated many times. In real-time data processing, this is often difficult. On the contrary, time averages are much easier to come by for time series. Hence, *ergodicity* property is of great practical importance. A process for which corresponding ensemble averages and time averages are equal is called *ergodic*. A stationary process is called *ergodic* if the following conditions are met:

(i) The time average is the same as the ensemble average for given time t; that is,

$$E[x(t)] = \lim_{T\to\infty} \frac{1}{2T}\int_{-T}^{T} x(t)dt = m \tag{8.51}$$

Or

$$E[x(t)] = \lim_{T\to\infty} \frac{1}{2T+1}\sum_{t=-T}^{T} x(t) = m \tag{8.52}$$

so that the variance of $\bar{x}(t)$ is zero as $T \to \infty$.

(ii) The autocorrelation function $R_x(\tau)$ (similar to (1.51)) can be expressed as a time average as well as the ensemble average

$$E[x(t)x(t+\tau)] = \lim_{T\to\infty} \frac{1}{2T}\int_{-T}^{T} x(t)x(t+\tau)dt = R_x(\tau) \tag{8.53}$$

Or

$$E[x(t)x(t+\tau)] = \lim_{T\to\infty} \frac{1}{2T+1}\sum_{t=-T}^{T} x(t)x(t+\tau) = R_x(\tau) \tag{8.54}$$

For complex signals, the autocorrelation function may be expressed by

$$R_x(\tau) = \lim_{T\to\infty} \frac{1}{2T}\int_{-T}^{T} x(t)x^*(t+\tau)dt \tag{8.55}$$

Or

$$R_x(\tau) = \lim_{T\to\infty} \frac{1}{2T+1}\sum_{t=-T}^{T} x(t)x^*(t+\tau) \tag{8.56}$$

where $x^*(t+\tau)$ is the complex conjugate of $x(t+\tau)$.

It should be noted that the class of ergodic processes is a proper subclass of the class of stationary processes. As such, an ergodic process could be strictly stationary but a stationary process does not have to be ergodic.

8.6 An overview of probability distributions

As earlier indicated, in radar systems, the signal received by the radar may be due to those reflected from clutter, or a combination of that from the target and surrounding surface. To achieve detection one must assume that some noise, or clutter, characteristics may appear at the radar receiver output. Careful analysis of the outcome should assist in minimizing the total probability of error. In practice, however, noise or clutter distribution patterns do not necessarily fit well into known distribution patterns. With experience, systems designers may modify such distributions and categorize them in a manner befitting recognizable patterns. Some well-known distributions are discussed in this section because they approximate physical problems and satisfy normal laws relating to the independence and randomness of physical quantities.

8.6.1 Uniform distribution

A continuous or discrete random variable that is equally likely to take on any value within a given interval is said to be uniformly distributed. If the random variable were continuous, its probability density function would be a series of equally weighted-impulse functions. If one allows the discrete random variable type to be a rectangular function, as shown in Figure 8.1, its mean value can be written as

$$E(x) = \frac{1}{2}(a + b) \tag{8.57a}$$

And its standard deviation by

$$\sigma_x = \frac{b - a}{2\sqrt{3}} \tag{8.57b}$$

Note that the height of the probability density function must be selected to give a unit area.

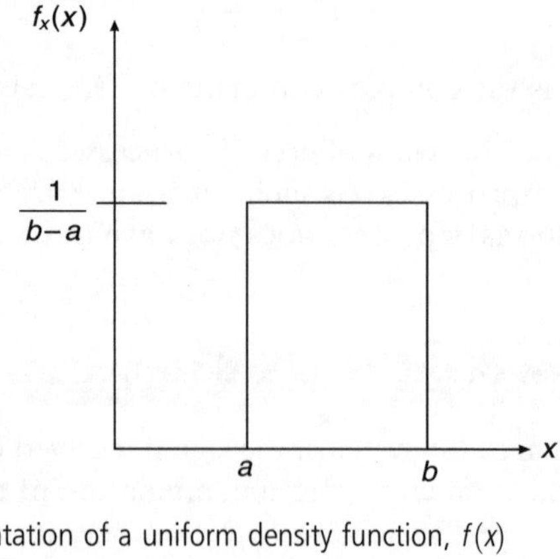

Figure 8.1 A representation of a uniform density function, $f(x)$

8.6.2 Normal or Gaussian distribution

A random variable x is said to be normally or Gaussianly distributed if its probability law has a density $f_x(x)$ that satisfies the normal or Gaussian law (see Figure 8.2). Specifically,

$$\frac{1}{\sqrt{2\pi}\sigma_x} e^{-[(x-\mu)^2/2\sigma_x^2)]} \tag{8.58}$$

The parameter μ is the mean of the variable x and the variance σ_x^2 is the second-order moment of the centred random variable $(x - \mu)$, where σ_x is called the standard deviation. Like (8.40), the function of a mean-centred (i.e. $\mu = 0$), the Gaussian characteristic function[2] may be written as

$$\Im(x) = e^{-x^2/2\sigma_x^2} \tag{8.59}$$

Using the series expansion of (8.41), the nth moment of the variable is written as

$$E(x^n) = \begin{cases} m_{2k} = \frac{2k!}{2^k k!}\sigma_x^{2k} & n = 2k \\ 0 & n = 2k+1 \end{cases} \quad k = 0, 1, \ldots, n \tag{8.60}$$

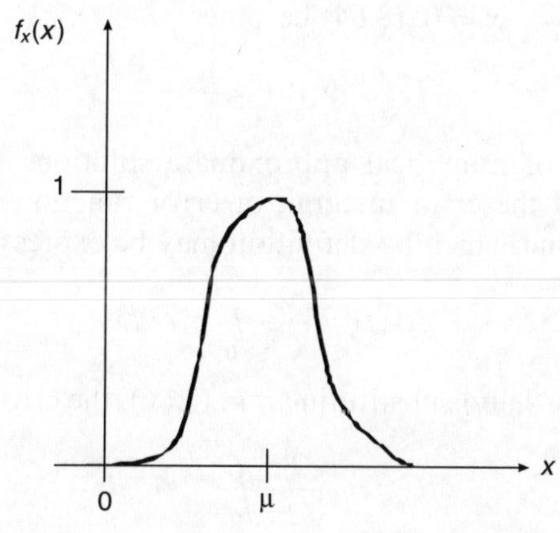

Figure 8.2 A Gaussianly distributed function

[2] A characteristic function is also defined for a real random variable x (say, chosen at time t from a process) (Adomian 1983):

$$\Im(\lambda) = \langle e^{j\lambda x} \rangle = \int_{-\infty}^{\infty} e^{j\lambda x} p(x) dx$$

where λ is real. It follows that the inverse Fourier transform of the characteristic function uniquely determines the distribution function $p(x)$; that is,

$$p(x) = \int_{-\infty}^{\infty} e^{-j\lambda x} \Im(\lambda) d\lambda$$

which is one of the principal reasons for the usefulness of the concept of the characteristic function.

For k-dimensional Gaussian variable $X(x_1, x_2, \ldots, x_k)$, the characteristic function is written as

$$\Im(u_1, u_2, \ldots, u_n) = \exp\left(-\frac{1}{2}\sum_{i=1}^{k}\sum_{j=1}^{k} m_{ij}u_j u_i\right) \tag{8.61}$$

where $m_{ij} = E[x_i x_j]$.

If a change of variable $y = x - \mu/\sigma_x$ is applied to (8.58), then

$$\int_{-\infty}^{\infty} f_x(x)dx = \frac{1}{\sqrt{2\pi}}\int_{-\infty}^{\infty} e^{-\frac{y^2}{2}}dy = 1 \tag{8.62}$$

This expression cannot be evaluated analytically. Instead a set of tables of numerical approximate solutions of $\Phi_x(x)$ is, by definition, given as (Abramowitz and Stegun 1968)

$$\Phi_x(x) = \frac{1}{\sqrt{2\pi}}\int_{-\infty}^{\infty} e^{-\frac{y^2}{2}}dy \tag{8.63}$$

which is the distribution function of a unit normal distribution. Since the integral is even, it follows that

$$\Phi_x(-x) = 1 - \Phi_x(x) \tag{8.64}$$

For the case of $x = -x = 0$, (8.64) becomes

$$\Phi_x(0) = \frac{1}{2} \tag{8.65}$$

Instead of tables of numerical approximate solutions to the distribution function, tables of the error integral, or error functions denoted by erf(t), are sometimes found which by definition may be expressed by

$$\text{erf}(t) = \frac{2}{\sqrt{\pi}}\int_{0}^{t} e^{-y^2}dy \tag{8.66}$$

By putting $y = x/\sqrt{2}$ and substituting it in (8.66), the error function becomes

$$\text{erf}(t) = \frac{2}{\sqrt{2\pi}}\int_{0}^{\sqrt{2}t} e^{-\frac{x^2}{2}}dx \tag{8.67}$$

It is evident from (8.63) and (8.64) that erf(t) is related to the normal distribution function from the perspective of unit normal distribution in the form

$$\text{erf}(t) = 2\Phi_x\left(\sqrt{2}t\right) - 1 \tag{8.68}$$

Using the previous definitions of conditional probability, the probability distribution function can thus be defined as

$$\begin{aligned} f_x(x\,|\,y) &= \frac{f_{xy}(x, y)}{f_x(x)} \\ f_y(y\,|\,x) &= \frac{f_{xy}(x, y)}{f_y(y)} \end{aligned} \tag{8.69}$$

Before closing the discussion on normal or Gaussian distribution, it is worth considering that the two constraints underlying the formulation of the previous expressions are seldom true in reality. For instance, mean zero and identical variance for multiple source data is seldom achieved in practice. The mean zero is an issue of choice of origin, which is easily accommodated in (8.61). For the variance, a group of components could be added in a way that would all have roughly the same variance. An additionally important feature of the Gaussian distribution is its behaviour under convolution. When two normal distributions are convoluted, the result is still a normal distribution whether the components have zero means or otherwise.

8.6.3 Bivariate Gaussian distribution

Suppose that two random variables X and Y have corresponding density functions $f_X(x)$ and $f_Y(y)$, see Figure 8.3. If they are independent, then, by applying rule 2 of section 8.2, their joint density function may be expressed as

$$f_{X,Y}(x, y) = f_X(x) f_Y(y) \tag{8.70}$$

Suppose that X and Y have corresponding mean values μ_x and μ_y, and variances σ_x^2 and σ_y^2, their distribution function may be written as

$$f_{X,Y}(x, y) = \frac{1}{2\pi\sigma_x\sigma_y} \exp\left\{-\frac{1}{2}\left[\left(\frac{x-\mu_x}{\sigma_x}\right)^2 + \left(\frac{y-\mu_y}{\sigma_y}\right)^2\right]\right\} \quad -\infty < x, y < \infty \tag{8.71}$$

In general, when variables X and Y are not independent, they become joint normal, or joint Gaussian, having joint density function

$$f_{xy}(x, y) = \kappa_{xy} e^{-\zeta_{xy}(x,y)} \quad -\infty < x, y < \infty \tag{8.72}$$

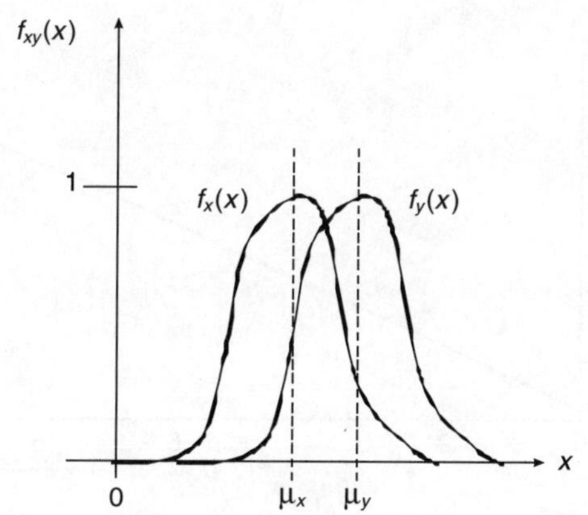

Figure 8.3 A bivariate gaussianly distributed function

where

$$k_{xy} = \frac{1}{2\pi\sigma_x\sigma_y\sqrt{1 - R_{xy}^2}} \tag{8.73a}$$

$$\zeta_{xy}(x, y) = \frac{1}{2\left(1 - R_{xy}^2\right)}\left[\left(\frac{x - \mu_x}{\sigma_x}\right)^2 + 2R_{xy}\left(\frac{x - \mu_x}{\sigma_x}\right)\left(\frac{y - \mu_y}{\sigma_y}\right) + \left(\frac{y - \mu_y}{\sigma_y}\right)^2\right] \tag{8.73b}$$

where R_{xy} is the *cross-correlation* coefficient of two functions. As discussed earlier in Chapter 1, section 1.3.4, when $R_{xy} = 0$, the functions are uncorrelated and independent. Hence (8.72) is the same as (8.71). However, when $R_{xy} = \pm 1$, equation (8.72) is meaningless because x and y are thus linearly related and are said to have a singular normal distribution. The joint density function $f_{x,\,y}(x, y)$ has non-zero values only on the line

$$\frac{x - \mu_x}{\sigma_x} = \pm\frac{y - \mu_y}{\sigma_y} \tag{8.74}$$

By rearranging (8.74), a linear relationship is then established between x and y as

$$y = \mu_y + \frac{\sigma_y(x - \mu_x)}{\sigma_x} \tag{8.75}$$

For $\mu_x = 0.1$, $\mu_y = 0.5$, and $\sigma_x = \sigma_y = 2$, the linear relationship between x and y where the joint density function $f_{x,\,y}(x, y)$ has non-zero values is demonstrated by Figure 8.4.

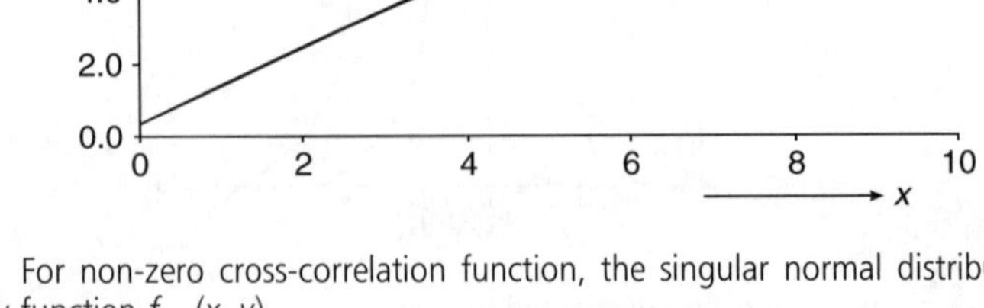

Figure 8.4 For non-zero cross-correlation function, the singular normal distribution of the joint density function $f_{x,\,y}(x, y)$

8.6.4 Rayleigh distribution

The Rayleigh distribution arises from the theory of post-detection noise. Let, for example, x and y be the Cartesian coordinate of a vector quantity, each satisfying the Gaussian distribution. Let the distribution functions of x and y have a representation described by (8.71). If the distribution of the modulus of the vector is required, a simple way of doing this is by transforming the quantities from one frame to another; that is, from (x, y) to (r, θ). So, the differentials of the coordinates may be expressed as

$$dxdy = rdrd\theta \tag{8.76}$$

If the mean of the distribution functions are assumed zero and their variances equal; that is, $\mu_x = \mu_y = 0$ and $\sigma_x = \sigma_y = \sigma$, then using (8.76) in (8.71) yields

$$\frac{1}{2\pi\sigma^2}e^{-\left(\frac{x^2+y^2}{2\sigma^2}\right)}dxdy = \frac{1}{2\pi\sigma^2}e^{-\left(\frac{r^2}{2\sigma^2}\right)}rdrd\theta \tag{8.77}$$

Since coordinates x and y are separable quantities, r and θ are also separable. To secure normalization, the radial density with respect to θ must be $1/2\pi$, noting that there is no dependence on θ. Hence, (8.77) resolves to

$$f(r) = \frac{r}{\sigma^2}e^{-\frac{r^2}{\sigma^2}} \tag{8.78}$$

which is the *Rayleigh* distribution with two degrees of freedom. Its generalized expectance can be shown to be

$$E(x^n) = \begin{cases} \frac{2^{\frac{n}{2}}}{\sigma^n} & n = \text{even} \\ \sqrt{\frac{2}{\pi}}1.3.5\ldots n\sigma^n & n = \text{odd} \end{cases} \tag{8.79}$$

Due to the statistical nature of radar received signals, the Rayleigh distribution is particularly useful in radar signal processing to characterize noise in the receiver prior to demodulation (detection) and certain types of clutter distribution across measurement domains, such as range, Doppler and azimuth. The expressions in (8.78) and (8.79) generally apply to radar noise. For post-detection signals, detected noise is called *video* noise. The video noise has a different probability distribution to the noise prior to detection.

The Rayleigh distribution has also been used to characterize clutter, particularly sea clutter, which is stochastic in nature arising perhaps from superposition of many processes or events. A simple Rayleigh clutter can be characterized as

$$f(v_e) = \frac{v_e}{\sqrt{\varphi_0}}e^{-\frac{v_e^2}{2\varphi_0}} \quad v_e > 0 \tag{8.80}$$

with its fluctuating input amplitude proportional to the mean, where $\sqrt{\varphi_0}$ is the clutter mean and v_e clutter amplitude or threshold voltage.

A sea clutter has been characterized by the *Weibull* distribution, which is the limiting case of the Rayleigh distribution, written as

$$f(v_e) = \alpha\vartheta^{\alpha-1}e^{-\vartheta^{\alpha}} \quad v_e > 0 \tag{8.81}$$

where α, β are constants and

$$\vartheta = \log_e(2v_e/\beta) \tag{8.82}$$

As demonstrated in Chapter 5, section 5.4.1, a simple Rayleigh description of sea clutter is insufficient because a number of multivariate components are required to accurately describe sea clutter. If, however, a quick estimate of sea clutter is required, the Rayleigh function tends to overestimate the range of values obtained from real clutter.

8.6.5 Poisson distribution

A random variable x is called a *Poisson* random variable if at $x = k$,

$$P(x = k) = \frac{e^{-\lambda_p}\lambda_p^k}{k!} \quad k = 0, 1, 2, \ldots, \infty \tag{8.83}$$

where λ_p is average intensity of the variable x, $\lambda_p \geq 0$ noting that $0! = 1$. The Poisson distribution function can therefore be given by

$$f_x(x) = \sum_{k=0}^{\infty} \frac{e^{-\lambda_p}\lambda_p^k}{k!} = 1 \tag{8.84}$$

The expectance of the distribution may be expressed as

$$\begin{aligned} E(x) &= \sum_{x=0}^{\infty} x\frac{e^{-\lambda_p}\lambda_p^x}{x!} \\ &= \lambda_p \sum_{x=1}^{\infty} \frac{e^{-\lambda_p}\lambda_p^{x-1}}{(x-1)!} \end{aligned} \tag{8.85}$$

If $(x - 1)$, in this expression, is replaced with y, then

$$E(x) = \lambda_p \sum_{y=0}^{\infty} \frac{e^{-\lambda_p}\lambda_p^y}{y!} = \lambda_p \tag{8.86}$$

since $\sum_{y=0}^{\infty} e^{-\lambda_p}\lambda_p^y/y! = 1$

The variance of the distribution is obtained by

$$\begin{aligned} \sigma_x^2 = E(x(x-1)) &= \sum_{x=0}^{\infty} x(x-1)\frac{e^{-\lambda_p}\lambda_p^x}{x!} \\ &= \lambda_p^2 \sum_{x=2}^{\infty} \frac{e^{-\lambda_p}\lambda_p^{x-2}}{(x-2)!} \end{aligned} \tag{8.87}$$

Replacing $(x-2)$ with y, to have

$$E(x(x-1)) = \lambda_p^2 \sum_{y=0}^{\infty} \frac{e^{-\lambda_p}\lambda_p^y}{y!} = \lambda_p^2 \tag{8.88}$$

Hence, the variance

$$\sigma_x^2 = E(x(x-1)) = \lambda_p^2 \tag{8.89}$$

Example 8.4 A surveillance station provides the statistics of system breakdowns per day as follows.

Breakdown/day	0	1	2	3	4	5	6
Frequency	340	121	53	30	12	4	0

Find the mean of the distribution. Using this mean, show that the distribution follows approximately a Poisson distribution. What is the probability that there are no breakdowns?

Solution

Let's denote breakdowns/day by x, and frequency by f. The expectance λ_p is defined by

$$\lambda_p = \frac{\sum_i f_i x_i}{\sum_i f_i} = 0.6875$$

Using (8.83), the following probability values are obtained and plotted as in Figure 8.5:

(i) No breakdown, $P(0) = e^{-\lambda_p} = 0.5028$
(ii) 1 breakdown, $P(1) = e^{-\lambda_p}\lambda_p = 0.3457$

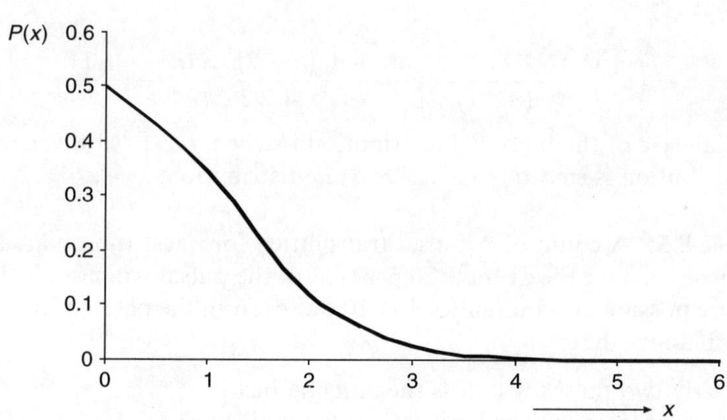

Figure 8.5 Probability of breakdowns/day

(iii) 2 breakdowns, $P(2) = \frac{e^{-\lambda_p}\lambda_p^2}{2!} = 0.1188$

(iv) 3 breakdowns, $P(3) = \frac{e^{-\lambda_p}\lambda_p^3}{3!} = 0.0272$

(v) 4 breakdowns, $P(4) = \frac{e^{-\lambda_p}\lambda_p^4}{4!} = 0.0047$

(vi) 5 breakdowns, $P(5) = \frac{e^{-\lambda_p}\lambda_p^5}{5!} = 0.0006$

(vii) 6 breakdowns, $P(6) = \frac{e^{-\lambda_p}\lambda_p^6}{6!} = 0.00007$

Using (8.84), $f_x(x) = \sum_{k=0}^{\infty} e^{-\lambda_p}\lambda_p^k/k! = 0.99999 \approx 1$, which demonstrates that the distribution is Poisson.

8.6.6 Binomial distribution

The binomial distribution is frequently used for multiple-pulse detection scenario. For instance, the possible outcomes from n-pulse received signals can be determined by writing

$$(p+q)^n \tag{8.90}$$

where $0 < p < 1$ is the probability of occurrence and $q = (1-p)$, not occurring. Following the binomial theorem

$$(p+q)^n = \sum_{r=0}^{N} \binom{n}{p} p^r q^{n-r} \tag{8.91}$$

The probability of r outcomes out of n pulses may be expressed as

$$P(x=r) = \binom{n}{r} p^r q^{n-r} \tag{8.92}$$

and the probability distribution as

$$f_x(x) = \sum_{r=0}^{N} \binom{n}{r} p^r q^{n-r} \tag{8.93}$$

where

$$\binom{n}{r} = \frac{n!}{r!(n-r)!} = \frac{n(n-1)(n-2)\ldots(n-r+1)}{1.2.3\ldots r} \tag{8.94}$$

A special case of the binomial distribution is when $n = 1$. At this condition, the distribution is said to have a *Bernoulli* distribution.

Example 8.5 A count of N pulses transmitted for a test run on an antenna dish shows that on the average 20 per cent of the pulses will not hit the dish. If it were possible to randomly select 10 pulses from the batch of pulses, find the probability that

(i) exactly two pulses will miss the antenna dish;
(ii) two or more pulses will miss the antenna dish; and
(iii) more than five pulses will miss the antenna dish.

Solution
$N = 10$
Probability of miss, $q = 0.2$
Probability of pulses hitting the dish, $p = 1 - q = 0.8$
Using (8.92) the following probability values are obtained:

(i) Exactly two pulses missing target

$$P(x = 2) = \binom{10}{2} 0.8^2 0.2^8 = \left(\frac{10 \times 9}{2}\right) 0.8^2 0.2^8 = 0.0000737$$

(ii) Two or more pulses missing target

$$P(x \geq 2) = 1 - [P(0) + P(1)] = 0.9999958$$

$$P(0) = \binom{10}{0} p^0 q^{10} = 0.2^{10} = 0.0000001$$

$$P(1) = \binom{10}{1} p^1 q^9 = 10 \times 0.2^9 \times 0.8 = 0.0000041$$

(iii) More than five pulses missing target

$$P(x > 5) = \sum_{i=6}^{10} P(i) \quad \text{or} \quad P(x > 5) = 1 - \sum_{i=0}^{5} P(i)$$

$$P(x > 5) = 0.9672065$$

8.7 Summary

The distribution of all orders that characterize a process is frequently too complicated and in some instances represent more than is needed. Often simpler and necessarily less complete characterizations in the form of expectations or means, dispersions or variances, covariances, joint moments, correlations, etc. are considered. These characterizations are useful ways of measuring our knowledge of the processes. This chapter has certainly provided such tools, complemented with examples. It is this author's belief that enough probability theory has been presented to the reader to understand the subsequent chapters. Since models of noise-corrupted signal processes usually specify system statistics, the importance of probability theory becomes self-evident.

Problems

1. In the process of manufacturing several radar system components, the factory estimates that 0.2 per cent of its production is defective. These components are sold in packets of 200. What percentage of the packets contains one or more defectives?

2. If a machine produces defective products with a probability of 4 per cent. What is the expected number of defective items in a random sample of 500 taken from its output? What is the variance of the number of defective items?
3. Consider a sample of 9 values x_i, ($i = 1, 2, \ldots, 9$), of a random variable X, which is known to be normally distributed with unit variance and unknown mean. Write a probability expression that an observed value would lie with any given range of the distribution.
4. A number is drawn from a hat that contains the numbers 1, 2, 3, ..., 50. Every number has an equal chance of being drawn from the hat. What is the probability of drawing a number divisible by 4?
5. Suppose three containers numbered 1, 2 and 3 contain, respectively, one red and one black ball, two red and three black balls, and four red and two black balls. Consider an experiment consisting of the selection of a container followed by the draw of a ball from it. The container is not observed but a red ball is drawn, with all events considered mutually exclusive and exhaustive. The probability that a ball in container 1 is drawn is 0.45, in container 2 is 0.35 and container 3 is 0.2. What is the probability that a red ball was drawn from container 1, 2, or container 3?
6. A coin is flipped seven times. [If an ith event, defined by $A_i = \{\omega \mid \omega$ has heads in the ith position$\}$ for $i = 1, 2, 3, \ldots, 7$.] All events $A_1, A_2, \ldots, A_7$ are mutually independent events. Show that the probability of selecting a head in ith position is

$$\prod_{j=1}^{r} P(A_{i_j}) = \left(\frac{1}{2}\right)^r.$$

7. Describe three realistic cases where the use of binomial distribution is an appropriate model for characterizing a random variable.

9

Decision theory

The last chapter provided the basis of decision theory – which is statistical and the main discussion of this chapter – and the signal detection process (to be discussed in Chapter 10).

The basic concepts of decision theory are fundamentally important in all analyses. As noted in the previous chapters, there is no pure signal. Signals received by a radar system may contain clutter, the target, or the target and clutter. The dilemma is making a correct decision that the signal received comes from the target or not. A decision is sought from statistical tests on which a hypothesis could be tested that the returns are truly from the target. A hypothesis is a statement of a possible decision. If the hypothesis were correctly postulated, the outcome would minimize the total probability of error. Hypothesis testing involves comparing (Lehman 1959):

- critical value(s) with defined population parameter(s);
- probability of acceptance with set value(s); and
- test value(s) with the specified confidence level(s).

Several decision criteria have been postulated in the literature, which use a different amount of information and specification. The most popular are the Maximum Likelihood, Neyman–Pearson, Minimum Error Probability (or Maximum *a posteriori* probability) and Bayes minimum risk decision rules. The basic ideas behind these criteria are discussed in this chapter. Examples are included to show how these rules are applied.

Before going into the main discussion on decision criteria, the author considers introducing the concept of the test of significance and the connection between error probabilities and decision criteria by an example of a binary detection problem. It is hoped that this approach will enable the reader to follow the flow of each rule's development. No attempt will be made to delve too deeply into mathematical details that govern these rules. However, the discussion will be explicit enough for easy comprehension of each of the criteria basic characteristics.

9.1 Tests of significance

The test of significance is a mode of inference within the framework of the sampling distribution techniques. This test is concerned with deciding whether or not a hypothesis concerning statistical parameters is true. As an illustration, suppose that it is required to test whether a sample space Ω of certain observations $x_1, x_2, \ldots, x_n$ is compatible with the hypothesis that they come from a normal probability density function with specified values μ_0, σ_0^2, for the mean and variance.

The steps involved in setting up a significance test follow closely Galati (1993) and Jenkins and Watts (1968).

(a) Assume a form for the probability density function associated with the samples is

$$f_{1,2,\ldots,n}(x_1, x_2, \ldots, x_n : \mu, \sigma^2) = \frac{1}{\left(\sqrt{2\pi\sigma^2}\right)^n} \exp\left\{-\frac{1}{2\sigma^2}\sum_{i=1}^{n}(x_i - \mu)^2\right\} \quad (9.1)$$

where the samples' mean μ and variance σ^2 may or may not be known. And set up a null hypothesis, H_0, that the samples are distributed normally with the mean μ_0 but unknown variance σ^2.

(b) Decide a set of alternative hypotheses. For example, it would be natural to take these to be $\mu > \mu_0$, meaning that a set of samples would be rejected if the mean were too high.

(c) Decide on the best function of the observations or *statistic* to test the null hypothesis. If the variance σ^2 is known, it is possible to show that the best statistic is the mean μ. When the variance is unknown, the best statistic is

$$t_v = \frac{\sqrt{n}(\mu - \mu_0)}{\sigma} \quad (9.2)$$

where 't' is a sampling distribution with subscript 'v' denoting its degrees of freedom. The probability density function of the random variable t_v is called the *Student's t distribution with v degrees of freedom*. You might be interested to know that the name *Student* was a pseudonym for W.S. Gosset, a French statistician. He used 't' to denote the standardized Student variable given by (9.2).

(d) Derive the sampling distribution of the statistic under the null hypothesis. From (c), the sampling distribution may be taken as *chi-squared distribution with v degrees of freedom* or *Student's t distribution with v degrees of freedom*. In this sampling example, the degrees of freedom $v = n - 1$.

A quick review of these distributions is now given to broaden the knowledge of the reader. The sampling distribution of the mean involves the distribution of sums of random variables; e.g.

$$f_\mu(\mu) = \frac{1}{\sqrt{2\pi}\left(\frac{\sigma}{\sqrt{n}}\right)} \exp\left\{-\frac{n}{2}\left(\frac{\mu - \mu_0}{\sigma}\right)^2\right\} \quad (9.3)$$

The sampling distribution of the variance of normal random variables involves the sum of squares of random variables. For example, suppose there are n independent measurements from a normally distributed population of zero mean, unit variance $N(0,1)$ and it is required to find the sampling distribution of the random variable:

$$\chi_n^2 = x_1^2 + x_2^2 + \cdots + x_n^2 \tag{9.4}$$

This distribution χ_n^2 is called the *chi-squared distribution with v degrees of freedom*, and with probability density function

$$f_{\chi_n^2}(x) = \frac{1}{2^{\frac{v}{2}}\Gamma\left(\frac{v}{2}\right)} x^{\left(\frac{v}{2}\right)-1} e^{-\frac{x}{2}} \quad (0 \le x \le \infty) \tag{9.5a}$$

where $\Gamma(v/2)$ is the gamma function with argument $(v/2)$ defined by

$$\Gamma\left(\frac{v}{2}\right) = \int_0^\infty e^{-t} t^{\left(\frac{v}{2}\right)-1} dt \tag{9.5b}$$

The first two moments of χ_n^2 distribution, obtained from the (9.5a), are

$$\begin{aligned} E\left[\chi_v^2\right] &= v \\ \mathrm{Var}\left[\chi_v^2\right] &= 2v \end{aligned} \tag{9.5c}$$

Plots of $f_{\chi_n^2}(x)$ against x for $v = 1, 2, 3, 5, 7$ and 9 are shown in Figure 9.1. As observed in Figure 9.1, at $v = 2$ the function $f_{\chi_n^2}(x)$ is exponential, and afterwards ($v \ge 3$) the function $f_{\chi_n^2}(x)$ settles down to a unimodal form. For values of $0 \le v \le 1$, as shown in Figure 9.2, the function $f_{\chi_n^2}(x)$ has an infinite ordinate as x tends to zero but tends to zero as x tends to infinity. Usually, the observation's normal distribution is written as $N(\mu, \sigma^2)$. In the case of χ_n^2, it can be written as $N(\mu/\sigma, 1^2)$. For unknown variance σ^2

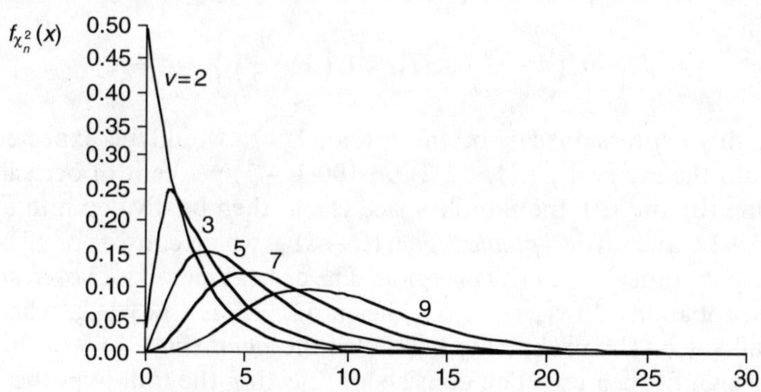

Figure 9.1 Chi-squared probability density function

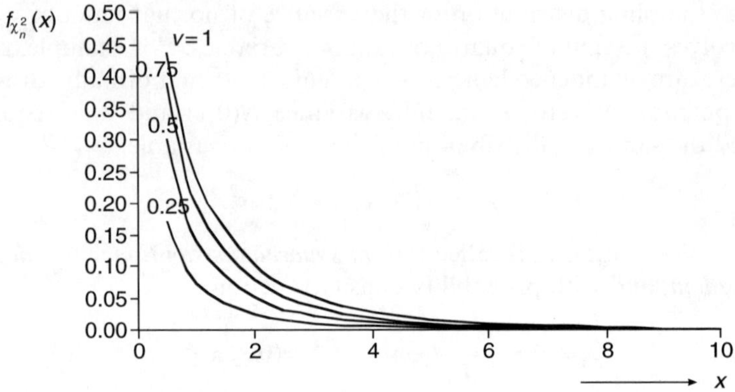

Figure 9.2 Chi-squared probability density function for $v \leq 1$

chi-squared probability density function for null variance σ_0^2, degrees of freedom v, the probability limits may be expressed in the form

$$P\left\{x_v\left(\frac{\xi}{2}\right) < \frac{v\sigma^2}{\sigma_0^2} \leq x_v\left(1 - \frac{\xi}{2}\right)\right\} = 1 - \xi \tag{9.6}$$

and may be obtained from statistical tables. Rearranging (9.6), it follows that the random variable satisfies

$$P\left\{\frac{v}{x_v\left(1 - \frac{\xi}{2}\right)} < \frac{\sigma^2}{\sigma_0^2} \leq \frac{v}{x_v\left(\frac{\xi}{2}\right)}\right\} = 1 - \xi \tag{9.7}$$

The Student's 't_v' distribution may be constructed on intervals $t_v(\xi/2)$ and $t_v(1 - \xi/2)$ in which t_v is allowed to lie on a proportion $(1 - \xi)$ of occasions. Since the Student's probability density function is symmetric $t_v(\xi/2) = -t_v(1 - \xi/2)$, the probability limits may be expressed as

$$P\left\{-t_v\left(1 - \frac{\xi}{2}\right) < T_v \leq t_v\left(1 - \frac{\xi}{2}\right)\right\} = 1 - \xi \tag{9.8}$$

So, this expression can be interpreted as t_v would be expected to lie within the interval $\pm t_v(1 - \xi/2)$ on $100(1 - \xi)$ per cent of occasions.

(e) Using (b) and (d), the sample space Ω can then be divided into a *critical region* Ω_c and an *acceptable region* $(\Omega - \Omega_c)$, which consists of all points in the space outside the critical region. The critical region is chosen such that the probability $P\{x_1, x_2, \ldots, x_n \text{ lies in } \Omega_c \mid H_0 \text{ is true}\} = \xi$, where ξ is a small value. The probability ξ is called the *significance level* of the test.

(f) The significance test then consists of rejecting the null hypothesis if the observed sample $x_1, x_2, \ldots, x_n$ falls in Ω_c and not rejecting if it falls in $(\Omega - \Omega_c)$. Given that there is a small probability that the sample point falls in Ω_c when H_0 is true, any cases when this happens are taken as evidence against the null hypothesis.

Since $P\{t_v > t_v(1-\xi)\} = \xi$, following (9.2), the critical region is defined by

$$t_v = \frac{\sqrt{n}(\mu - \mu_0)}{\sigma} > t_{n-1}(1-\xi) \tag{9.9}$$

Alternatively

$$\mu = \mu_0 + \frac{\sigma t_{n-1}(1-\xi)}{\sqrt{n}} \tag{9.10}$$

If the observed value μ does not lie in the critical region, the null hypothesis is not rejected at the ξ significance level. This approach is called a *one-sided significance test*.

Another situation might arise where $\mu > \mu_0$ and $\mu < \mu_0$ are of equal importance. For example, if the mean of a sample has to conform to the specified mean μ_0. In such a case, it would be reasonable to define the critical region as

$$t > t_{n-1}\left(1 - \frac{\xi}{2}\right), \quad t < -t_{n-1}\left(1 - \frac{\xi}{2}\right) \tag{9.11}$$

In this situation, following (9.8) to write the probability limits, the mean test may be expressed as

$$\mu > \mu_0 + \frac{\sigma t_{n-1}\left(1 - \frac{\xi}{2}\right)}{\sqrt{n}}, \quad \mu < \mu_0 - \frac{\sigma t_{n-1}\left(1 - \frac{\xi}{2}\right)}{\sqrt{n}} \tag{9.12}$$

If the observed value μ does not lie in the critical region limits, the null hypothesis would not be rejected at the ξ significance level. This approach is called a *two-sided significance test* as opposed to the one-sided test given by (9.10).

9.2 Error probabilities and decision criteria

Suppose a binary source produces possible signals $x_i\{x_0, x_1\}$ with respective probabilities $p_i\{p_0 = P(x_0), p_1 = P(x_1)\}$. The received signals $y_i(=x_i + \text{noise})$ reached the observer in a deteriorated form because of the signals' contamination by various random distances. Knowing the binary nature of the source, the observer can set two hypotheses about the signal identity on the basis of the observer's continuous, or discrete, observation of the received signals y_i. For this, the observer must apply a decision criterion. The hypothesis testing, in this case, is the problem of deciding which hypothesis is correct based on a single measurement, y, from the observation space, $\Omega\,\{\Omega \in y_i\}$. That is, a decision of ascertaining whether 'a target is' or 'a target is not' present in Ω. Denoting the two outcomes by d_0 and d_1 respectively as 'a target is not' and 'a target is' in the desired observation space Ω. This becomes a binary detection problem.

However, if the radar returns from the surveillance area or observation space contain a set of M hypotheses H_i, where $i = 0, 1, 2, \ldots, M - 1$, then the sequence would have M-ary detection problem.

The next step is to partition the observation space into two decision regions Y_0 and Y_1. When y lies in Y_0, d_0 is taken as the correct hypothesis and whenever y lies in Y_1, d_1 is taken as the correct hypothesis. The question is: how then does one choose from these regions to minimize probability of error?

To begin with a suitable criterion for the observation space to test which of the hypotheses is true is written as

$$P(d_i \mid y) \quad i = 0, 1 \tag{9.13}$$

which means that the probability that d_i is the true hypothesis given a particular value of y. With this formulation, it is possible to decide whether the true hypothesis is the one corresponding to the larger of the two possibilities.

The error probabilities can be defined as either of the first kind α (also called Type I) or second kind β (also called Type II). A Type I error may be expressed as

$$\begin{aligned} \alpha &= p(d = d_1 \mid y = y_0) \\ &= p(d_1 \mid x_0) \end{aligned} \tag{9.14}$$

which is the probability of deciding on event x_1 when x_0 actually happened (and hence measurement y_0 was generated). This type of error is similar to the probability of false alarm, P_{fa}.

A Type II error may similarly be written as

$$\begin{aligned} \beta &= p(d = d_0 \mid y = y_1) \\ &= p(d_0 \mid x_1) \end{aligned} \tag{9.15}$$

which is the probability of deciding on event x_0 when x_1 actually happened (and hence measurement y_1 was generated). This type of error is similar to the probability of *miss* detection, denoted by $\bar{P}_D$, which is the same as $(1 - P_D)$, where P_D is the probability of detection.

At this junction, decisions are made on the basis of:

If y exists in region $Y_0 (y \in Y_0)$, d_0 is decided;
If y exists in region $Y_1 (y \in Y_1)$, d_1 is decided.

The error probabilities can be defined using the conditional probability density functions $p(y \mid x_0)$ and $p(y \mid x_1)$. Thus, the probability of making an incorrect decision can be defined for each type of error:

$$\alpha = p(d_1 \mid x_0) = \int_{y_1} p(y \mid x_0) dy \tag{9.16}$$

$$\beta = p(d_0 \mid x_1) = \int_{y_0} p(y \mid x_1) dy \tag{9.17}$$

Table 9.1 Error probabilities and decision criteria

	Events	
Decision	x_0	x_1
d_0	Correct decision $1-\alpha$	Error Type II β
d_1	Error Type I α	Correct decision $1-\beta$

Similarly, the probability of making the correct decisions is:

$$p(d_0 \mid x_0) = \int_{y_0} p(y \mid x_0)dy = 1 - P_{fa} \tag{9.18}$$

$$p(d_1 \mid x_1) = \int_{y_1} p(y \mid x_1)dy = P_D \tag{9.19}$$

It is obvious that (9.18) equates to $(1 - \alpha)$ while (9.19) equates to $(1 - \beta)$. It follows therefore from (9.16) and (9.18) that

$$p(d_0 \mid x_0) + p(d_1 \mid x_0) = 1 \tag{9.20}$$

And also from (9.17) and (9.19)

$$p(d_1 \mid x_1) + p(d_0 \mid x_1) = 1 \tag{9.21}$$

From (9.20) and (9.21) a decision and error probabilities table can be developed as in Table 9.1.

The functional relationships formalized in (9.16) through to (9.21) are used in the next sections.

9.3 Maximum likelihood rule

The *maximum likelihood rule* (MLR) is a decision based on most likely causal. It requires that the conditional probability density function of the observation is given and that every possible event is known. This statement can be formalized as $P(y \mid x_i)$, where y is the observation and x_i represents possible events. The decision rule is formed by choosing:

$$d(y) = \begin{cases} d_0 & \text{if} \quad p(y \mid x_0) > p(y \mid x_1) \\ d_1 & \quad\;\;\; p(y \mid x_1) > p(y \mid x_0) \end{cases} \tag{9.22}$$

Alternatively a *likelihood ratio* test can be used:

$$\Lambda(y) = \frac{p(y \mid x_1)}{p(y \mid x_0)} \tag{9.23}$$

The decision can be written concisely as

$$\Lambda(y) \underset{d_0}{\overset{d_1}{\gtrless}} 1 \tag{9.24}$$

Thus, the decision test consists of comparing the ratio $\Lambda(y)$ with a constant, called the threshold. The threshold in this instance is unity. This kind of decision process is called the *likelihood ratio test* (LRT).

In summary, the maximum likelihood is a simple decision rule. Its drawback is that it may not represent adequately practical problems. Nonetheless, it is a powerful tool in estimation problems.

Example 9.1 Two observations Y_0 and Y_1 are related to events x_0 and x_1 respectively. Their probability density functions are described by

$$Y_0 : p(y \mid x_0) = \frac{e^{-\frac{y^2}{2}}}{\sqrt{2\pi}} \tag{9.25a}$$

$$Y_1 : p(y \mid x_1) = \frac{e^{-\frac{(y-\mu)^2}{2\sigma^2}}}{\sigma\sqrt{2\pi}} \tag{9.25b}$$

Apply the MLR and decide which event truly comes from observation y_1.

Solution

From (9.23), calculate the likelihood ratio

$$\Lambda(y) = \frac{p(y \mid x_1)}{p(y \mid x_0)} = \frac{1}{\sigma} e^{\left\{\frac{y^2}{2} - \frac{(y-\mu)^2}{2\sigma^2}\right\}} \tag{9.26}$$

From (9.24), the decision test is

$$\exp\left\{\frac{y^2}{2} - \frac{(y-\mu)^2}{2\sigma^2}\right\} \underset{d_0}{\overset{d_1}{\gtrless}} \sigma \tag{9.27}$$

Rearranging after taking $\log_e$ of both sides to have

$$(\sigma^2 - 1)y^2 + 2y\mu - (\mu^2 + 2\sigma^2 \log_e \sigma) \underset{d_0}{\overset{d_1}{\gtrless}} 0 \tag{9.28}$$

Equation (9.28) is a quadratic equation having solutions:

$$y \underset{d_0}{\overset{d_1}{\gtrless}} -\left(\frac{\mu}{\sigma^2 - 1}\right) \mp \sqrt{\frac{2}{\sigma^2 - 1}(\mu^2 + \sigma^2 \log_e \sigma)} \tag{9.29}$$

From (9.29), the following decisions are made

$$\text{If} \begin{cases} y < -\left(\frac{\mu}{\sigma^2-1}\right) - \sqrt{\frac{2}{\sigma^2-1}(\mu^2 + \sigma^2 \log_e \sigma)} & d_1 \quad \text{decided} \\ y > -\left(\frac{\mu}{\sigma^2-1}\right) + \sqrt{\frac{2}{\sigma^2-1}(\mu^2 + \sigma^2 \log_e \sigma)} & d_0 \quad \text{decided} \end{cases} \tag{9.30}$$

If $\mu = 0$, and $\sigma^2 > 1$,

$$|y| \underset{d_0}{\overset{d_1}{\gtrless}} \sqrt{\frac{2\sigma^2 \log_e \sigma}{\sigma^2 - 1}} \tag{9.31}$$

For a case of $\sigma^2 = 2$,

$$|y| \underset{d_0}{\overset{d_1}{\gtrless}} 1.18 \tag{9.32}$$

In conclusion, for two observations with normally distributed probability density functions with a spread of two or more, it can be suggested that the signals are truly coming from observation y_1 if the likelihood ratio is greater than 1.18, otherwise they come from observation y_0.

9.4 Neyman–Pearson rule

The Neyman–Pearson rule (NPR) is a problem of constrained optimization that uses the Lagrange multiplier λ. The rule can be expressed as

$$\max_{y_1} \Gamma$$

Alternatively in the functional form

$$\Gamma = p(d_1 \mid x_1) - \lambda[p(d_1 \mid x_0) - \alpha_0] \tag{9.33}$$

where α_0 is a desired value. From this expression, it can be said that the Neyman–Pearson rule expresses the desire to estimate, or set constant, the probability of false alarm, P_{fa}, at a given value α_0 while maximizing the probability of detection, P_D. The optimum value of $P(d_1 \mid x_1)$ is a function of λ, where λ itself is a function of the desired value α_0. So, in the likelihood terms,

$$\Lambda(y) \underset{d_0}{\overset{d_1}{\gtrless}} \lambda \tag{9.34}$$

It should be noted that the explicit computation of λ is not necessary. If $\lambda = 1$, the Neyman–Pearson rule will be identical to the MLR depicted by (9.24).

The Neyman–Pearson rule is particularly suited to radar applications owing to the concept of 'P_{fa} threshold' to be *fixed a priori* while maximizing P_D (Galati 1993).

Example 9.2 Using Example 9.1 as the basis for this problem but with a unity spread, apply the Neyman–Pearson rule to estimate the following variables:

(a) probability of false alarm;
(b) probability of detection;
(c) probability of miss; and finally
(d) optimum threshold.

Solution
Given $\sigma^2 = 1$.

$$Y_0: p(y \mid x_0) = \frac{e^{-\frac{y^2}{2}}}{\sqrt{2\pi}} \tag{9.35a}$$

$$Y_1: p(y \mid x_1) = \frac{e^{-\frac{(y-\mu)^2}{2}}}{\sqrt{2\pi}} \tag{9.35b}$$

The likelihood ratio test produces

$$\Lambda(y) = e^{\mu\left(y-\frac{\mu}{2}\right)} \underset{d_0}{\overset{d_1}{\gtrless}} \lambda \tag{9.36}$$

Rearranging the terms after taking $\log_e$ of both sides of (9.36)

$$\Lambda(y) \underset{d_0}{\overset{d_1}{\gtrless}} \lambda \underset{d_0}{\overset{d_1}{\gtrless}} \frac{\log_e \lambda}{\mu} + \frac{\mu}{2} \tag{9.37}$$

This decision is graphically shown in Figure 9.3.

(a) The probability of false alarm:

$$P_{fa} = p(d_1 \mid x_0) = \int_{\frac{\log_e \lambda}{\mu}+\frac{\mu}{2}}^{\infty} p(y \mid x_0)dy$$

$$= \int_{\frac{\log_e \lambda}{\mu}+\frac{\mu}{2}}^{\infty} \frac{e^{-\frac{y^2}{2}}}{\sqrt{2\pi}} dy \tag{9.38}$$

$$= \mathrm{erf}c\left[\frac{\log_e \lambda}{\mu} + \frac{\mu}{2}\right]$$

where $\mathrm{erf}c[\cdot]$ denotes the complementary error function of $[\cdot]$.

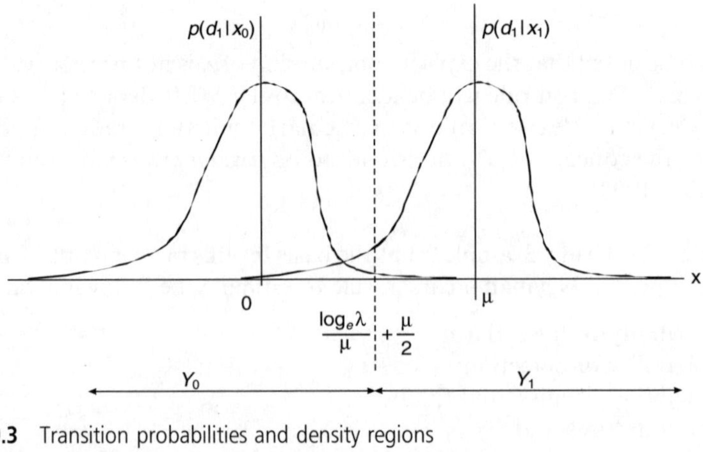

Figure 9.3 Transition probabilities and density regions

(b) The probability of detection:

$$
\begin{aligned}
P_D &= \int_{\frac{\log_e \lambda}{\mu}+\frac{\mu}{2}}^{\infty} \frac{e^{-\frac{(y-\mu)^2}{2}}}{\sqrt{2\pi}} dy \\
&= \operatorname{erf}c\left[\frac{\log_e \lambda}{\mu} - \frac{\mu}{2}\right]
\end{aligned}
\tag{9.39}
$$

(c) The probability of a miss:

$$
\begin{aligned}
\bar{P}_D &= 1 - P_D \\
&= 1 - \operatorname{erf}c\left[\frac{\log_e \lambda}{\mu} - \frac{\mu}{2}\right]
\end{aligned}
\tag{9.40}
$$

(d) The optimum threshold is obtained when the probability of false alarm equals the probability of a miss; that is, $P_{fa} = \bar{P}_D$. Hence, (9.38) = (9.40):

$$
\operatorname{erf}c\left[\frac{\log_e \lambda}{\mu} + \frac{\mu}{2}\right] = 1 - \operatorname{erf}c\left[\frac{\log_e \lambda}{\mu} - \frac{\mu}{2}\right]
\tag{9.41}
$$

In several radar applications, there is the tendency to have a pre-assigned value as acceptable threshold for which the false alarm probability value can be tolerated. If the desired threshold value is denoted by γ_{fa}, then a miss over and above the threshold can be attained:

$$
\gamma_{fa} = \frac{\log_e \lambda}{\mu} + \frac{\mu}{2}
\tag{9.42}
$$

With prior knowledge of γ_{fa} and μ, the Lagrange multiplier λ can be solved. However, if there is no preferred value of γ_{fa} the use of (9.41) is appropriate.

9.5 Minimum error probability rule

The *minimum error probability* (MEP) rule is often referred to as the ideal observer. For simplicity, the MEP rule is defined by using two case events x_0 and x_1 where the total error probability P_e is written as

$$
P_e = p(x_0)p(d_1 \mid x_0) + p(x_1)p(d_0 \mid x_1)
\tag{9.43}
$$

where $p(x_0)$ and $p(x_1)$ are the a *priori* probabilities of events x_0 and x_1 occurring respectively.

Following (9.21),

$$
p(d_0 \mid x_1) = 1 - p(d_1 \mid x_1)
\tag{9.44}
$$

Substituting (9.44) in (9.43), the total error probability can be written as

$$
\begin{aligned}
P_e &= p(x_0)p(d_1 \mid x_0) + p(x_1)[1 - p(d_1 \mid x_1)] \\
&= p(x_1) + \{p(x_0)p(d_1 \mid x_0) - p(x_1)p(d_1 \mid x_1)\}
\end{aligned}
\tag{9.45}
$$

The conditional probabilities $p(d_1 \mid x_0)$ and $p(d_1 \mid x_1)$ have been described as the probabilities of false alarm and detection in (9.16) and (9.19) respectively. The goal is to minimize the total probability of error. So, it is necessary to make the terms in the curly bracket $\{\cdot\}$ in (9.45) negative.

Mathematically, a decision rule can be instituted as follows. Substitute (9.16) and (9.19) in (9.45) and then consider only the $\{\cdot\}$ terms, which will be the integrand to be made negative.

$$\begin{aligned} \Lambda(y) &= \frac{p(y \mid x_1)}{p(y \mid x_0)} \underset{d_0}{\overset{d_1}{\gtrless}} \frac{p(x_0)}{p(x_1)} \\ &= \frac{p(x_1 \mid y)}{p(x_0 \mid y)} \underset{d_0}{\overset{d_1}{\gtrless}} 1 \end{aligned} \tag{9.46}$$

If the *a priori* probabilities are equal, that is, $p(x_0) = p(x_1)$, then the MEP decision rule coincides with the MLR.

Instead of using the initial, or prior, probability concerning the occurrence of some event as described above, a similar procedure can be obtained in terms of an amended, or posterior, probability. This procedure is known as the maximum *a posteriori* probability (MAP). This is left to the reader to verify, using the Bayes' theorem discussed in Chapter 8 as a guide.

The above discussion can be extended to situations where a decision between two hypotheses d_0 and d_1 is based on multiple observations, n. Let the successive measurements of several parameters or combinations thereof be denoted by $y_1, y_2, \ldots, y_n$. These observations can be described by the conditional density function $p(y_i \mid x_1)$ and $p(y_i \mid x_0)$ where $i = 1, 2, \ldots, n$. As indicated earlier in Chapter 8, multiple observations are easily expressed in vector format as $\mathbf{y} = [y_1, y_2, \ldots, y_n]^T$ where superscript T denotes transposition. In such a case, the decision rule can be written by considering the observations as a point in the n-dimensional space:

$$\begin{aligned} \Lambda(\mathbf{y}) &= \frac{p(\mathbf{y} \mid x_1)}{p(\mathbf{y} \mid x_0)} \\ &= \frac{p(y_1, y_2, \ldots, y_n \mid x_1)}{p(y_1, y_2, \ldots, y_n \mid x_0)} \underset{d_0}{\overset{d_1}{\gtrless}} \gamma_a \end{aligned} \tag{9.47}$$

If a bias value (threshold value) γ_{fa} is given as the acceptable probability of false alarm, then γ_a can be determined from

$$\int_{\gamma_a}^{\infty} p(\Lambda(\mathbf{y}) \mid x_0) d\mathbf{y} = \gamma_{fa} \tag{9.48}$$

Example 9.3 Estimate the minimum probability of error, given the following conditional probability densities:

$$
\begin{aligned}
p(y \mid x_1) &= \frac{e^{-\frac{(y-\mu_1)^2}{2\sigma^2}}}{\sqrt{2\pi}\sigma} \\
p(y \mid x_2) &= \frac{e^{-\frac{(y-\mu_2)^2}{2\sigma^2}}}{\sqrt{2\pi}\sigma}
\end{aligned} \tag{9.49}
$$

with $\sigma_1 = \sigma_2 = \sigma$ and $\mu_1 - \mu_2 \neq 0$.

Solution

Error decision:

$$
\Lambda(y) \underset{d_1}{\overset{d_2}{\gtrless}} \gamma_a \tag{9.50a}
$$

It is known that the sum of *a priori* probabilities is unity; that is,

$$
p(x_1) + p(x_2) = 1 \tag{9.50b}
$$

The *a priori* probabilities can thus be written in terms of the decision threshold γ_a:

$$
\begin{aligned}
p(x_1) &= \frac{\gamma_a}{1+\gamma_a} \\
p(x_2) &= \frac{1}{1+\gamma_a}
\end{aligned} \tag{9.51a}
$$

noting that

$$
\gamma_a = \frac{p(x_1)}{p(x_2)} \tag{9.51b}
$$

Taking the likelihood ratio of densities of (9.49):

$$
\begin{aligned}
\Lambda(y) = \frac{p(y \mid x_2)}{p(y \mid x_1)} &= \exp\left[\frac{(y-\mu_1)^2 - (y-\mu_2)^2}{2\sigma^2}\right] \\
&= \exp\left[\frac{2y(\mu_2-\mu_1) + \mu_1^2 - \mu_2^2}{2\sigma^2}\right]
\end{aligned} \tag{9.52}
$$

Taking $\log_e$ of both sides, and replacing the likelihood ratio with the decision threshold γ_a:

$$
y \underset{d_1}{\overset{d_2}{\gtrless}} \sigma\left\{\frac{\sigma \log_e \gamma_a}{\mu_2 - \mu_1} + \frac{1}{2}\left(\frac{\mu_2 + \mu_1}{\sigma}\right)\right\} \tag{9.53}
$$

Note that by factorization, $\mu_2^2 - \mu_1^2 = (\mu_2 - \mu_1)(\mu_2 + \mu_1)$. If the right-hand terms of (9.53) are replaced by λ, that is, $\lambda = \sigma\{(\sigma \log_e \gamma_a)/(\mu_2 - \mu_1) + (1/2)(\mu_2 + \mu_1)/\sigma\}$, then the next task is to define the error probabilities.

First, similar to the definition in (9.41), the false alarm probability is

$$P_{fa} = p(d_2|x_1) = \int_{\lambda}^{\infty} \frac{e^{-\frac{(y-\mu_1)^2}{2\sigma^2}}}{\sqrt{2\pi}\sigma} dy$$
$$= \text{erf}c\left[\frac{\lambda - \mu_1}{\sigma}\right] = \text{erf}c\left[\frac{\sigma \log_e \gamma_a}{\mu_2 - \mu_1} + \frac{1}{2}\left(\frac{\mu_2 - \mu_1}{\sigma}\right)\right] \tag{9.54}$$

The ratio $(\mu_2 - \mu_1/\sigma)$ is often referred to as the *signal-to-noise ratio* (SNR). Second, the miss probability is obtained as

$$\bar{P}_D = p(d_1 \mid x_2) = \int_{-\infty}^{\lambda} p(y \mid x_2) dy$$
$$= \text{erf}c\left[\frac{\mu_2 - \mu_1}{2\sigma} - \frac{\sigma \log_e \gamma_a}{\mu_2 - \mu_1}\right] \tag{9.55}$$

Third, given (9.51a), (9.54) and (9.55), estimate the total probability of error P_e from (9.43):

$$P_e = p(x_1)\text{erf}c\left[\frac{\mu_2 - \mu_1}{2\sigma} + \frac{\sigma \log_e \gamma_a}{\mu_2 - \mu_1}\right] + p(x_2)\text{erf}c\left[\frac{\mu_2 - \mu_1}{2\sigma} - \frac{\sigma \log_e \gamma_a}{\mu_2 - \mu_1}\right]$$
$$= \frac{\gamma_a}{1 + \gamma_a}\text{erf}c\left[\frac{\mu_2 - \mu_1}{2\sigma} + \frac{\sigma \log_e \gamma_a}{\mu_2 - \mu_1}\right] + \frac{1}{1 + \gamma_a}\text{erf}c\left[\frac{\mu_2 - \mu_1}{2\sigma} - \frac{\sigma \log_e \gamma_a}{\mu_2 - \mu_1}\right]$$
$$= \frac{1}{1 + \gamma_a}\left\{\gamma_a\text{erf}c\left[\frac{\mu_2 - \mu_1}{2\sigma} + \frac{\sigma \log_e \gamma_a}{\mu_2 - \mu_1}\right] + \text{erf}c\left[\frac{\mu_2 - \mu_1}{2\sigma} - \frac{\sigma \log_e \gamma_a}{\mu_2 - \mu_1}\right]\right\}$$
$$= \frac{1}{1 + \gamma_a}\left\{\gamma_a\text{erf}c\left[\frac{\text{SNR}}{2} + \frac{\log_e \gamma_a}{\text{SNR}}\right] + \text{erf}c\left[\frac{\text{SNR}}{2} - \frac{\log_e \gamma_a}{\text{SNR}}\right]\right\} \tag{9.56}$$

It is easily seen in this expression that the minimum probability of error is dependent on the numeric values of γ_a and SNR. Increasing the value of SNR for a given threshold γ_a reduces the probability of error P_e.

In summary, the *minimum error probability* (MEP) decides which is the most likely event for a set of observations particularly for events with the greater *a posteriori* probability. However, by deduction, the *maximum a posteriori probability* (MAP) decision rule ensures that comparison is made between the probabilities of the *causes*, having observed the *effects*. These decision rules (MAP and MEP) are directly suited to radar applications owing to the difficulties in evaluating the *a priori* probabilities.

9.6 Bayes minimum risk rule

The Bayes minimum risk rule is based on defining a cost for each conditioned decision. To choose an optimum decision rule, one must first assign costs to the decision through a cost function C_{ij}. Literally, C_{ij} is the cost in deciding d_i when x_j is true. A practical example is in radar detection problem in which the parameter x_j may be related to target position and velocity. A suggestion of a cost structure might be to assign higher premium to minimizing close or fast moving targets than missing slower, more distant targets.

For n observations, the Bayes risk, denoted by B, can be represented by the average cost for all decisions:

$$\begin{aligned} B &= \sum_{i=0}^{n-1}\sum_{j=0}^{n-1} C_{ij}p(x_j)p(d_i \mid x_j) \\ &= \sum_{i=0}^{n-1}\sum_{j=0}^{n-1} C_{ij}p(x_j)\int_Y p(y \mid x_j)dy \end{aligned} \tag{9.57}$$

Note that

$$\int_{Y_i} p(y \mid x_i)dy + \int_{Y_j} p(y \mid x_j)dy = 1 \tag{9.58}$$

If the number of observations is restricted to 2 for simplicity sake, the Bayes risk becomes

$$\begin{aligned} B &= C_{00}p(x_0) + C_{01}p(x_1) \\ &+ \int_{Y_1} \{[(C_{10} - C_{00})p(x_0)p(y \mid x_0)] - [(C_{01} - C_{11})p(x_1)p(y \mid x_1)]\}dy \end{aligned} \tag{9.59}$$

where C_{00} and C_{11} are direct costs for making incorrect and right decisions respectively. And the average cost can be written as

$$\begin{aligned} B_{av} &= E\{C_{ij}\} \\ &= E\{C_{ij} \mid x_0\}p(x_0) + E\{C_{ij} \mid x_1\}p(x_1) \\ &= B_0p(x_0) + B_1p(x_1) \end{aligned} \tag{9.60}$$

where B_0 and B_1 are called the *conditional costs*. Following (9.20) and (9.21) the conditional costs may be written as

$$\begin{aligned} B_0 &= C_{00} + (C_{10} - C_{00})p(d_1 \mid x_0) \\ B_1 &= C_{01} + (C_{01} - C_{11})p(d_1 \mid x_1) \end{aligned} \tag{9.61}$$

The goal is to minimize the average cost. Like the case of the *minimum error probability* (MEP) rule, the integrand in (9.59) needs to be made negative. So, the cost decision would be

$$\Lambda(y) = \frac{p(y \mid x_1)}{p(y \mid x_0)} \underset{d_0}{\overset{d_1}{\gtrless}} \frac{p(x_0)(C_{10} - C_{00})}{p(x_1)(C_{01} - C_{11})} \tag{9.62}$$

This expression is the hypothesis for which conditional risk is the minimum. As in the previous two decision rules, the Bayes decision rule also involves knowledge of the events' *a priori* probabilities. If there are no known direct costs associated with making incorrect or correct decisions, it would be appropriate to put $C_{00} = C_{11} = 0$ and $C_{01} = C_{10} = 1$. By choosing $C_{10} - C_{00} = C_{01} - C_{11}$ the Bayes decision rule of (9.62) will be the same as the 'maximum *a posteriori* probability' (MAP) decision rule:

$$\Lambda(y) = \frac{p(y \mid x_1)}{p(y \mid x_0)} \underset{d_0}{\overset{d_1}{\gtrless}} \frac{p(x_0)}{p(x_1)} \tag{9.63}$$

In summary, it may not be practical to directly apply the *Bayes minimum risk* rule to radar applications owing to the difficulties in evaluating the *a priori* probabilities and defining and/or obtaining the Bayes costs C_{ij}.

Example 9.4 Determine the Bayes rule associated with the following conditional probabilities

$$\begin{aligned} p(y \mid x_0) &= e^{-\frac{|y|}{2}} \\ p(y \mid x_1) &= e^{-2|y|} \end{aligned} \tag{9.64}$$

given the costs: $C_{11} = C_{00} = 0$; $C_{01} = 1 \quad C_{10} = 2$
and *a priori* probability: $P(x_1) = 0.75$.

Solution
The likelihood ratio:

$$\Lambda(y) = \frac{p(y \mid x_1)}{p(y \mid x_0)} = 2e^{-|y|} \tag{9.65}$$

From (9.62), consider the Bayes cost decision

$$2e^{-|y|} \underset{d_0}{\overset{d_1}{\gtrless}} \frac{p(x_0)(C_{10} - C_{00})}{p(x_1)(C_{01} - C_{11})} \underset{d_0}{\overset{d_1}{\gtrless}} \frac{0.25(2 - 0)}{0.75(1 - 0)} \tag{9.66}$$

in which the value of y is readily obtained as

$$|y| \underset{d_0}{\overset{d_1}{\gtrless}} -\log_e\left(\frac{0.5}{0.75}\right) = 1.10 \tag{9.67}$$

noting that $p(x_0) + p(x_1) = 1$. To estimate the Bayes risk, the false alarm and detection probabilities must be determined. Hence, the false alarm probability is

$$p(d_1 \mid x_0) = p(y \mid x_0) = \int_{-1.1}^{1.1} e^{\frac{-|y|}{2}} dy = 0.67 \tag{9.68}$$

And the detection probability is

$$p(d_1 \mid x_1) = p(y \mid x_1) = \int_{-1.1}^{1.1} e^{-2|y|} dy = 0.89 \tag{9.69}$$

By substituting (9.68) and (9.69) in (9.61) the following cost values are obtained

$$\begin{aligned} B_0 &= 1.34 \\ B_1 &= 0.11 \\ B &= 0.42 \end{aligned} \tag{9.70}$$

9.7 Summary

In this chapter, the basic criteria for developing decision rules have been discussed. The decision rules discussed are the Maximum Likelihood, Neyman–Pearson, Minimum Error Probability, and Bayes minimum risk decision rules. These decision rules are essentially a measure of the comparison between a function of observations, called likelihood ratio, with a suitable constant, whose value is a characteristic for each of the rules. The main difference between the decision rules is in the way the threshold is chosen. The suitability of each rule to radar problems was discussed.

Problems

1. Suppose that three sensors taken at random from a batch were found to have lifetimes of 2.8, 1.9 and 1.6 hours respectively. Write an expression of the sensors' likelihood function. Find the value that maximizes the likelihood function.

2. Consider a multiple hypothesis-testing problem for five known signals. All the signals are positive and are equally likely. Suppose the hypotheses are defined by

$$H_0: y = -2x + w$$
$$H_1: y = -x + w$$
$$H_2: y = w$$
$$H_3: y = x + w$$
$$H_4: y = 2x + w$$

where x is the signal and w denotes Gaussian noise of zero mean with variance σ^2. If the boundaries of the distributions are given as

$$H_0: -\infty < \alpha < -\frac{3x}{2}$$
$$H_1: -\frac{3x}{2} \leq \alpha < -\frac{x}{2}$$
$$H_2: -\frac{x}{2} \leq \alpha < \frac{x}{2}$$
$$H_3: \frac{x}{2} \leq \alpha < \frac{3x}{2}$$
$$H_4: \frac{3x}{2} \leq \alpha < \infty$$

where α is the abscissa of the hypotheses distribution curves. If the cost of a correct decision is 0, and the costs of all incorrect decisions are 1, determine which hypothesis to accept by choosing the H_i with the largest *a posterior* hypothesis. Can you extend your result to an M observation case?

3. Develop a computer program that tests a simple binary hypothesis problem with variable signal $s(t)$ and Gaussianly distributed noise $v_i(t)$ with variable mean and variance. If the hypothesis is defined by

$$\begin{aligned} H_0 &: y_i = s(t) + v_i(t) \\ H_1 &: y_i = v_i(t) \end{aligned} \qquad i = 1, 2, \ldots, n$$

4. If the cost of a correct decision is 0.12, and the costs of all incorrect decisions are 0.88 in Example 9.4, calculate the *a priori* probabilities.

10

Signal-peak detection

Typically the radar problem may be divided into two parts: detection of the presence of target and estimation of target parameters of interest; for example, target range, target speed, and target bearing. Detection will only be possible if adequate processing is done on the received signals.

Signal processing techniques used for conventional radar may differ from that of the skywave radar. However, there are some processing techniques that are commonly applicable to each of the radar systems. Signal processing is performed for the purpose of accomplishing certain functions. The functions include signal enhancement, clutter suppression or data conditioning, sidelobe suppression, radio frequency interference suppression, target detection or extraction, target classification estimation and imaging. Most of these processing techniques have been discussed in Chapters 1 and 3 (e.g. Fourier analysis, spectral correlation, weighting, and sidelobe suppression), and Chapter 7, section 7.2.5 (e.g. data conditioning, CFAR). Paradoxically, useful signals and noise have many common features, and to some extent, follow similar classification. What is discussed in this chapter, however, is the general description of radar signal processing operations that have not been discussed previously in section 7.2.5.

Decision-testing rules are useful tools in solving signal-peak detection problems. Detection of a signal in noise is a question of statistical hypothesis testing – already discussed in Chapter 9. Detection is an essential stage that lends itself to analysis; a subsequent objective is tracking – the formation of tracks. There is no way that tracking can be successfully performed unless the prior stages hand over a reliable input stream of detected peaks. Track formation on detected peaks is the central theme of discussion in Chapter 12. As already demonstrated in Chapter 9, as one attempts to reduce the probability of error (false alarm) one increases the likelihood of another signal being missed. The vesting question is: what can be done to maximize the chances of peak detection? The obvious suggestion will be to ensure that the signal-to-noise ratio is maximized at the receiver output; for instance, by enabling the input noise bandwidth to match with the signal noise bandwidth. This is achieved with matched filtering; more is said about matched filtering later in this chapter. Where there are fluctuations in target cross-sections, the matched filtering technique may not be adequate. Consequently, the probability density function and correlation properties over time must be known for the target and its trajectory. Unfortunately, these properties are usually difficult to obtain in a particularly target-scintillating case, but, with reasonable assumptions, it is possible to propose models that are credible and that closely represent physical characteristics of the target. Such models include those already discussed in Chapter 5, section 5.4, thresholding, already discussed in Chapter 9.

10.1 Signal processing

Signal processing plays a large part in radar operations, since signals contain information transmitted or propagated from sources to receivers, and they take different forms. It is computationally demanding because, as in the case of skywave radar, the received signal environment contains clutter that may originate as Earth surface backscatter or as Doppler-smeared ionospheric backscatter. The power of this clutter may distort target echoes by multiple folds. As such, a first distinction is between useful signals and noise – non-white or non-stationary phenomena. In reality, noise sources are always present. So any received signal will contain noise, and a significant part of signal processing operations is aimed at removing the noise. A significant process of reducing, or eliminating, noises and other biases had been discussed in Chapter 7, section 7.2.5. A brief discussion of some of the other techniques as well as expanding on others discussed previously follows.

10.1.1 Processes for detection

The method of signal processing depends on the prevailing environmental conditions under which the signal becomes available. A commonly used set of procedures includes preprocessing, prewhitening or data conditioning, and interpolation with smoothing.

10.1.1.1 Preprocessing

Preprocessing is a method of conditioning the signal into a form suitable for analysis. For instance, the presence of large amplitude, slowly fluctuating trends, which may prevent effective analysis of small rapid changes in the signal by restricting the usable dynamic range to a function of that of the trend (Beauchamp 1973). Preprocessing may include identification of the calibration signal, dominant clutter signatures, and calibration technique used and subsequent removal of such biases or trends. Some of the techniques used during signal preprocessing include ‘classical methods’ such as the average slope method, least squares method, decimation, truncation of record length, and reduction to zero mean.

Decimation is a process of data reduction involving the selection of, say, m samples of the digital data at uniformly spaced intervals throughout the data sequence. Such data reduction may be necessary because too high a rate of digitization may have been originally used. It could simply have been constrained by the limited resolution requirements of the analysis. The implication of this is demonstrated by the ensuing example.

Consider a sequence of data points x_i, spaced at equal interval p. Under this distribution, only constituent frequencies may be represented by up to $1/2p$ (Hz). If every mth point is retained, then the new sampling interval p' equals mp. In this instance, only frequencies up to $1/2mp$ (Hz) can be represented. Unless frequencies higher than this are filtered out from the data they will be effectively translated (aliased or folded) into the band 0 to $1/2mp$ (Hz) and thus distort the baseband signal. The concept of aliasing has been discussed in Chapter 1.

In essence, it is important to know the physical characteristics of the system under study, together with the preprocessing conditions and recording format, for a realistic interpretation of the results to be achieved.

10.1.1.2 Prewhitening or data conditioning

Prewhitening is a means of bringing the spectrum of the signal close to that of white noise; that is, rejecting any unwanted data from the signal before analysis starts. White noise is defined as having constant spectral density. By prewhitening, one attempts to make the rate of change of power spectral density with frequency relatively small. Hence, prewhitening is particularly valuable where intermodulation distortion is encountered.

The whitening technique does not restore the received signal-to-noise ratio (SNR) but prevents local false detections. Following the discussion in Chapter 7, impulsive phenomena, ionospheric biases, and other non-white or non-stationary phenomena can be located by their signatures, and be localized to one or more cells. Recognition always implies the excision of these biases, correlating their signatures against known templates, and then forming stable or smoothed estimates of the bias in each of the affected cells.

10.1.1.3 Data points interpolation (with smoothing)

Some interpolation may be necessary to compensate for the dispersion of signal energy over adjacent resolution cells when attempting to formulate stable or smoothed estimates. Interpolation involves weighting and summing of sets of three[1] adjacent values to form a new series of, say, N data points. This method is analogous to *low-pass* filtering, which represents a moving average form of the digital filter. A number of smoothing or interpolating algorithms are available for this purpose. A well-known example is that of Blackman's, which may be represented linearly as

$$y_i = \beta_0 y_{i-1} + \beta_1 y_i + \beta_2 y_{i+1} \tag{10.1}$$

[1] Depending on the level of accuracy expected, the set could be more than three using the Blackman–Harris constants in Table 1.2 of Chapter 1.

where βs are the weightings obtainable from the Blackman constants discussed in Chapter 1, Table 1.2, and y_i is the interpolate which weights adjacent data.

Filters are used in signal processing for a number of reasons; some of which are smoothing of data, event detection, bandwidth selection, bandwidth limitation, and signal-to-noise ratio enhancement. Filters are particularly useful in the preprocessing stage of peak detection. For further reading on filter design, the reader is advised to consult among many books Orfanidis (1996). Three broad types of techniques used in practice for signal enhancement are briefly discussed:

- Correlation technique, which is aimed at identifying and retrieving signals in noise – already discussed in Chapter 1, section 1.3.4.
- Filtering technique is used to reduce the effects of noise components that lie in a different part of the spectrum to the signal. This technique may be limited in its application if signal and noise components overlap in frequency.
- Coherent time averaging technique involves the summation of successive repetitions of a signal in such a way that the time signal reinforces itself, while the noise tends to cancel out. This technique includes other methods, such as the equal weight summation, sliding window average, and exponentially weighted running average.

10.2 Peak detection

After preprocessing, a threshold is applied to isolate the target returns from the residual power present in various range-angle-Doppler resolution cells. To detect peaks in the signal, a simple threshold test is applied to the signal mostly of variable amplitude by comparing a pre-assigned (threshold) value, say γ_a, to the magnitude, M, of the processed (whitened) data. As in Chapter 9, the decision rule may be written as

$$|M| \underset{no-target}{\overset{target}{\gtrless}} \gamma_a \tag{10.2}$$

This rule is interpreted with the aid of Figure 10.1 as follows. If the amplitude of the processed data exceeds γ_a, then a target is declared to be present. Otherwise a decision of no-target is made. Cases 1, 2, and 3 in Figure 10.1 are probably false targets, which could trigger a false alarm. The average noise floor is represented by Φ_0. The abscissa of Figure 10.1 could be range (in km), azimuth (in radians), or time (in seconds).

If the noise spectrum is characterized as Gaussian, and knowing threshold γ_{th}, then the

- probability of false alarm P_{fa}, that is, the probability of interference greater than the threshold,
- probability of detection P_D, the probability of signal plus interference greater than the threshold,

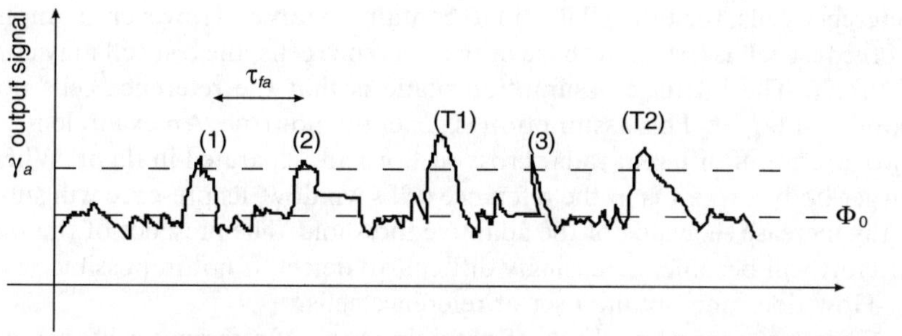

Figure 10.1 Signal response of a detector. Peaks (1), (2) and (3) are probably false, while (T1) and (T2) are target echo peaks

can be evaluated using the definitions given in Chapter 9 for these variables. The error probabilities are *signal-to-noise ratio* (SNR) dependent; demonstrated by equation (9.56). The average time τ_{fa} between false target peaks can be computed as

$$\tau_{fa} = \frac{1}{B_n P_{fa}} \tag{10.3}$$

where B_n is the noise bandwidth in Hertz.

In a real-time scenario the residual power is usually higher than those occurring in a noise-only environment. Even in a noise-only environment, as demonstrated by (9.56), the probability of error is dependent on the threshold as well as the SNR, and increasing the value of SNR for a given threshold γ_a could increase the false-alarm rate. If a constant threshold is applied, the false-alarm rate will increase and the saturation of the processor will result. To prevent saturation, a method of preventing the false-alarm rate increasing is mandatory. False-alarm rate increase is unacceptable because target detectability is significantly reduced, which is also unsatisfactory. Similar observations hold for the clutter-plus-noise environment, which is often the case in real life, and a method of preventing the false-alarm rate increase is again mandatory. Clearly, a form of adjustable, or adaptive, threshold will be required to monitor the residual power while maintaining *constant false-alarm rate* (CFAR) and increasing the chances of target detectability.

10.2.1 CFAR detection

CFAR is used in automatic detection systems to keep the false-alarm rate as the noise level at the receiver. It also prevents concealment of detectable targets by weaker clutter by maintaining the clutter output from the receiver at a constant value well below the saturation level of the display. The basic concept of a CFAR technique is that the amplitude of a test cell is compared to that of a set of reference cells. If the test cell is identical to those of the reference cells, the test cell is said to contain no target. However, if amplitude of the test cell is less than those of the reference cells, the test cell may contain a target. The intrinsic assumption made is that the reference cells do not contain a target. This assumption is generally not true. An example is where two aircraft of different radar cross-sections are separated in flight. When the larger body aircraft is in the reference cell's window, its presence will substantially increase the value of the adaptive threshold. The presence of the smaller aircraft will become increasingly difficult to detect, if not impossible.

How does one obtain a set of reference cells?

There are several methods of obtaining a set of reference cells; prominent among these methods include cell averaging and clutter mapping. However, the appropriate method employed, in a particular situation, depends on the type of propagation environment. The development of any of these methods is based on the assumption that the interference characteristics do not change over the periods the measurements are taken. It should be noted that there is no definitive method available that caters for all environmental conditions: this calls for the jurisprudence of the radar operators.

A common technique is the cell averaging CFAR, which computes an adaptive threshold to maintain a constant false-alarm rate, see Figure 10.2. The signal passes through a tapped shift register with the centre point being the sample of video under consideration at any instant. The centre tap is reduced by the greater of the average value of the taps proceeding or succeeding the centre point. The resultant signal should, in principle, be reduced to a near noise-like signal, which then passes through the threshold arrangement. The applicable threshold is set based on the preceding average value.

The clutter mapping technique uses data from previous sweeps to estimate the clutter power for every resolution cell. The clutter mapping can be

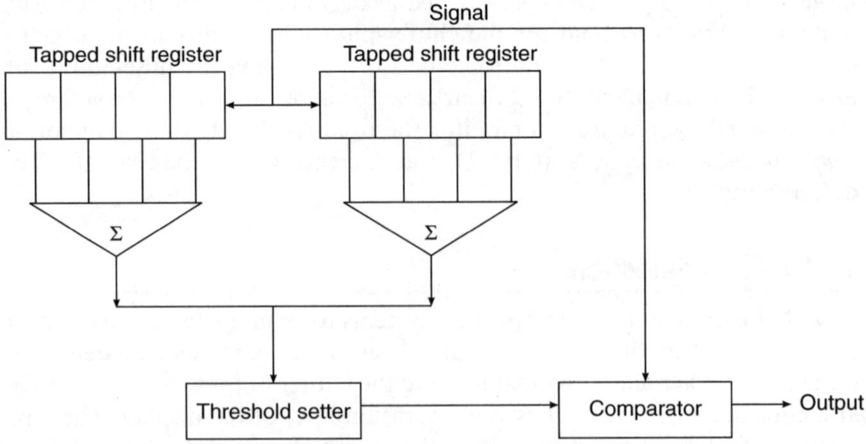

Figure 10.2 CFAR processor using cell averaging technique

instantaneous on-line processing, where the clutter floor can be estimated from the on-line scan-to-scan samples, using a suitable application of statistical principles. From the instantaneous map, the clutter intensities are sorted in a descending order. The sorting allows picturing of an instantaneous profile of the target environment and setting the interference bandwidth to achieve optimum filtering. As the target enters any of the resolution cells, an increase in the reflected power will indicate the presence of the target at that instant and allows its detection. The clutter map information is used to set the threshold for target tracking.

10.3 Matched filter

If the input noise bandwidth matches the noise bandwidth of the radar receiver, its performance will be optimum. This assertion can be investigated by the following example. Suppose a simple linear process can be represented by Figure 10.3.

If the input signal $s_i(t)$ can be defined within a finite time T, then the impulse of the optimum linear filter $h_{op}(t)$ can be estimated by running the signal backwards $x(-t)$ in time from the instant, t_m, at which the maximum signal-to-noise ratio (SNR) has occurred. This type of filter is normally referred to as *a matched filter*. This description is depicted by Figure 10.4. Mathematically, the output response $s_o(t)$ can be expressed by

$$s_o(t) = \left[s_i(t) \otimes h_{op}(t)\right] = \int_0^T s_i(\tau)x(\tau + t)d\tau \qquad (10.4)$$

This neatly expresses the relationships between convolution correlation and filtering. Zadek and Ragazzini (1952) gave the classical solution[2] that demonstrated a sufficient and necessary condition for obtaining the impulse response $h_{op}(t)$ of the optimum filter in the form

$$\int_0^T h_{op}(\tau)\delta(t+\tau)d\tau = \frac{N_0}{2}h_{op}(t) = ks_i(t_m + \tau) \qquad (10.5)$$

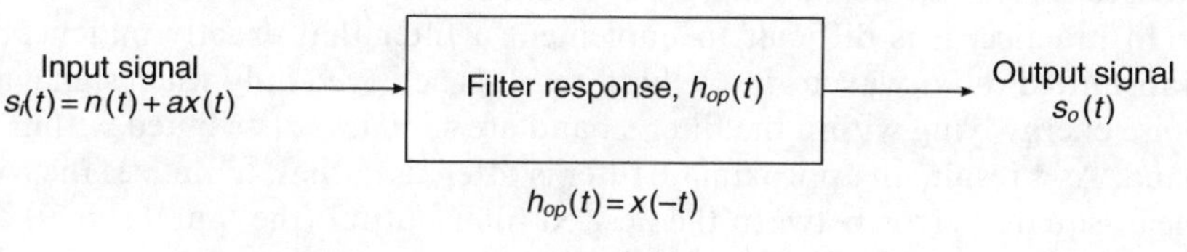

Figure 10.3 A representation of a matched filter

[2] A complete treatment of the solution can be found in their paper. Deductions are only drawn to explain how the optimum filter approach is used to solve the matched filtering problem and to draw upon theories developed in previous chapters.

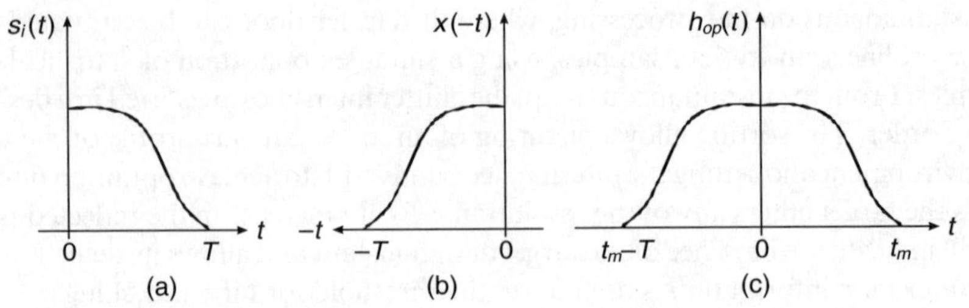

Figure 10.4 A schematic diagram of (a) signal $s_i(t)$ of finite period T run backwards as in (b) from an instant t_m with an impulse filter's response $h_{op}(t)$ as in (c)

where $\delta(t)$, N_0, and k are, respectively, the Dirac function, input-noise spectral power, and constant of proportionality. From this expression, an impulse response can be written as a function of input signal and noise:

$$h_{op}(t) = \frac{2k}{N_0} s_i(t_m + \tau) \tag{10.6}$$

By Fourier transformation, the optimum filter's frequency response can be expressed as

$$H_{op}(f) = \frac{2k}{N_0} S_i^*(f) e^{-j2\pi f t_m} \tag{10.7}$$

This expression is sometimes referred to as a conjugate filter, where superscript (*) indicates complex conjugation. Certain requirements need to be met when designing a matched filter: namely, that

- the filter must be at least as wide as the signal spectrum, otherwise it will reject signal and reduce SNR, and
- the frequencies where the signal is strong must be emphasized and conversely where the noise is strong the associated frequencies must be de-emphasized.

For a rigorous development of matched filter requirements, the reader is advised to consult Berkowitz (1965) and Cook and Bernfeld (1967).

In practice, it is difficult to implement a filter that exactly matches the transmitted radar waveform. A 'best' result is achieved only when signal and noise energy lying within the filter's band are similarly distributed within the band. As a result, an approximate filter is often used that minimizes the root-mean-squared error between the desired filter output (the 'pure' signal) and the actual output (filtered signal-plus-noise).

Example 10.1 Design a matched filter for a finite radio frequency pulse train, depicted by Figure 10.5. For simplicity, take the interpulse period T as a multiple of the pulse width and the pulse amplitude as unity.

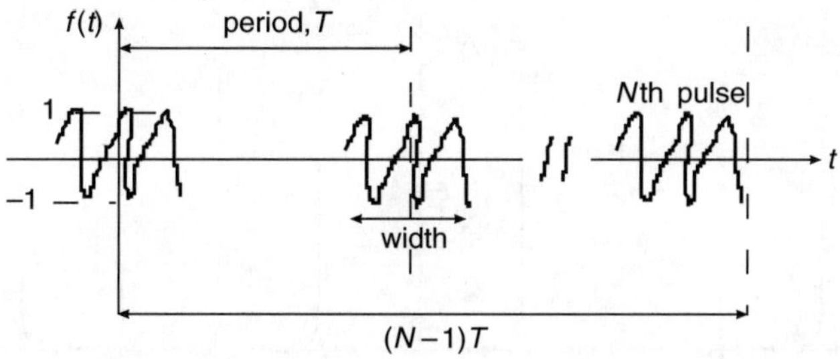

Figure 10.5 A finite pulse train

Solution

If the receiver received sequentially the pulse train, the 1st pulse of the train can be denoted by $f(t)$. Supposing the 1st pulse is taken as the phase centre, then the 2nd pulse train will be advanced by T and denoted by $f(t - T)$. The 3rd pulse will be advanced by $2T$ and denoted by $f(t - 2T)$. Following this thread, the Nth pulse train can be denoted as $f(t - (N - 1)T)$. So, the total pulse train received by the receiver is expressed by

$$f_T(t) = f(t) + f(t - T) + f(t - 2T) + \cdots + f(t - (N - 1)T) \qquad (10.8)$$

If the 1st pulse of the train can be Fourier transformed, and be represented by $F_\Delta(\omega)$, then the Fourier transform of all the pulse train can be expressed as

$$F_T(\omega) = F_\Delta(\omega)\left[1 + e^{-j\omega T} + e^{-j2\omega T} + \cdots + e^{-j\omega(N-1)T}\right] \qquad (10.9)$$

Multiplying (10.9) by $e^{-j\omega T}$ to yield

$$F_T(\omega)e^{-j\omega T} = F_\Delta(\omega)\left[e^{-j\omega T} + e^{-j2\omega T} + e^{-j3\omega T} + \cdots + e^{-j\omega NT}\right] \qquad (10.10)$$

This expression has a geometric progression with ratio $e^{-j\omega T}$. Now subtract (10.10) from (10.9) and divide by $1 - e^{-j\omega T}$ to get

$$F_T(\omega) = F_\Delta(\omega)\frac{1 - e^{-j\omega NT}}{1 - e^{-j\omega T}} \qquad (10.11)$$

Using known geometric series expansion

$$F_T(\omega) = F_\Delta(\omega)\frac{\sin\left(\frac{\omega NT}{2}\right)}{\sin\left(\frac{\omega T}{2}\right)}e^{-\frac{j}{2}\omega(N-1)T} \qquad (10.12)$$

A plot of (10.12) is shown in Figure 10.6. The response of the pulse train is comprised of large spikes resembling the teeth of a comb, whose centres are separated in frequency by $2\pi/T$. The matched filter of the pulse train is called a *comb filter*, whose transfer function $H_{op}(\omega)$ will be proportional to the complex conjugate of (10.12).

It is worth noting that a comparison of the pulse-train response (10.12) and the linear array of n isotropic radiators (of Chapter 4, equation (4.6a)),

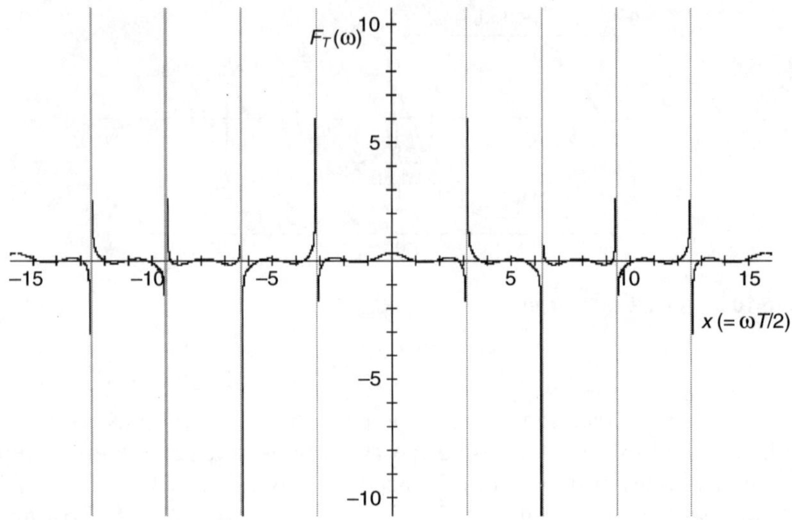

Figure 10.6 The response of a comb filter

shows that, for synthetic purposes, a linearly configured array of isotropic radiators can be represented by a comb filter approximation.

The idealized approximation to the matched filter, called the uniform comb filter, and its performance has been the subject of many studies. A typical classical performance analysis of a uniform comb filter has been given by George and Zamanakos (1954).

10.4 Summary

In this chapter, the basic extraction process for discerning the presence of target returns in a noisy environment has been discussed. This process is called *detection*. Detection is only possible if adequate processing is done on the received signals. The commonly used sets of procedures for peak detection, the concept of adaptive thresholding and CFAR as well as matched filtering were discussed.

An important application area of signal processing is target estimation and tracking, an area that forms the central theme of Part IV.

Problems

1. Consider an RF pulse of duration T (sec) and unity energy with a rectangular envelope described by

$$s(t) = \begin{cases} \sqrt{\frac{2}{T}}\cos\omega_c t & 0 \le t \le T \\ 0 & \text{elsewhere} \end{cases}$$

Design a filter that matches the pulse. What will happen if the pulse is sampled at a rate lower than ω_c/π?

2. During observation, the noise was seen to be completely uncorrelated between successive versions of the response. If the frequency bandwidth of the noise is the same of the signal, will there be an improvement in the signal-to-noise ratio? Under what condition is an improvement in the signal-to-noise ratio optimum?
3. Write a computer program that sequentially

   - samples and sorts radar data into descending order of intensities
   - selects the average values of background noise or clutter windows across the range-angle-Doppler resolution cells.

   The program should be robust to extract target peaks above threshold values. Observe what happens if more than one target enters the reference cell. Does the program need re-evaluation of the threshold value(s)? Explain your reasons.
4. Your radar system is tasked to function detect targets in air and sea environments. Will a simple threshold be sufficient for such operational environments? Why?
5. Why does the concept of matched filter occupy a rather central role in signal theory?

Part IV

Estimation and Tracking

Part IV is structured into two chapters – 11 and 12 – covering parameter estimation and radar tracking. Having discussed in the previous chapter the process of detecting the presence of a particular target signal, among other candidate signals in a noisy or clutter environment, attention now turns to how to estimate some characteristics of the target signal that is assumed to be present. This process ensures that the signal-reception problem is decoupled into two distinct domains: detection and estimation. Detection is the first type of optimization problem, which has been studied in Chapter 10. Estimation is the second type of optimization problem and exploits the several parallels with the decision theory of Chapter 9. Three estimation procedures are considered in Chapter 11, namely, maximum likelihood, *a posteriori*, and linear estimation.

Tracking is the central theme of Chapter 12 and it brings to the fore all the concepts discussed in previous chapters. For example, target tracking now turns the tentative decision statistics, discussed in Chapters 9 and 11, into more highly refined decision statistics. The probability theory discussed in Chapter 8 is expanded to solve the problem of uncertainty in track initiation and establishment as well as data association.

11

Parameter estimation and filtering

Systems are often described by equations in which the independent variable is time. A system may operate on discrete, or continuous, data with the defined equations either differential or difference in nature.

Parameter is a term used to name a scalar, or vector-valued, quantity. An *estimator* is a formula or a procedure for deriving from a sample or set of observations to generate an *estimate.* In essence, the estimator is the parallel of the decision rule discussed in Chapter 9, while the estimate is the parallel of the decision. Parameter estimation refers to the computation of the numerical values of the parameters, which invariably enter into system equations. For example, the detection of a target is generally followed by the estimation of related quantities, such as range, bearing, Doppler frequency or speed.

After the basic principles of parameter estimation, especially for a system acted upon by random inputs, have been introduced, the criteria used for selecting estimators are discussed. Following the two popular estimators, such as maximum likelihood and Bayesian, linear estimators are discussed. Linear estimators form the basis of linear filtering and prediction.

11.1 Basic parameter estimator

The process of estimation may be defined as a process of making a decision concerning the appropriate value of certain unknown parameters when the decision is influenced, or weighted, by all available information. As an illustration, let β be a parameter: a state vector unknown to an observer that one wishes to estimate using observations or data corrupted by *random error* or noise, $\in$. Let Y be a random variable such that $Y = \beta + \in$. The random error $\in$ is assumed to be an uncorrelated random variable with zero mean and known variance. Suppose a random sample $y_1, y_2, y_3, \ldots, y_n$ of size n has been taken. It would be natural to take an estimate of the unknown parameter β to be

$$\hat{y} = \frac{1}{n}\sum_{i=1}^{n} y_i \tag{11.1}$$

which is the arithmetic mean of Y. In this expression, the hat on 'y' means that the quantity is an estimate. It should be emphasized that this arithmetic mean $\hat{y}$ is also called an *estimate* of the parameter β.

In practical estimation problems, particularly as they applied to real-time tracking, the parameters are time dependent and the exact knowledge of samples distribution is not possible. In such situations, a less detailed description of the estimator is required preferably in terms of its lower-order moment. Also the sequence in which the measurements are observed must be preserved in finding the required estimate(s) of the system's parameters. The above illustration has been used to give some meaning to the parameter β that has sought an estimator $\hat{y}$, which has a similar meaning with respect to the sample.

11.1.1 Choice of an estimator

To choose between different estimators, it is important to define an optimality criterion. The philosophy of these estimation procedures is closely related to those discussed in Chapter 9. Basically the choice of an estimator and the design of its associated estimation procedure is partly dependent on the type of data, the availability of the data's *a priori* statistics and importantly a matter for the user. A wide variety of methods exists, which lead to a number of estimators such as the maximum likelihood, Bayes estimator, least squares, minimum or linear minimum variance, estimators *a posteriori* including *maximum a posteriori* (MAP) and recursive linear estimators. The most important of these is the mean square error criterion, which is discussed in section 11.4 as part of linear estimators. A class of estimators, which have smallest mean square errors for a large sample size, is the *maximum likelihood estimators*, to be discussed in section 11.2. Another class in which a prior knowledge amounts to less than a prior distribution of parameter values is called the estimators *a posteriori*, which is discussed in section 11.3.

11.2 Maximum likelihood estimator

The maximum likelihood estimator relates to choosing from among the possible values for the parameter; that is, the value that maximizes the probability of obtaining the sample that was obtained. As an illustration, suppose vector $\mathbf{x}$ contains a set of n independent random samples described by $(x_1, x_2, x, \ldots, x_n)$ and characterized by a probability density function denoted by $f_n(\mathbf{x}; \beta)$ where β is a parameter of the distribution. If L is introduced as a likelihood function, then the likelihood function of $\mathbf{x}$ can be expressed as

$$L(\mathbf{x}; \beta) = f_n(\mathbf{x}; \beta) \tag{11.2}$$

Some authors use L for the logarithm of likelihood. Since the samples are independent, the likelihood function can be written as

$$L(\mathbf{x};\beta) = f(x_1;\beta)f(x_2;\beta)f(x_3;\beta)\ldots f(x_n;\beta)$$
$$= \prod_{i=1}^{n} f(x_i;\beta) \tag{11.3a}$$

If the data set consists of discrete elements, (11.3a) is simply written as

$$L(\mathbf{x};\beta) = \prod_{i=1}^{n} p_i(\beta) \tag{11.3b}$$

where $p_i(\beta)$ is the probability associated with the ith sample.

The goal of maximum likelihood is selecting (estimating) $\hat{\beta}$ for β that will maximize L. Maximization implies differentiating L (or logarithmic function of L) with respect to the variable(s) to be estimated, in this case β, and equating the resultant to zero; i.e.,

$$\frac{\partial}{\partial\beta}\log_e(L) = 0 \tag{11.4}$$

Since $\log_e(L)$ is a monotonic function, it attains its maximum when L is a maximum. Equation (11.4) is usually called the *likelihood equation*. Any solution of $\hat{\beta}$ for β that satisfies (11.4) is called a *maximum likelihood estimate* of β. Naturally, as in algebra, a second derivative of (11.3) is sought to ensure that a maximum estimate has been obtained. If the second derivative is negative, a maximum has been obtained.

In essence, the maximum likelihood estimator is extremely useful because of its simplicity and requires a minimum amount of statistical information for its implementation.

Example 11.1 If a sample is normally distributed with mean μ and standard deviation, σ can be functionally defined by

$$f_n(x) = \frac{1}{\sigma^n\sqrt{(2\pi)^n}}\exp\left(-\sum_{i=1}^{n}\frac{(x_i-\mu)^2}{2\sigma^2}\right)$$

Find the maximum likelihood estimators of μ and standard deviation, σ, for distribution.

Solution

Define the distribution's likelihood function by

$$L(x) = \frac{1}{\sigma^n\sqrt{(2\pi)^n}}\exp\left(-\sum_{i=1}^{n}\frac{(x_i-\mu)^2}{2\sigma^2}\right) \tag{11.5}$$

It is easier to work with the natural logarithm of the likelihood than just L. So (11.5) can be written as

$$\log_e(L(x)) = -n\log_e \sigma - \frac{n}{2}\log_e(2\pi) - \sum_{i=1}^{n}\frac{(x_i-\mu)^2}{2\sigma^2} \quad (11.6)$$

Differentiating (11.6):

$$\frac{\partial \log_e(L)}{\partial \mu} = \frac{1}{\sigma^2}\sum_{i=1}^{n}(x_i-\mu) = 0 \quad (11.7a)$$

$$\frac{\partial \log_e(L)}{\partial \sigma} = \frac{1}{\sigma}\left\{-n + \frac{1}{\sigma^2}\sum_{i=1}^{n}(x_i-\mu)^2\right\} = 0 \quad (11.7b)$$

$$\frac{\partial^2 \log_e(L)}{\partial \mu^2} = -\frac{n}{\sigma^2} = 0 \quad (11.7c)$$

$$\frac{\partial^2 \log_e(L)}{\partial \sigma^2} = \frac{1}{\sigma^2}\left\{-n + \frac{3}{\sigma^2}\sum_{i=1}^{n}(x_i-\mu)^2\right\} = 0 \quad (11.7d)$$

The mean estimate $\hat{\mu}$ of x_i and its standard deviation estimate $\hat{\sigma}$, which make the first derivative equal to zero, maximize $\log_e(L)$. So, from (11.7a) and (11.7b), the maximum likelihood estimates of μ and σ can be obtained as follows:

$$\underbrace{\sum_{i=1}^{n}\mu}_{=n\hat{\mu}} = \sum_{i=1}^{n}x_i \quad (11.8a)$$

Hence,

$$\hat{\mu} = \frac{1}{n}\sum_{i=1}^{n}x_i \quad (11.8b)$$

From (11.7b),

$$\hat{\sigma} = \sqrt{\frac{1}{n}\sum_{i=1}^{n}(x_i-\hat{\mu})^2} \quad (11.9)$$

Since the second derivative of μ is always negative, the second derivative of σ at the point where the two first derivatives are zero is

$$\frac{n}{\hat{\sigma}^2} - \frac{3n\hat{\sigma}^2}{\hat{\sigma}^4} = -\frac{2n}{\hat{\sigma}^2} \quad (11.10)$$

which again is negative; hence the maximum likelihood estimators.

Before leaving the maximum likelihood estimation problem, one of the issues in the radar estimation problem is the multiple nature of the observations. How then does one obtain an optimum estimate from multi-dimensional data? This approach is discussed next.

11.2.1 Maximum likelihood estimators of multiple observations

Suppose a linear function $L = L(\mathbf{x}; \beta_i)$ has joint probability density functions that depend on unknown parameters $\beta_i = (\beta_1, \beta_2, \beta_3, \ldots, \beta_p)$. The estimate of each of the unknown parameters is a linear function of each of the unknown estimates. For example,

$$\begin{aligned}\hat{\beta}_1 &= \beta_1(x_1, x_2, x_3, \ldots, x_n)\\ \hat{\beta}_2 &= \beta_2(x_1, x_2, x_3, \ldots, x_n)\\ &\vdots\\ \hat{\beta}_p &= \beta_p(x_1, x_2, x_3, \ldots, x_n)\end{aligned} \tag{11.11}$$

The function of $\mathbf{x}$ can then be maximized whenever $\beta_i = (\beta_1, \beta_2, \beta_3, \ldots, \beta_p)$ is replaced by $\hat{\beta} = (\hat{\beta}_1, \hat{\beta}_2, \hat{\beta}_3, \ldots, \hat{\beta}_p)$. Hence $\hat{\beta}$ is called the maximum likelihood estimator of β_i.

As in Example 11.1, there is a unique set of $\hat{\beta}$ that maximizes the likelihood function L via partial differentiation of the function(s). An illustration is given as follows.

Example 11.2 Consider a random, Gaussianly distributed function $\mathbf{x}$ of order p with mean μ and variance v. If the likelihood function can be expressed by

$$L(\mathbf{x}) = \frac{1}{(2\pi)^{\frac{np}{2}} v^{\frac{n}{2}}} \exp\left\{-\frac{1}{2}\sum_{i=1}^{n}(x_i - \mu)^T v^{-1}(x_i - \mu)\right\} \tag{11.12}$$

where superscript T implies transposition, then estimate the parameters $\hat{\mu}$ and $\hat{v}$ that maximize the likelihood function.

Solution

The likelihood function can be maximized by setting the partial derivative of equation (11.12) with respect to the variables, namely μ and v, to be estimated to zero to obtain their maximum likelihood estimate(s). Specifically,

$$\begin{aligned}\frac{\partial \log_e(L)}{\partial\mu} &= -\frac{np}{2}\log_e(2\pi) - \frac{n}{2}\log_e|v| - \frac{1}{2}\sum_{i=1}^{n}(x_i - \mu)^T v^{-1}(x_i - \mu)\\ &= -\frac{np}{2}\log_e(2\pi) - \frac{n}{2}\log_e|v| - \frac{1}{2}\sum_{i=1}^{n}(x_i - \hat{x})^T v^{-1}(x_i - \hat{x})\\ &\quad -\frac{n}{2}(\hat{x} - \mu)^T v^{-1}(\hat{x} - \mu)\end{aligned} \tag{11.13}$$

It should be noted that logarithmic function is maximized if the quadratic form is positive definite

$$\frac{n}{2}(\hat{x}-\mu)^T v^{-1}(\hat{x}-\mu)=0 \tag{11.14}$$

providing $n \neq 0$, $v \neq 0$, or $(\hat{x}-\mu)=0$ implying that $\hat{x}=\mu$. Consequently, if $\hat{x}=\hat{\mu}$ is the maximum likelihood estimator of μ, it may be deduced that the maximum likelihood estimate of function $\mathbf{x}$ is $\hat{\mu}$.

The next stage is to find the estimator $\hat{v}$. This involves taking the partial derivative of (11.12) with respect to v and equating the result to zero; that is,

$$\frac{\partial \log_e(L)}{\partial v} = -\frac{np}{2}\log_e(2\pi) - \frac{n}{2}\log_e|v| - \frac{1}{2}\mathrm{tr}\left\{ v^{-1}\sum_{i=1}^{n}(x_i-\hat{x})^T(x_i-\hat{x})\right\}$$
$$-\frac{n}{2}(\hat{x}-\hat{\mu})^T v^{-1}(\hat{x}-\hat{\mu}) \tag{11.15}$$

where $\mathrm{tr}\{\cdot\}$ denotes the *trace* of $\{\cdot\}$. The trace of a square matrix is by definition the sum of its diagonal matrix. As demonstrated above for the case of estimator $\hat{\mu}$, if $\hat{x}=\hat{\mu}$ then the right-hand term of (11.15), that is, $-(n/2)(\hat{x}-\hat{\mu})^T v^{-1}(\hat{x}-\hat{\mu})$, must be equal to zero. Thus, the problem now is that of maximizing the remainder of (11.15); that is,

$$-\frac{np}{2}\log_e(2\pi) - \frac{n}{2}\log_e|v| - \frac{1}{2}\mathrm{tr}\left\{ v^{-1}\sum_{i=1}^{n}(x_i-\hat{x})^T(x_i-\hat{x})\right\} = 0 \tag{11.16}$$

The variance v is maximized if $\hat{x}=\hat{\mu}$ and

$$\hat{v}=\frac{1}{n}\sum_{i=1}^{n}(x_i-\hat{x})^T(x_i-\hat{x}) \tag{11.17}$$

Thus the maximum likelihood estimator of v is $\hat{v}$.

11.3 Estimators a *posteriori*

If a prior knowledge of the value or values a set of parameters is likely to have is known, one would be prepared to obtain better estimates by using the available information. If the parameter were a random variable, a value of which has been selected in agreement with its distribution, this value being an unknown constant throughout the experiment, the use of Bayes' theorem would enable the result of the experiment to be obtained that incorporates prior information. In a situation where a prior knowledge is insufficient, it may be useful to form one or two hypothetical prior distributions and see what estimators are suggested by the distributions.

For instance, suppose a prior distribution of parameters $\mathbf{x}$ is available. If the prior distribution has a density function $g(\theta)$ and the conditional probability density function of observations given θ is

$f(\mathbf{x} \mid \theta) = f(x_1, x_2, x_3, \ldots, x_n|\theta)$, then by Bayes' theorem the posterior density function of $\mathbf{x}$ may be written as

$$g(\theta \mid x_1, x_2, x_3, \ldots, x_n) = \frac{g(\theta)f(x_1, x_2, x_3, \ldots, x_n \mid \theta)}{\int_{-\infty}^{\infty} g(u)f(x_1, x_2, x_3, \ldots, x_n \mid u)du} \qquad (11.18)$$

If a prior knowledge of the probability function of the parameter is known, a sum will appear in place of the integral. For some applications of (11.18), it is convenient to note that θ appears only in the numerator of the right hand of the equation. So, a suitable *a priori* is found among functions $g(\theta)$ that is proportional to the likelihood function. In such a case, $g(\theta \mid x_1, x_2, x_3, \ldots, x_n)$ will be proportional to a possible likelihood function.

Example 11.3 Suppose a family of Bernoulli distribution is to be investigated, which is likely to be near 2/3 with the probability density falling off to zero at $\theta = 0$ and $\theta = 1$, and with an expected value of about 0.6. A prior probability density function is given as $12\theta^2(1 - \theta)$. For x successes in n trials, find the posterior density function.

Solution

Given a prior probability density function:

$$g(\theta) = 12\theta^2(1 - \theta) \quad 0 \leq \theta \leq 1 \qquad (11.19)$$

The conditional probability density function $\mathbf{x}$ for n trials given θ can be written as

$$f(x_1, x_2, x_3, \ldots, x_n \mid \theta) = \theta^x(1 - \theta)^{n-x} \quad x = 0, 1, 2, \ldots, n \qquad (11.20)$$

Using (11.19) and (11.20), the denominator of (11.18) can be expressed:

$$\int_0^1 12\theta^2(1 - \theta)\theta^x(1 - \theta)^{n-x}d\theta = \int_0^1 12\theta^{x+2}(1 - \theta)^{n-x+1}d\theta \qquad (11.21)$$

This is a classical integral problem, which has a known solution in gamma $\Gamma(\cdot)$ form:

$$\int_0^1 12\theta^{x+2}(1 - \theta)^{n-x+1}d\theta = \frac{12\Gamma(x + 3)\Gamma(n - x + 2)}{\Gamma(n + 5)} \qquad (11.22)$$

Thus, the posterior density function:

$$g(\theta \mid x_1, x_2, x_3, \ldots, x_n) = \frac{\Gamma(n + 5)}{12\Gamma(x + 3)\Gamma(n - x + 2)} 12\theta^{x+2}(1 - \theta)^{n-x+1} \qquad (11.23)$$

A plot of (11.23) is shown in Figure 11.1 for $x = \theta = 2/3$.

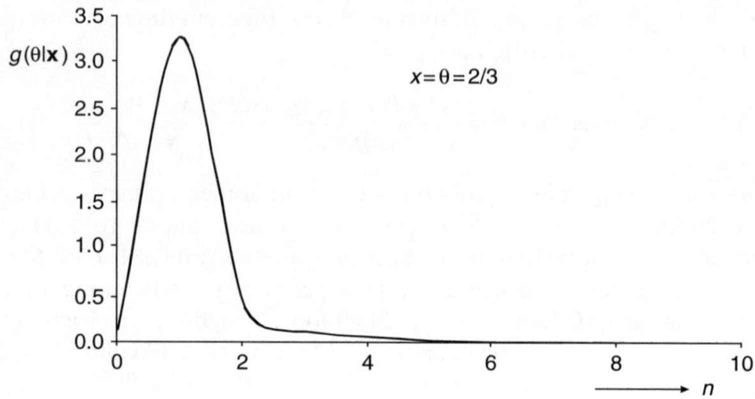

Figure 11.1 The posterior density function of equation (11.23)

11.4 Linear estimators

Linear estimators are recursive, or sequential, estimators. These estimators use procedures for each new observation in time, be it discrete or continuous, to refine the previous estimate(s). In them the latest measurement (or observation) is approximately weighted to determine its contribution to the estimate. The contribution is then combined with the previous estimate to yield an updated estimate. The procedure is repeated for each new data point of the observed sequence. For the case of continuous estimation, where the observations are a continuous function of time of the estimates, the corresponding estimates are updated continuously.

A simple example of recursive estimation is explained as follows. Consider the estimate of a scalar constant x from noise corrupted observations y_i, where

$$y_i = x + v_i \quad i = 0, 1, 2, \ldots, N-1 \tag{11.24}$$

The notation v_i is the noise measurement on the observations, assumed to be uncorrelated random variables with zero mean and variance σ_v^2. For N observations, an unbiased minimum variance estimate of scalar constant x is the average value of the measurements:

$$\hat{x}_N = \frac{1}{N}\sum_{i=0}^{N-1} y_i \tag{11.25}$$

The expectance of $\hat{x}_N$ is x; that is, $E\{\hat{x}_N\} = x$. If a new observation y_N is made, the new estimate of x can be expressed as

$$\begin{aligned} \hat{x}_{N+1} &= \frac{1}{N+1}\sum_{i=0}^{N} y_i \\ &= \frac{1}{N+1}\left[y_N + \sum_{i=0}^{N+1} y_i\right] \\ &= \frac{1}{N+1}[y_N + N\hat{x}_N] \end{aligned} \tag{11.26}$$

This expression may be rearranged as

$$\begin{aligned} \hat{x}_{N+1} &= \frac{N\hat{x}_N}{N+1} + \frac{y_N}{N+1} \\ &= \hat{x}_N - \frac{\hat{x}_N}{N+1} + \frac{y_N}{N+1} \\ &= \hat{x}_N + \frac{1}{N+1}(y_N - \hat{x}_N) \end{aligned} \tag{11.27}$$

This is the *recursive linear estimator* for the scalar constant x. Equation (11.27) shows that a new estimate is given by the prior estimate plus an appropriately weighted contribution of the difference between it and the most recent measurement y_N. Following from (11.27) if an element called *error* (or residual) denoted by 'e' can be introduced, then the difference between the estimate and the actual can equally be defined as

$$e_N = \hat{x}_N - x \tag{11.28}$$

The ultimate objective is to make the absolute value of the error as small as possible. In view of (11.27), expression (11.28) may be written, in terms of measurement noise, as

$$e_{N+1} = e_N + \frac{1}{N+1}(v_N - e_N) \tag{11.29}$$

The $(N+1)$th error variance P_{N+1}, may be written as

$$P_{N+1} = \frac{N^2}{(N+1)^2}P_N + \frac{\sigma_v^2}{(N+1)^2} \tag{11.30}$$

where the components of error are considered to have equal dispersion σ_v^2; that is,

$$P_N = E\left[e_N^2\right] = \sigma_v^2 \tag{11.31}$$

Equation (11.30) is the *propagation equation* of the variance of the estimator.

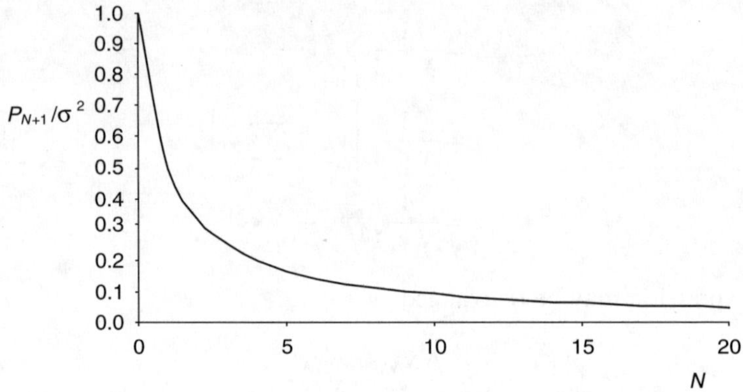

Figure 11.2 Normalized propagation equation

The initial condition is when $N = 0$. From (11.30), when $N = 0$, $P_1 = \sigma_v^2$. Recursively, the variance propagation equation for successive iterations $(n \geq 0)$ can be written as

$$P_{N+1} = \frac{\sigma_v^2}{(N+1)} \tag{11.32}$$

The normalized propagation equation of (11.32) is plotted in Figure 11.2.

Following the above simple linear scalar estimator discussion, vector notations can be introduced to the estimation procedure. The vector notations are used for convenience and in subsequent discussions.

Define a random vector $\mathbf{x}$ whose estimate $\hat{\mathbf{x}}$ is sought. If the error vector is denoted by $\mathbf{e}$, like (11.28), the error vector as the difference between the estimate and the actual can be defined as

$$\mathbf{e} = \hat{\mathbf{x}} - \mathbf{x} \tag{11.33}$$

If the expectance of the error vector is zero; that is,

$$E(\mathbf{e}) = \mathbf{0} \tag{11.34a}$$

or the expectance of the estimate to be equal to the actual; i.e.

$$E(\hat{\mathbf{x}}) = \mathbf{x} \tag{11.34b}$$

then $\hat{\mathbf{x}}$ is said be an *unbiased* estimate of $\mathbf{x}$.

Like (11.31), the covariance error matrix can be expressed as

$$P = E\left[\mathbf{e}\mathbf{e}^T\right] = E\left[(\hat{\mathbf{x}} - \mathbf{x})(\hat{\mathbf{x}} - \mathbf{x})^T\right] \tag{11.35}$$

An estimator designed in this fashion is called the *minimum mean square estimator*. This estimator is unbiased if the covariance error matrix equals

$$P = E\left[(\hat{\mathbf{x}} - E(\hat{\mathbf{x}}))(\hat{\mathbf{x}} - E(\hat{\mathbf{x}}))^T\right] \tag{11.36}$$

The objective is to minimize the error matrix, P. The diagonal terms of P are the variances of the components of the estimate. For this reason, the estimator is called the *minimum variance unbiased estimator*. If the estimates are obtained by performing linear operations on the measurements, the estimator is called the *linear minimum variance unbiased estimator*.

Perhaps the most important part of studying a radar-tracking problem is to determine a good model that reasonably describes the target dynamics and orientation. Models frequently take on the form of dynamic systems. A dynamic system is a mathematical description of a quantity that evolves over time. A brief introduction to a dynamic system is discussed in the next section. This is intended to enhance the capability of the reader to follow the development of prediction and filtering techniques as well as tracking algorithms.

11.4.1 An overview of a dynamic system

Assume that a class of single-input, single-output dynamic system can be described by an nth order of ordinary differential equation

$$\frac{d^n y}{dt^n} + a_{n-1}\frac{d^{n-1}y}{dt^{n-1}} + a_{n-2}\frac{d^{n-2}y}{dt^{n-2}} + \cdots + a_1\frac{dy}{dt} + a_0 y = u(t) \tag{11.37}$$

This expression can be reduced to the form of n first-order state equations by redefining the differentials as follows

$$x_1 = y, \quad x_2 = \frac{dy}{dt}, \quad x_3 = \frac{d^2y}{dt^2}, \quad \ldots, \quad x_n = \frac{d^{n-1}y}{dt^{n-1}} \tag{11.38}$$

Consequently, the $n-1$th first-order differential equations may be expressed as

$$\dot{x}_1 = x_2, \quad \dot{x}_2 = x_3, \quad \cdots, \quad \dot{x}_{n-1} = x_n, \quad \dot{x}_n = \frac{d^n y}{dt^n} \tag{11.39}$$

In view of (11.38) and (11.39), the expression (11.37) is concisely rewritten as

$$\dot{x}_n = -a_0 x_1 - a_1 x_2 - a_2 x_3 - \cdots - a_{n-1} x_n + u(t) \tag{11.40a}$$

$$y = x_1 \tag{11.40b}$$

which in matrix terms

$$\dot{\mathbf{x}} = \underbrace{\begin{bmatrix} 0 & 1 & 0 & \ldots & 0 \\ 0 & 0 & 1 & \ldots & 0 \\ \vdots & \vdots & \vdots & \ldots & \vdots \\ 0 & 0 & 0 & \ldots & 1 \\ -a_0 & -a_1 & -a_2 & \ldots & -a_{n-1} \end{bmatrix}}_{\Phi} \mathbf{x} + \underbrace{\begin{bmatrix} 0 \\ 0 \\ 0 \\ \ldots \\ 1 \end{bmatrix}}_{B} \mathbf{u} \tag{11.41a}$$

$$\mathbf{y} = \underbrace{[1 \quad 0 \quad 0 \quad \ldots \quad 0]}_{H} \mathbf{x} \tag{11.41b}$$

Concisely further as

$$\dot{\mathbf{x}} = \Phi\mathbf{x} + B\mathbf{u} \tag{11.42a}$$

$$\mathbf{y} = H\mathbf{x} \tag{11.42b}$$

Equation (11.42a) is the 'state space' representation of a *continuous-time* linear stochastic system, where $\mathbf{x}$ is the state vector, $\dot{\mathbf{x}}$ its time derivative, $\mathbf{u}$ the input disturbance or process noise, and Φ and B are matrices. Equation (11.42b) is the 'measurement' vector of the system.

A solution to (11.42a) can be found by multiplying both sides of (11.42a) by $e^{-\Phi t}$, and integrating it between t_0 and t, to obtain

$$\mathbf{x}(t) = e^{\Phi(t-t_0)}\mathbf{x}(t_0) + \int_{t_0}^{t} e^{\Phi(t-\tau)} B\mathbf{u}(\tau)d\tau \tag{11.43}$$

This expression is the sum of the motions due to the initial condition and those due to the forcing function. The motion due to the forcing function depends on B. If $\mathbf{u}(\tau) = \mathbf{u}(t_0) =$ constant, the integrand component of (11.43) becomes

$$\int_{t_0}^{t} e^{\Phi(t-\tau)} B\mathbf{u}(\tau)d\tau = B(t - t_0) \tag{11.44}$$

If the input variable is maintained constant within a time interval, $t_1 - t_0 = t_2 - t_1 = \cdots = t_{k+1} - t_k = T$, and the fundamental matrix notation is defined by $A(T) = e^{\Phi T}$, then in view of (11.44), the discrete form of the state equation (11.43) may be written as

$$\mathbf{x}((k+1)T) = A(T)\mathbf{x}(kT) + B(T)\mathbf{u}(kT) \tag{11.45}$$

This expression holds so long as the interest is only in the solution at the instants of sampling.

If a white noise is assumed with zero mean for the forcing function, its covariance matrix can be defined as

$$E\left[\mathbf{u}(k)\mathbf{u}^T(j)\right] = \mathbf{Q}(k)\delta_{kj} \tag{11.46}$$

where δ_{kj} is the Kronecker symbol, which equals to unity when $k = j$, and zero otherwise. In a simplified index-only time notation, the discrete-time dynamic model of (11.45) is written as

$$\mathbf{x}(k+1) = A\mathbf{x}(k) + B\mathbf{Q}(k) \tag{11.47}$$

In real life, a system's measurement components contain some noise. If a random measurement noise vector, $\mathbf{v}$, with zero mean is introduced to the output equation (11.42b), the discrete-time measurement equation can be written as

$$\mathbf{y}(k) = H\mathbf{x}(k) + \mathbf{R}(k) \tag{11.48a}$$

where the covariance matrix **R**; defined by

$$E\left[\mathbf{v}(k)\mathbf{v}^T(j)\right] = \mathbf{R}(k)\delta_{kj} \tag{11.48b}$$

Since the sampling interval is held constant, matrices A and H do not depend on k. If the covariances **Q** and **R** are also independent of the sampling interval k, then the discrete-time system will be completely time invariant.

By iterative process, it is easy to find the solution to the discrete equations (11.47) and (11.48a). For example, assume that **Q** and **R** are known, the solution to the input states can be found iteratively when sampled at discrete time intervals, such as

$$k = 0; \quad \mathbf{x}(1) = A\mathbf{x}(0) + B\mathbf{Q}(0) \tag{11.49a}$$

$$\begin{aligned} k = 1; \quad \mathbf{x}(2) &= A\mathbf{x}(1) + B\mathbf{Q}(1) \\ &= A^2\mathbf{x}(0) + AB\mathbf{Q}(0) + B\mathbf{Q}(1) \end{aligned} \tag{11.49b}$$

Following same procedure, the kth term can be written as

$$\mathbf{x}(k) = A^k\mathbf{x}(0) + \sum_{i=1}^{k} A^{k-1}B\mathbf{Q}(i-1) \tag{11.49c}$$

Substituting (11.49c) in (11.48a), the solution to the output variable can be written as

$$\mathbf{y}(k) = HA^k\mathbf{x}(0) + \left[H\sum_{i=1}^{k} A^{k-1}B\mathbf{Q}(i-1)\right] + \mathbf{R}(k) \tag{11.49d}$$

For an unbiased state vector the initial value is set at $\mathbf{x}(0) = \mathbf{0}$.

If **Q** and **R** are not known, the components of these vectors may be considered to have equal dispersion and hence

$$\begin{aligned} \mathbf{Q}(k) &= \sigma_x^2 I \\ \mathbf{R}(k) &= \sigma_y^2 I \end{aligned} \tag{11.49e}$$

where

I = the identity matrix
σ_x^2 = system noise (plant) variance
σ_x^2 = measurement noise variance.

With the preceding explanation, a system state's algorithm can be expressed for discrete-time state and measurement equations. The state vectors, in practical terms, may comprise several kinematic variables; for example, range, azimuth or bearing, velocity (Doppler or range rate), and elevation or direction cosines. For the case of over-the-horizon radar (OTHR), the measurements of interest are often slant range, azimuth, Doppler and signal-to-noise ratio (SNR) because of a peak selection and interpolation process, which is intended to compensate for the dispersion of signal energy over adjacent resolution cells. Thus, it could be said that the order of the state vector depends on the target kinematic variables of interest.

The theory of the dynamic system is rich and fascinating; it is only employed in this section to explain the linear estimation theory in the particular form of the Kalman estimator, which is discussed next.

11.4.2 Kalman estimator

Originally, the Kalman estimator was designed as an optimal Bayesian technique to estimate state variables at a time $t + \Delta t$ from indirect noise measurements at time t, assuming that the statistical correlation between variables and time is known (Gauss 1963). The previous sections have provided the components necessary to develop a mathematical model for the Kalman filter. To summarize, suppose that the model for the discrete dynamic system is defined recursively, step by step, by *state equations* given by (11.47):

$$\mathbf{x}(k+1) = A\mathbf{x}(k) + B\mathbf{u}(k) \tag{11.50}$$

with $\mathbf{x}(k+1)$ the variables at step $(k+1)$, A (the system transition matrix), B (the matrix that relates external input noise $\mathbf{u}(k)$ to the state $\mathbf{x}(k)$), and $\mathbf{u}(k)$ (the system process noise vector) with step k. For simplicity, the process noise has been assumed to be a Gaussian random vector with zero mean and covariance matrix Q. The Gaussian random signals are assumed to remain Gaussian after passing through a linear system. The state transition specifies how a form of the state is transformed into another as time passes. The state equations $\mathbf{x}(k)$ are linearly related to measurements $\mathbf{y}(k)$ by other recursive equations, called *measurement equations*, given by (11.48a)

$$\mathbf{y}(k) = H\mathbf{x}(k) + \mathbf{v}(k) \tag{11.51}$$

with H (the measurement transition matrix) and $\mathbf{v}$ (observation or measurement noise vector) with step k. The measurement noise vector $\mathbf{v}$ is still Gaussian with zero mean and covariance matrix, R.

If the optimal, unbiased estimate of the system state is $\hat{\mathbf{x}}$ and the estimate's corresponding covariance $P(k)$ is defined from the error vector like (11.36) as

$$P = E\left[(\hat{\mathbf{x}} - E(\hat{\mathbf{x}}))(\hat{\mathbf{x}} - E(\hat{\mathbf{x}}))^T\right] \tag{11.52}$$

Given the dependence of prediction on available observation, what the Kalman filter computes is the best, linear, unbiased estimate of $\mathbf{x}$ at time k given measurements $\mathbf{y}_0, \mathbf{y}_1, \mathbf{y}_2, \ldots, \mathbf{y}_k$. This statement introduces 'conditionality'. As such, the estimate can be written concisely as $\hat{\mathbf{x}}(k \mid k)$, where the first k in the notation refers to which variable is being estimated, while the second refers to which measurements are being used for the estimate. Thus, in general, $\hat{\mathbf{x}}(i \mid j)$ is the estimate of the value that $\mathbf{x}$ assumes at time

i given the first $j+1$ measurements $\mathbf{y}_0, \mathbf{y}_1, \mathbf{y}_2, \ldots, \mathbf{y}_j$. The system state variables, estimates and their associated covariances would conditionally depend on the measurements.

11.4.2.1 Measurement update stage

The optimal prediction of the next system state value, in the absence of any new observation, is based on the current estimate $\hat{\mathbf{x}}(k \mid k)$ and is given by

$$\hat{\mathbf{x}}(k+1 \mid k) = A\hat{\mathbf{x}}(k \mid k) \tag{11.53}$$

while the observation estimate is given by

$$\hat{\mathbf{y}}(k \mid k) = H\hat{\mathbf{x}}(k \mid k) \tag{11.54}$$

The above equations are often called single-stage optimal prediction. The prediction properties of the transition matrix are employed, which link the current states of the system to states at the next time instant. It should be noted that prediction is a stated expectation about a given attribute that may be verified by subsequent observation.

The optimal unbiased initial state estimate $\hat{\mathbf{x}}(0 \mid 0)$ is a Gaussian random vector with zero mean:

$$E\{\hat{\mathbf{x}}(0 \mid 0)\} = \mathbf{x}(0) = 0 \tag{11.55a}$$

and with an error covariance matrix $\mathbf{P}(0)$ defined as

$$E\{\mathbf{x}(0 \mid 0)\mathbf{x}^T(0 \mid 0)\} = \mathbf{P}(0 \mid 0) = \mathbf{P}(0) \tag{11.55b}$$

The covariance matrix $P(k \mid k)$ must be computed to keep the Kalman filter running. Following (11.35), at time k given measurements $\mathbf{y}_0, \mathbf{y}_1, \mathbf{y}_2, \ldots, \mathbf{y}_k$, the Kalman filter computes the covariance matrix $P(k|k)$ from the error $\mathbf{e}(k \mid k)$. Computation happens according to the phases of updates and propagation. The next phase is to incorporate the new measurement $\mathbf{y}_k$ into the estimate that is progressing from $\hat{\mathbf{x}}(k \mid k)$ to $\hat{\mathbf{x}}(k+1 \mid k)$. Given $\mathbf{y}_k$, consider the *residue*

$$\mathbf{r}_\mathrm{k} = \mathbf{y}_k - \hat{\mathbf{y}}(k \mid k) \tag{11.56a}$$

which, from (11.54), is equal to

$$\mathbf{r}_\mathrm{k} = \underbrace{\mathbf{y}_k - H\hat{\mathbf{x}}(k \mid k)}_{\tilde{y}} \tag{11.56b}$$

The gain matrix

$$\kappa_k = P(k \mid k)H^T R^{-1} \tag{11.57}$$

The gain matrix is usually called the *Kalman gain* matrix because it specifies the amount by which the residue must be multiplied or amplified to obtain the correction term that transforms the old estimate from $\hat{\mathbf{x}}(k \mid k)$ to $\hat{\mathbf{x}}(k+1 \mid k)$.

If the residue is zero, it means that the initial estimate is exact, otherwise there is a need to correct the estimate $\hat{\mathbf{x}}(k\,|\,k)$ so that the new prediction of the measurement is very close to the old prediction. The reader might ask how much correction should one introduce to the state estimate? This requires some mechanism for comparing the quality of the new measurement $\mathbf{y}_k$ with the old estimate $\hat{\mathbf{x}}(k\,|\,k)$. The uncertainty of the new measurements arises from the covariance of the measurement error, $\mathbf{R}(k)$, while that of the states is $\mathbf{P}(k\,|\,k)$. The update stage of Kalman filter uses $\mathbf{R}(k)$ and $\mathbf{P}(k\,|\,k)$ to weigh past estimate and new measurements. Also, the uncertainty value of $\mathbf{P}(k\,|\,k)$ must be updated so that it is available for the next step $(k+1)$. Propagation then accounts for the development of the system state, as well as increasing uncertainty. Before attempting to write the next state update expressions, let us examine the recursive nature of (11.50) in obtaining an expression for the state uncertainty $P(k\,|\,k)$. Following (11.49c), the error vector

$$\mathbf{e}(k+1\,|\,k) = A\mathbf{e}(k\,|\,k) + \sum_{i=1}^{k} A^{k-i} B\mathbf{Q}(i-1) \tag{11.58}$$

From (11.58) it is apparent that $\mathbf{e}(j\,|\,k)$, where $j = k+1, k+2, \ldots$ is a Gaussian discrete process with a zero mean since for $\mathbf{u}(i-1)$ equation (11.45) holds. Following similar considerations that applied to (11.58)

$$\mathbf{e}(k+1\,|\,k) = A\mathbf{e}(k\,|\,k) + B\mathbf{Q}(k) \tag{11.59}$$

from which it is apparent that the process considered is a Markov process. So, the state covariance can be derived from the expectance of (11.58) to form the relation

$$P(k+1\,|\,k) = AP(k\,|\,k)A^T + BQ(k)B^T \tag{11.61}$$

The next stage is to update the previous state estimate. The updated state parameter estimate may be written as

$$\begin{aligned} \hat{\mathbf{x}}(k+1\,|\,k) &= \hat{\mathbf{x}}(k\,|\,k) + \kappa_k \mathbf{r}_k \\ &= \hat{\mathbf{x}}(k\,|\,k) + \kappa_k \tilde{\mathbf{y}} \end{aligned} \tag{11.62}$$

where $\tilde{\mathbf{y}}$, defined by (11.56b), is called the *innovations sequence*. It provides an easy check for the optimality of the Kalman filter. And the updated covariance matrix may be expressed by

$$P(k+1\,|\,k) = P(k\,|\,k)[I - \kappa_k H] \tag{11.63}$$

where I is the identity matrix. This expression describes how the error variance propagates.

The updated Kalman filter gain matrix

$$\begin{aligned} \kappa_{k+1} &= P(k+1\,|\,k)H^T \underbrace{\left[HP(k+1\,|\,k)H^T + R\right]}_{S}{}^{-1} \\ &= P(k+1\,|\,k)H^T S^{-1} \end{aligned} \tag{11.64}$$

is effectively the ratio between the uncertainty in the state estimates and the uncertainty in the measurements, where S is called the *residual covariance matrix*.

In applying the linear filter to a specific system, matrices A, B, H and noise statistics Q and R must be specified. The initial estimates, $\hat{\mathbf{x}}(0)$ and $\mathbf{P}(0)$, are assumed *a priori* of the target's position at the beginning of the navigation period. In the absence of *a priori* data, the best estimate of $\mathbf{x}(0)$, given no observations, may be expressed as

$$\hat{\mathbf{x}}(1\,|\,0) = A\hat{\mathbf{x}}(0\,|\,0) = \mathbf{x}(0) \tag{11.65}$$

This implies that $\hat{\mathbf{x}}(1\,|\,0)$ is the estimate of $\mathbf{x}(1)$, given observation up to and including $k = 0$. Since the observations do not start until $k = 1$, there are no observations, and hence $\hat{\mathbf{x}}(1\,|\,0)$ is the *a priori* value. Therefore, the initial covariance value may be expressed by

$$\begin{aligned}\mathbf{P}(1\,|\,0) &= \mathrm{var}\{\hat{\mathbf{x}}(1\,|\,0) - \mathbf{x}(0)\}\\ &= \mathrm{var}\{A\hat{\mathbf{x}}(0\,|\,0) - \mathbf{x}(0)\}\\ &= A\mathbf{P}(0\,|\,0)A^T + B\mathbf{Q}(0)B^T\\ &= \mathbf{P}(0)\end{aligned} \tag{11.66}$$

where 'var' means variance. The covariance matrix $\mathbf{P}(0)$ will be a diagonal matrix with large values to ensure relatively fast convergence by the Kalman filter. Of course, if $\hat{\mathbf{x}}(1\,|\,0)$ and $\mathbf{P}(1\,|\,0)$ are specified, rather than $\hat{\mathbf{x}}(0\,|\,0)$ and $\mathbf{P}(0\,|\,0)$, then they may be used directly as the initial conditions for the problem. Targets are often tracked in an environment that is cluttered, and tracks may be initiated on clutter. This will provide an additional problem of tracks' initialization where *a priori* information would be handy for effective initialization. For easy reference, the preceding algorithms, which describe the Kalman estimator one-stage prediction, are collated in Table 11.1.

The real merit of the Kalman algorithms, in their application to filtering and prediction problem, lies in the fact that not only is a solution obtained, but that the solution directly specifies practical implementation of the results. The algorithms can handle both *stationary* (fixed values of noise statistics Q and R) and *non-stationary* data (Q, R time varying).

An issue that should be considered regarding the validity of the Kalman estimator is that precise models need to be postulated for both the target-state and the measurement process. If these underlying models are not accurate, the Kalman filter will not perform optimally. The Kalman filtering approach also implies that the residual covariance matrix S, is adaptively and optimally 'matching' the target and measurement characteristics. Any deviation from this implied notion indicates the imperfection in the models and undermines the stability of the system. By comparing the residual's statistics (e.g. mean, variance and autocorrelation function) to preset threshold(s), the presence of a model mismatch can be inferred and corrected. This may be achieved at a cost of an added delay for forming a sliding window average (Bolgler 1990).

Table 11.1 Summary of discrete one-stage Kalman filter predictor algorithms

System model	$\mathbf{x}(k+1) = A\mathbf{x}(k) + B\mathbf{u}(k)$	(11.50)
Measurement model	$\mathbf{y}(k) = H\mathbf{x}(k) + \mathbf{v}(k)$	(11.51)
Prior statistics	$E\{\hat{\mathbf{x}}(0\,|\,0)\} = \mathbf{x}(0) = 0$	(11.55a)
	$E\{\mathbf{x}(0\,|\,0)\mathbf{x}^{\mathrm{T}}(0\,|\,0)\} = \mathbf{P}(0\,|\,0) = \mathbf{P}(0)$	(11.55b)
Extrapolation:	$\hat{\mathbf{x}}(k+1\,|\,k) = A\hat{\mathbf{x}}(k\,|\,k)$	(11.53)
Predictor algorithms	$\hat{\mathbf{y}}(k\,|\,k) = H\hat{\mathbf{x}}(k\,|\,k)$	(11.54)
	$P(k+1\,|\,k) = AP(k\,|\,k)A^T + BQ(k)B^T$	(11.61)
Update:	$\hat{\mathbf{x}}(k+1\,|\,k) = \hat{\mathbf{x}}(k\,|\,k) + \kappa_k\tilde{\mathbf{y}}$	(11.62)
New estimation algorithm	$\tilde{\mathbf{y}} = \mathbf{y}_k - H\hat{\mathbf{x}}(k\,|\,k)$	(11.56b)
	$\kappa_{k+1} = P(k+1\,|\,k)H^T S^{-1}$	(11.64)
Gain algorithm	$S = HP(k+1\,|\,k)H^T + R$	(11.64)
Error variance algorithm	$P(k+1\,|\,k) = P(k\,|\,k)[I - \kappa_k H]$	(11.63)
	$\hat{\mathbf{x}}(1\,|\,0) = A\hat{\mathbf{x}}(0\,|\,0) = \mathbf{x}(0)$	(11.65)
Initial conditions	$\mathbf{P}(1\,|\,0) = \mathbf{P}(0)$	(11.66)
	If Q and R unknown	
	$\mathbf{Q}(k) = \sigma_x^2 I$	
	$\mathbf{R}(k) = \sigma_y^2 I$	(11.49e)
	$\mathbf{x}(0) = \mathbf{0}$	(11.49d)

11.4.2.2 Application of the Kalman estimator to engineering problems

The recursive Kalman filtering technique has been applied to various engineering problems including:

- Missile projectile monitoring and tracking
- Space navigation (Murtagh 1965)
- Plant control (Strejc 1981)
- Anti-submarine warfare (Laing 1967)
- Aircraft and maritime surveillance (Colegrove *et al.* 1986; Kolawole 1994)

An application of the technique to the problem of estimating a missile trajectory is illustrated by Example 11.4. This problem allows the reader the opportunity of seeing some of the technical issues involved in setting up the dynamic equations for the system and the subsequent use of the preceding algorithms.

Example 11.4: A video sensor is attached to a warship, depicted in Figure 11.3, which spots the oncoming missile threat. The variation in intensity of the blobs appearing on the ship's sensory screen estimates how close the threat is. The missile is spotted at about 30 km away from the ship and released at about 500 m above the sea surface. The ship's sensor estimates the missile's radial speed of about 0.601 km/s, at about 9.46° azimuth. Plot the missile's true and estimated speed trajectory as well as its position against time. Determine also whether there is sufficient time to launch an evasive action to counter the missile threat.

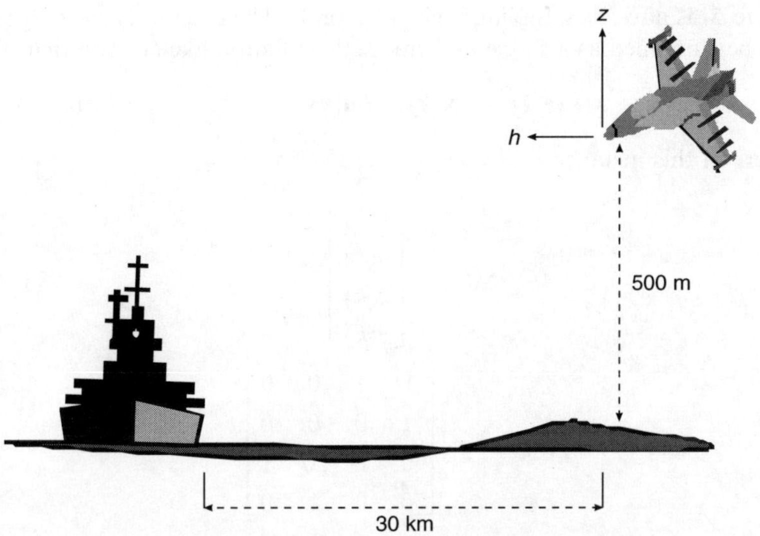

Figure 11.3 An illustration of a missile released by a fighter

Solution
A prior knowledge of the type of missile used is assumed to be known as well as the sensor's parameters. It is also assumed that the ship sensor keeps track of the missile's location within its designated window. At a particular point in space, the missile's coordinates could be measured.

Let the missile's horizontal coordinate and vertical coordinate be represented by h and z respectively. So, the dots of these coordinates will denote their derivatives. If air drag is very small as to affect the course of the missile and be considered negligible, at any time t, the missile dynamic state equations and transition matrix can be established following the laws governing the ballistic trajectory motion.

$$h(t) = h(0) + \dot{h}(0)t \tag{11.67a}$$

$$z(t) = z(0) + \dot{z}(0)t - \frac{g}{2}t^2 \tag{11.67b}$$

where g denotes acceleration due to gravity ($= 9.8\,\text{m/s}^2$). Due to the differential nature of the above equations, they are continuous. In practice, measurements are sampled at a given time interval, say k. Since the interest is only in the solution at the instants of sampling (11.67) can be rewritten as follows:

$$h(k+1) = h(k) + \dot{h}(k)\Delta t \tag{11.68a}$$

$$z(k+1) = z(k) + \dot{z}(k)\Delta t - \frac{g}{2}(\Delta t)^2 \tag{11.68b}$$

where Δt is now the sampling time, in seconds. The equations given by (11.68) can be combined as a single dynamic state equation like (11.50); that is,

$$\mathbf{x}(k+1) = A\mathbf{x}(k) + \mathbf{Bu}(k) \qquad \text{(same as 11.50)}$$

where in this instance

$$\mathbf{x}(k) = \begin{bmatrix} h(k) \\ \dot{h}(k) \\ z(k) \\ \dot{z}(k) \end{bmatrix} \tag{11.69a}$$

$$A = \begin{bmatrix} \Delta t & 1 & 0 & 0 \\ 1 & 0 & 0 & 0 \\ 0 & 0 & \Delta t & 1 \\ 0 & 0 & 1 & 0 \end{bmatrix} \tag{11.69b}$$

$$B = \begin{bmatrix} 0 \\ 0 \\ \frac{-(\Delta t)^2}{2} \\ \Delta t \end{bmatrix} \tag{11.69c}$$

The system process noise vector $\mathbf{u}(k)$ with step k is assumed a Gaussian random vector with components having equal dispersion. Hence, following (11.49e),

$$E\{\mathbf{uu}^T\} = \mathbf{Q} = \sigma_x^2 I \tag{11.69d}$$

where I is the identity matrix. Due to limited knowledge of the state's covariance matrix, the dispersion is made small, that is, $\sigma_x^2 = 0.1$. So

$$Q = 0.1 \begin{bmatrix} 1 & 0 & 0 & 0 \\ 0 & 1 & 0 & 0 \\ 0 & 0 & 1 & 0 \\ 0 & 0 & 0 & 1 \end{bmatrix} \tag{11.69e}$$

Having obtained the state dynamic equations, the next phase is to establish the measurement equations. From *a prior* information of the missile, the elevation and blob size in pixels at any point in space can be expressed. If missile elevation is represented by e_m and blob size by s_m, then

$$e_m = m\frac{z}{h} \tag{11.70a}$$

Similarly, the size of the blob, in pixels,

$$s_m = \frac{m}{\sqrt{h^2 + z^2}} \tag{11.70b}$$

where m is the proportionality constant obtainable from the video parameters. For more information on how depth from image pixels or sequences can be estimated, the reader is advised to consult Broida *et al.* (1990). The task is to express (11.70) as linear approximations, using Taylor series, around the current estimates to obtain a change in coordinates measurements. For instance,

$$\begin{aligned} e_m &= m\frac{z}{h} \\ &\approx m\left\{\frac{\hat{z}}{\hat{h}} - \frac{\hat{z}-z}{\hat{h}} + \frac{\hat{z}}{\hat{h}}\left(\hat{h}-h\right)\right\} \end{aligned} \tag{11.71a}$$

from which a change in elevation measurement as a function of the missile coordinates is

$$\begin{aligned} \Delta e_m &= e_m - m\frac{\hat{z}}{\hat{h}} \\ &\approx m\left\{-\frac{\hat{z}}{\hat{h}^2}h + \frac{1}{\hat{h}}z\right\} \end{aligned} \tag{11.71b}$$

Similarly, from (11.70b), the blob size is expressed by

$$\begin{aligned} s_m &= \frac{m}{\sqrt{h^2+z^2}} \\ &\approx m\left\{\frac{1}{\sqrt{\hat{h}^2+\hat{z}^2}} + \frac{\hat{h}\left(\hat{h}-h\right)}{\left(\hat{h}^2+\hat{z}^2\right)^{3/2}} + \frac{\hat{z}(\hat{z}-z)}{\left(\hat{h}^2+\hat{z}^2\right)^{3/2}}\right\} \end{aligned} \tag{11.72a}$$

from which a change in image depth measurement as a function of the missile coordinate is

$$\begin{aligned} \Delta s_m &= s_m - \frac{m}{\sqrt{\hat{h}^2+\hat{z}^2}} \\ &\approx m\left\{\frac{\hat{h}}{\left(\hat{h}^2+\hat{z}^2\right)^{3/2}}h + \frac{\hat{z}}{\left(\hat{h}^2+\hat{z}^2\right)^{3/2}}z\right\} \end{aligned} \tag{11.72b}$$

The changes depicted by (11.71b) and (11.72b) constitute the elements of the measurement transition matrix H. Specifically

$$H = \begin{bmatrix}\Delta e_m \\ \Delta s_m\end{bmatrix} \approx -m\begin{bmatrix} \frac{\hat{z}}{\hat{h}^2} & 0 & -\frac{1}{\hat{h}} & 0 \\ \frac{\hat{h}}{\left(\hat{h}^2+\hat{z}^2\right)^{3/2}} & 0 & \frac{\hat{z}}{\left(\hat{h}^2+\hat{z}^2\right)^{3/2}} & 0 \end{bmatrix} \tag{11.73}$$

Like (11.51), the measurement equation contains a noise vector $\mathbf{v}(k)$ with step k with components having equal dispersion. So, following (11.49e),

$$E\{\mathbf{v}v^T\} = \mathbf{R} = \sigma_y^2 I \tag{11.74}$$

where I is the identity matrix. Due to limited knowledge of the state's covariance matrix, it is reasonable to put $\sigma_y^2 = m$.

From the available information, the initial state estimate of the projectile is

$$\hat{\mathbf{x}}(0) = \begin{bmatrix} \hat{h}(0) \\ \hat{\dot{h}}(0) \\ \hat{z}(0) \\ \hat{\dot{z}}(0) \end{bmatrix} = \begin{bmatrix} 30 \\ -0.601\cos(9.46) \\ 0.5 \\ 0.601\sin(9.46) \end{bmatrix} = \begin{bmatrix} 30 \\ -0.6 \\ 0.5 \\ 0.1 \end{bmatrix} \tag{11.75}$$

Having collated the necessary expressions for the state and measurement vectors, the Kalman filtering equations can now be applied to study the behaviour of the missile and consider what evasive action can be implemented before it reaches its intended destination. The other task is to simulate the missile trajectory using the state and measurement equations, for each time step k, taking uniform sample time $\Delta t = 100$ ms. For brevity, $m = 1000$. Plots of the missile trajectories are shown in Figures 11.4 and 11.5.

The missile trajectories should be read from right to left; that is, trajectories start from the right inwardly. The difference between the true and estimated trajectories is shown in the figures. The trajectories in Figure 11.4 are essentially the missile tracks (more is said of tracks in Chapter 12).

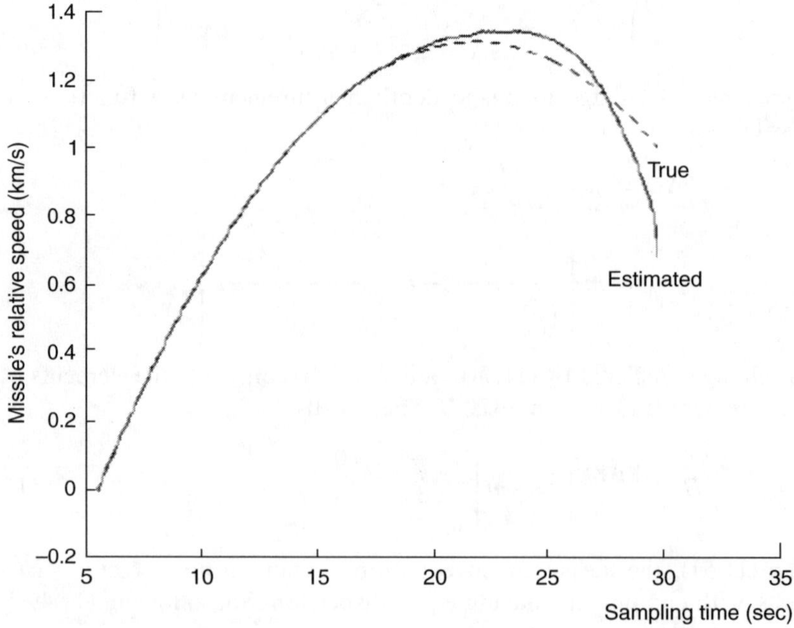

Figure 11.4 The missile's true and estimated trajectories versus time

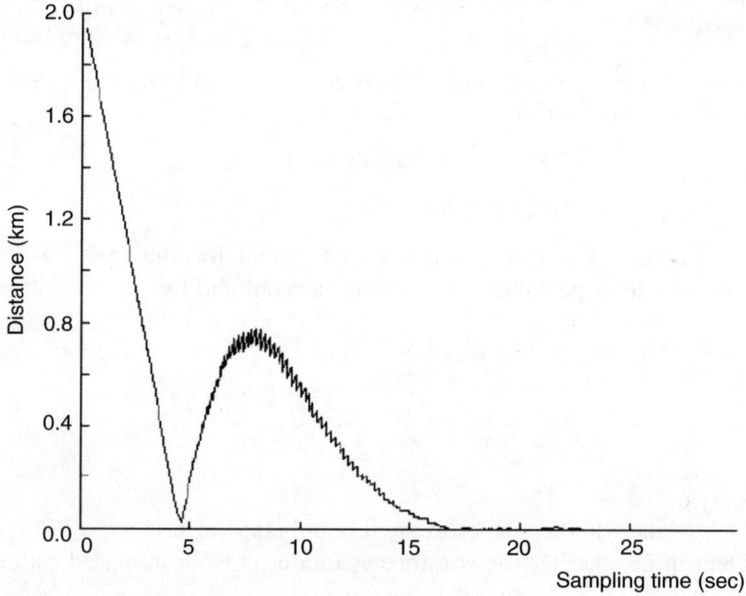

Figure 11.5 Distance between estimated and true missile position against time

The true and estimated target positions converge readily and quickly to zero, as shown in Figure 11.5, at about 20 seconds. This allows sufficient time for any evasive action to be implemented; that is, shooting the missile down before it reaches its intended destination, or abandoning the warship.

11.5 Summary

Estimation, the second type of optimization problem, has been discussed in this chapter. It exploited the several parallels with the decision theory discussed in Chapter 9. The dynamic nature of the parameter in signal estimation adds a new dimension to the statistical modelling of estimation problems. For example, the dynamic properties of the signal, such as how fast and in what manner the signal can change, must be modelled at least statistically to obtain meaningful signal estimation techniques.

The concept of filtering, interpolation (data smoothing) and prediction, encapsulated in the linear estimation procedure, leads to the Kalman estimator, which allows the study of system causation from the past to the future. A typical real-time missile example was given that provided the opportunity of seeing how the dynamic equations for the system were formulated and the subsequent use of the preceding algorithms.

Problems

Suppose a sequence $x_1, x_2, x_3, \ldots, x_n$ is given as a random variable such that its expectance and variance are defined as

$$E(x_i) = \mu \quad \text{where } i = 1, 2, 3, \ldots, n$$
$$\mathrm{var}(x_i) = \sigma^2$$

Suppose that $x_i - \mu$ and $x_j - \mu$ are orthogonal for the case $i \neq j$. If the estimates of the expectance and variance are defined by

$$\hat{\mu} = \frac{1}{n}\sum_{i=1}^{n} x_i$$
$$\hat{\sigma}^2 = \frac{1}{n-1}\sum_{i=1}^{n} (x_i - \hat{\mu})^2$$

(a) Determine whether the estimate $\hat{\mu}$ is unbiased or not.
(b) Determine whether the variance estimate $\hat{\sigma}^2$ is an unbiased estimate of σ^2 or not.
(c) Prove that $n(\hat{\mu} - \mu) = \sum_{i=1}^{n} (x_i - \mu)$.
(d) Prove that $\left[\sum_{i=1}^{n} (x_i - \mu)\right]^2 = \sum_{i=1}^{n} (x_i - \mu)^2 + \sum_{i \neq j}^{n}\sum_{i \neq j}^{n} (x_i - \mu)(x_j - \mu)$.

12

Tracking

The previous chapters have presented preliminary materials on radar principles, estimation and applicable probability theories. Radar tracking is an important application area of signal processing. A radar system repeatedly scans a geographical area and produces data from which can be inferred the location, speed, and size of the objects detected. As noted in Chapter 10, tracking can only be successfully performed when signal processing is capable of producing a reliable input stream of detected peaks (also called *detections*). For each of, the radar scans, a myriad of data points, or *returns*, are produced. The returns correspond to reflections of the radar beam from real targets of interest including that from other objects in the vicinity of the real targets and clutter. Clutter refers to returns from the Earth's surface, electromagnetic interference, meteor, lightning and even other objects in the vicinity of the target(s) of interest.

In this chapter, a discussion on the basic principles of radar tracking is presented. The reader may ask: what is tracking? Simply, tracking is a process of determining the speed and direction of targets and which enables monitoring of the target throughout the radar cover area. Technically, tracking is parameter estimation. The missile trajectory problem (of Example 11.4, Chapter 11) is an example of tracking. However, the real-time stage-by-stage observation of tracks established on returns from a target, as detected peaks, separates 'parameter estimation technique' from 'tracking'. For clarity, it is necessary to distinguish between a target and a track. A target is a physical object that can produce sensor measurements while a track is the symbolic representation of a target, formed from successive detected positions. In general, the determination of tracks means the development of a mathematical model that represents target structure, the parameter values and, if needed, the values of dependent variables such as state variables.

The tracking process involves filtering, interpolation and prediction: a process with behaviour dependent on variables past parameter values as well as the current values of parameters. Armed with the knowledge of the present state, the future state may be predicted. Measurements of the present state may include noise, errors and inaccuracies. By preprocessing the data to remove some, or all, of the clutter prior to tracking reduces the level of the uncertainties. Despite this, the problem of identifying whether the measurements actually originate from the target, or are due to false alarms, still remains. It is because of these errors that tracking and smoothing theory are needed. The procedure used to imply the origin of measurement uncertainties is called *data association.*

This chapter discusses two commonly used tracking methods: the fixed tracking coefficients (commonly called $\alpha\beta$ filter or $\alpha\beta\gamma$ filter) method and adaptive coefficients via the Kalman filtering method. As demonstrated in Chapter 11, section 11.4.2, the Kalman filter computes the parameters of posterior distributions for certain kinds of stochastic process, characterized by linear transformations and additive Gaussian noise. The Gaussian random signals are considered to remain Gaussian after passing through a linear filtering system. In reality the $\alpha\beta$, $\alpha\beta\gamma$ and Kalman are of the fading memory type, which are implemented recursively.

A brief description of $\alpha\beta$, $\alpha\beta\gamma$ and Kalman filters will be given in section 12.2. Before developing each of the algorithms that allow tracking to be achieved it is necessary to reflect on what these algorithms were meant to achieve. The goal of a tracking algorithm is to provide a best estimate of quantities of interest. However, actual signals received are corrupted by disturbances, which are random in nature. As such, the goal of obtaining the best estimate provides the disposition to create an optimum system or algorithms, which will produce a minimum mean-square error between the actual and the desired output. The minimum mean-square error criterion may not be ideal for all systems; however, the criterion leads to the Kalman filter theory.

For completeness, more recently another modelling idea, the *hidden Markov technique* (Xie and Evans 1991), was introduced. The Hidden Markov technique formulates the tracking problem in terms of a hidden Markov model and produces track estimates via the Viterbi algorithm. Like the Kalman filter, the hidden Markov model deals with stochastic processes in which the hidden states and measurements are continuous random variables. The Markov's nature of computation changes in a sense: instead of dealing in explicit probability distributions over a finite state space, the means and variance are dealt with. Detailed discussion and derivation of this approach are not considered in this book; however, the reader is advised to consult Xie and Evans (1991) for more details.

12.1 Basic tracking process

When a target is viewed remotely from a point or points of reference, the determination of the object's transfer functions, as evident in classical control theory, is not achievable. Instead, the returns (i.e. the reflected or backscattered signals) from such a target are processed. The processing technique has been discussed in Chapter 10. As an aid to understanding the sensing and tracking process, Figure 12.1 is provided.

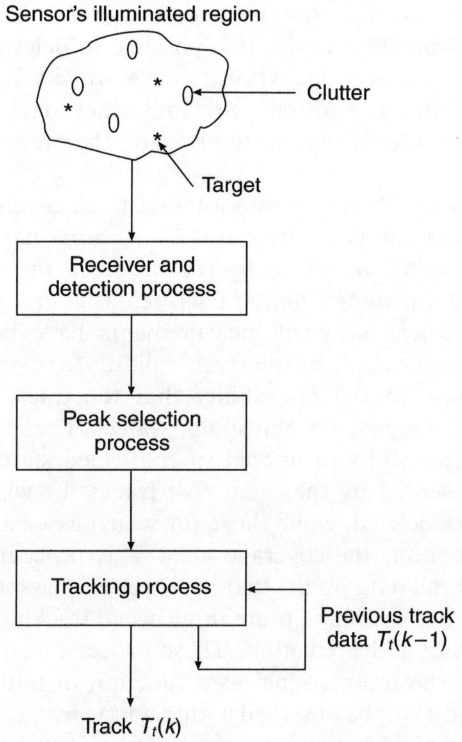

Figure 12.1 Basic sensing and tracking process

The radar operates in a track-while-scan mode. A surveillance area (or region) is defined, and the radar beams are pointed in a particular direction for a few seconds, denoted as a dwell. Scanning is performed from left to right at regular intervals until the entire area is covered, after which the process is repeated. During each scan, a large volume of data points, or *returns*, are produced. These returns also contain information that is not from the targets. During initial scans little information is available to assist in discerning which returns are due to targets and which are noise. Based on the strength of the processed signals and after thresholding target peaks are selected from that of clutter. The peaks are then associated with the target.

The reader might wonder whether selected peaks actually belong to targets since returns from other objects within the vicinity of the target might have comparable strength. But it is understood from the basic physics principles that real targets travel in short time intervals in physically realizable paths, for example straight lines or near smooth curves. With this knowledge it is highly likely that in the next radar scan there will be a return present at a location that extrapolates out from the position of each return in the previous return that corresponds to the target. The converse is also true; it is unlikely that a return from clutter will be positioned in a regular sequence in the next scan. Hence, by associating returns in subsequent scans with returns from previous scans, it is possible to determine which returns are targets and which are not. At any time k, tracks $T_t(k)$ can be formed from successive returns' positions. Previous track data (or a set of track data) $T_t(k-1)$ are interrogated to improve on the present and future track position estimates.

Initial conversion of a return into a target track is called *track initiation.* When new tracks are initiated, they are of a *tentative* nature. Such tentative tracks are *updated,* or *smoothed,* by returns from the next scan that are within reasonable kinematics limits. Each tentative track is promoted to a *confirmed* track when sufficient measurements have been received from subsequent scans that confirm the track validity to a significant level. The validation is based on the probability that the track is a target return exceeding a prearranged, or threshold, value. The number of ensuing measurements successfully promoted to confirmed status is dependent on the environment sensed by the radar. All tracks for which no association could be made are deleted, while those for which associations are made are maintained throughout the coverage area. This procedure is repeated for each scan, and for all data points that have not been associated with a track.

The preceding paragraphs capture three broad tracking processes, namely smoothing, filtering and prediction. These processes can be formulated as follows. Suppose the return signal as a function of time is represented by $x(t)$. Also suppose it can be observed within a time frame $t_0 \leq t \leq t_k$. Then at time $t = t_j$, these processes can be expressed concisely:

- smoothing (interpolation) at $t_j < t_k$;
- filtering at $t_j = t_k$; and
- predicting at $t_j > t_k$.

These processes are closely connected and can be achieved within the general framework of dynamic system theory. A simple transformation process from a continuous-time linear stochastic system to a state space discrete time has been discussed in Chapter 11, section 11.4.1. The discrete-time system notation is followed throughout this chapter, allowing a system's behaviour to be described conveniently in the form of a vector matrix.

12.2 Filters for tracking

A large number of filtering algorithms have been developed for tracking including

- $\alpha\beta$; a two-point extrapolator filter,
- $\alpha\beta\gamma$; a three-point extrapolator filter; and
- Kalman filter, which is a multi-point extrapolator filter.

For each of the filters, the radar system's selection process takes each track $T_t(k)$ from the measurement data $y(k) = \{y_1(k), y_2(k), \ldots, y_n(k)\}$. The selection process also has access to the previous track data $T_t(k-1)$ up to the time $k-1$. For every track $T_t(k)$ in the scanned region at time k, there is a corresponding target. The track is assumed to contain the target's state estimate $\hat{x}(k)$. With this preamble, each filter's algorithm is discussed in the following sections under their appropriate heading.

12.2.1 αβ filter

The αβ trackers are a widely used two-dimensional class of time-invariant filters for estimating system states (e.g. position and velocity) having the form

$$\hat{x}(k) = x_p(k) + \alpha\left[y(k) - x_p(k)\right] \tag{12.1}$$

$$\hat{v}(k) = \hat{v}(k-1) + \frac{\beta}{T}\left[y(k) - x_p(k)\right] \tag{12.2}$$

$$x_p(k) = \hat{x}(k-1) + T\hat{v}(k-1) \tag{12.3}$$

where, on the k radar scans

α = position damping factor
β = velocity damping factor
$\hat{x}(k)$ = estimated or smoothed position
$x_p(k)$ = forecast or predicted position
$\hat{v}(k)$ = estimated or smoothed velocity
$y(k)$ = measured or plotted position
T = data interval (or sampling period).

From (12.1) to (12.3), the time-invariant αβ filter model for estimating target kinematics $\mathbf{x}(k)$ can concisely be written in the forms

$$\hat{\mathbf{x}}(k+1) = \mathbf{x}_p(k) + W\tilde{\mathbf{y}}(k) \tag{12.4}$$

$$x_p(k) = \hat{x}(k-1) + T\hat{v}(k-1) \tag{12.5}$$

where

$\tilde{\mathbf{y}}$ = residual (innovation) vector; being the difference between measured and predicted quantities
W = the weighting factor, or filter gain, which equates to

$$W = \begin{bmatrix} \alpha \\ \frac{\beta}{T} \end{bmatrix} \tag{12.6}$$

In practice, the gains (α, β) are selected and adjusted using a combination of intuition, experience and rules of thumb.

The time index k assumes integer values for a constant sampling period T. In the event that a true measurement is not detected for radar scans, for which no measurements are received, then set $\hat{x}(k) = x_p(k)$. The velocity

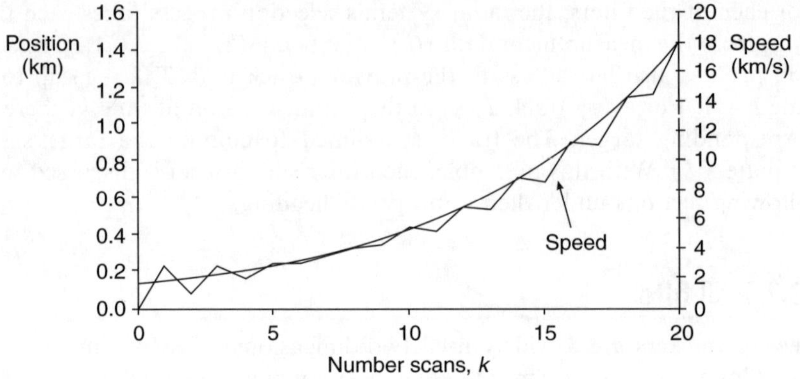

Figure 12.2 Estimated target track (position) and speed for successive scans with αβ tracker

equation remains unaltered and the prediction $x_p(k)$ to the next scan follows the same sequence as in (12.3).

Figure 12.2 shows the track (estimated position) of a target as well as its speed at sampling time T of 0.1 s for filter gains $\alpha = 0.25$ and $\beta = 1.15$, and initial conditions $\hat{x}(-1) = 0.01$ km and $\hat{v}(-1) = \hat{v}(0) = 0.0001$ km/s.

12.2.1.1 Coordinate system

Typically, equations (12.1) to (12.3) are applied separately in Cartesian coordinate systems, although polar and track-oriented coordinate systems have also been used. To understand the translation from one frame to another, the Cartesian coordinate system shown in Figure 12.3 is presented.

Assuming that the predicted range r_p and bearing θ_p are available, the following relationships are developed between these coordinate systems. The target predicted position is at (r_p, θ_p) and the measured position is at (r_m, θ_m). Let a Cartesian frame (ξ, Ψ) be constructed at an angle θ_p relative to north axis N such that ξ and Ψ lie respectively along the line of sight and

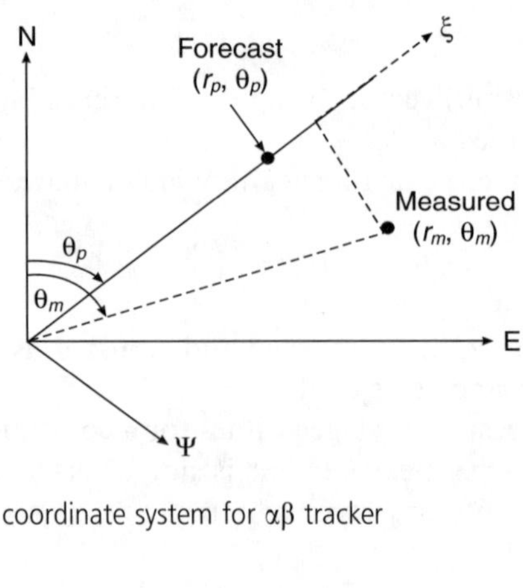

Figure 12.3 Cartesian coordinate system for αβ tracker

across the line of sight. In the (ξ, Ψ) frame, the predicted position can be written as

$$\begin{aligned} \xi_p &= r_p \\ \Psi_p &= 0 \end{aligned} \tag{12.7}$$

And, for the measurement position

$$\xi_m = r_m \cos(\theta_m - \theta_p) \tag{12.8}$$

$$\Psi_m = r_m \sin(\theta_m - \theta_p) \tag{12.9}$$

The $\alpha\beta$ filter equations (12.1) to (12.3) are applied as follows. For the next radar scan (i.e. ξ_p, Ψ_p), the new predicted range and bearing can be calculated as

$$r_p = \sqrt{\xi_p^2 + \Psi_p^2} \tag{12.10}$$

$$\theta_p = \theta_p + \partial\theta \tag{12.11}$$

$$\partial\theta = \tan^{-1}\left(\frac{\xi_p}{\Psi_p}\right) \tag{12.12}$$

The next step is to rotate $\partial\theta$ such that the across the line of sight term Ψ_p becomes zero and ξ_p assumes the predicted range r_p. If frame quantities ξ and Ψ are to be modified by rotation of axes through angle $\partial\theta = \theta' - \theta$ given ξ' such that Ψ' equals to zero, then

$$\overline{\xi}' = \overline{\xi}\cos\partial\theta + \overline{\xi}\sin\partial\theta \tag{12.13}$$

$$\overline{\Psi}' = \overline{\Psi}\cos\partial\theta - \overline{\Psi}\sin\partial\theta \tag{12.14}$$

Substituting $\cos\partial\theta = \Psi/r$ and $\sin\partial\theta = \xi/r$ in (12.13) and (12.14),

$$\overline{\xi}' = \frac{1}{r}(\overline{\xi}\xi + \overline{\Psi}\Psi) \tag{12.15}$$

$$\overline{\Psi}' = \frac{1}{r}(\overline{\Psi}\xi - \overline{\xi}\Psi) \tag{12.16}$$

where range r is determined by (12.10). Given that $(\theta_m - \theta_p)$ is nominally small, the trigonometric functions are simplified using a linear approximation.

12.2.1.2 Smoothing factor

The amount of smoothing applied to the system state estimates is determined by the selection of the values of α and β. According to Simpson (1963), in order to maintain stability, the limiting values of α and β are determined using the following relations for a least square fit to the incoming data:

$$\alpha = \frac{2(k-1)}{k(k+1)} \tag{12.17}$$

$$\beta = \frac{6}{k(k+1)} \tag{12.18}$$

where k is the measurement number. The effect of these damping factors is graphically shown in Figure 12.4.

Values of α (and β) approaching zero give heavy damping, whereas values approaching unity give light damping. Benedict and Bordner (1962) suggested that the $\alpha\beta$ tracker is optimized when

$$\beta = \frac{\alpha^2}{2 - \alpha} \tag{12.19}$$

Another method of estimating the values of α and β was given by Benedict and Bordner (1962) in terms of steady-state variance reduction ratio K_x. This reduction ratio is the ratio of the output variance of the smoothed position estimate $\sigma_{x_s}^2$ to the input measurement variance σ_m^2 as follows:

$$K_x = \frac{\sigma_{xs}^2}{\sigma_m^2} = \frac{2\alpha^2 + \beta(2 - 3\alpha)}{\alpha(4 - 2\alpha - \beta)} \tag{12.20}$$

The velocity reduction variance ratio K_v is measured as the velocity estimation output given only the noise input

$$K_v = \frac{\sigma_{xs}^2}{\sigma_{mnoise}^2} = \frac{2\beta^2}{\alpha T^2(4 - 2\alpha - \beta)} \tag{12.21}$$

Expressions (12.20) and (12.21) are valid only if the probability of detection is unity.

The $\alpha\beta$ approach is satisfactory for straight tracks, and needs some modification to cope with manoeuvres (i.e. when the target deviates from a straight-line constant velocity trajectory).

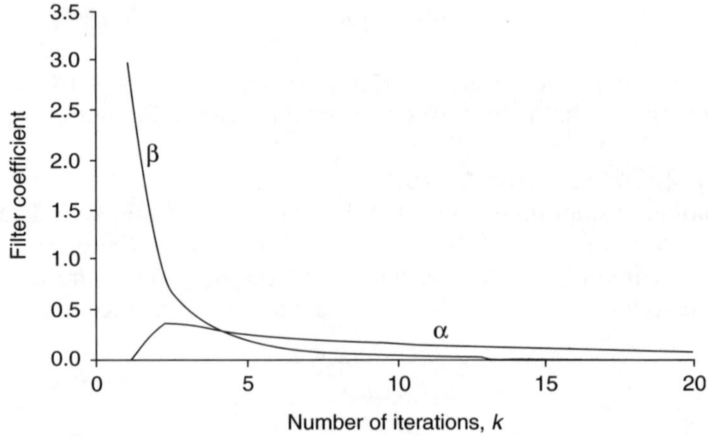

Figure 12.4 $\alpha\beta$ filter coefficient

12.2.2 αβγ filter

The αβγ tracker is a logical extension to the αβ type. It incorporates the estimates of acceleration, *a*, but hypothesizes constant acceleration. The αβγ filter equations are written as follows:

$$\hat{x}(k) = x_p(k) + \alpha[y(k) - x_p(k)] \qquad \text{(same as (12.1))}$$

$$\hat{v}(k) = \hat{v}(k-1) + \frac{\beta}{T}[y(k) - x_p(k)] \qquad \text{(same as (12.2))}$$

$$\hat{a}(k) = \hat{a}(k-1) + \frac{\gamma}{T^2}[y(k) - x_p(k)] \qquad (12.22)$$

$$x_p(k) = \hat{x}(k-1) + T\hat{v}(k-1) + \frac{1}{2}T^2\hat{a}(k-1) \qquad (12.23)$$

where

$\hat{a}(k)$ = the smoothed acceleration (m/s^2)
γ = the acceleration component damping coefficient, which is dimensionless
Other terms are the same as defined for the αβ type in section 12.2.1.

In essence, the above time-invariant αβγ filter model equations are concisely written:

$$\hat{\mathbf{x}}(k+1) = \mathbf{x}_p(k) + W\tilde{\mathbf{y}}(k) \qquad (12.24a)$$

$$x_p(k) = \hat{x}(k-1) + T\hat{v}(k-1) + \frac{1}{2}T^2\hat{a}(k-1) \qquad (12.24b)$$

where $\mathbf{x}(k)$ is the target kinematics vector and

$$W = \begin{bmatrix} \alpha \\ \frac{\beta}{T} \\ \frac{\gamma}{T^2} \end{bmatrix} \qquad (12.24c)$$

Figures 12.5(a) and 12.5(b) show the target track (position) as well as its velocity and acceleration profile using the αβγ tracker, a sampling time T of 0.1 s, filter gains $\alpha = 0.25$, $\beta = 1.15$, $\gamma = 3.0$, and initial conditions $\hat{x}(-1) = \hat{x}(0) = 0.01$ km, $\hat{v}(-1) = \hat{v}(0) = 0.0001$ km/s, $\hat{a}(-1) = \hat{a}(0) = \hat{a}(1) = 0.0$ km/s^2.

Following Simpson (1963), the coefficients are defined thus

$$\alpha = \frac{3(3k^2 - 3k - 2)}{k(k+1)(k+2)} \qquad (12.25)$$

$$\beta = \frac{18(2k-1)}{k(k+1)(k+2)} \qquad (12.26)$$

$$\gamma = \frac{30}{k(k+1)(k+2)} \qquad (12.27)$$

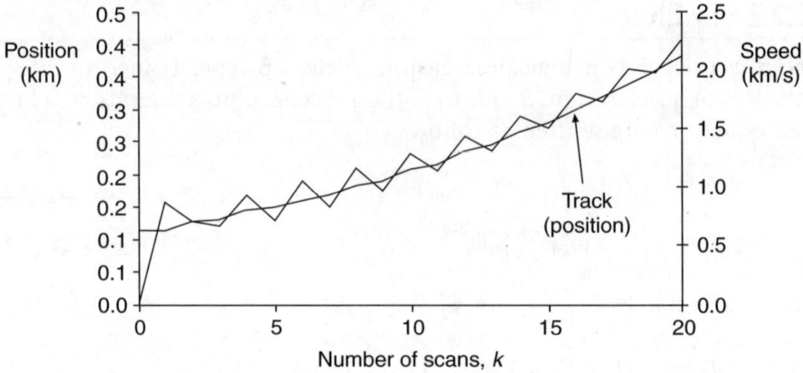

Figure 12.5(a) Estimated target track as well as speed for successive scans by αβγ tracker

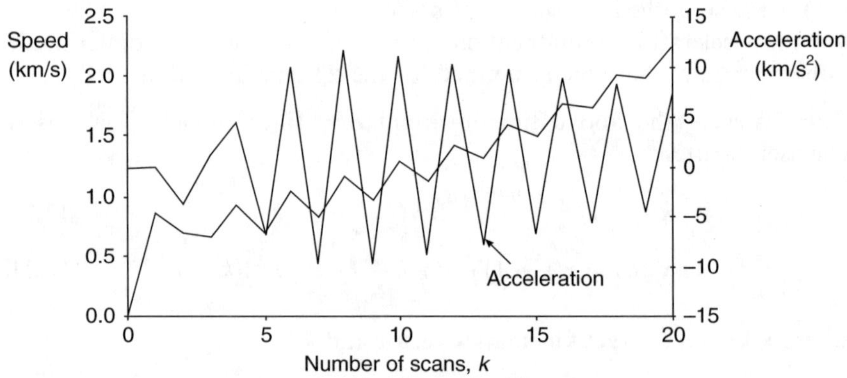

Figure 12.5(b) Estimated target speed and acceleration versus successive scans by αβγ tracker

And, in terms of variance reduction ratios:

$$K_x = \frac{2\beta(2\alpha^2 + 2\beta - 3\alpha\beta) - \alpha\gamma(4 - 2\alpha - \beta)}{(4 - 2\alpha - \beta)(2\beta\alpha + \alpha\gamma - 2\gamma)} \tag{12.28}$$

$$K_v = \frac{4\beta^2(\beta - \gamma) + 2\gamma^2(2 - \alpha)}{T^2(4 - 2\alpha - \beta)(2\alpha\beta + \alpha\gamma - 2\gamma)} \tag{12.29}$$

$$K_a = \frac{4\beta\gamma^2}{T^4(4 - 2\alpha - \beta)(2\alpha\beta + \alpha\gamma - 2\gamma)} \tag{12.30}$$

The values of the damping factors are graphically shown in Figure 12.6. Like α, β in Figure 12.4, values of α, β and γ approaching zero give heavy damping, whereas values approaching unity give light damping. Again by extension, when there are insufficient observations made, a low confidence premium is placed on the predicted target estimates.

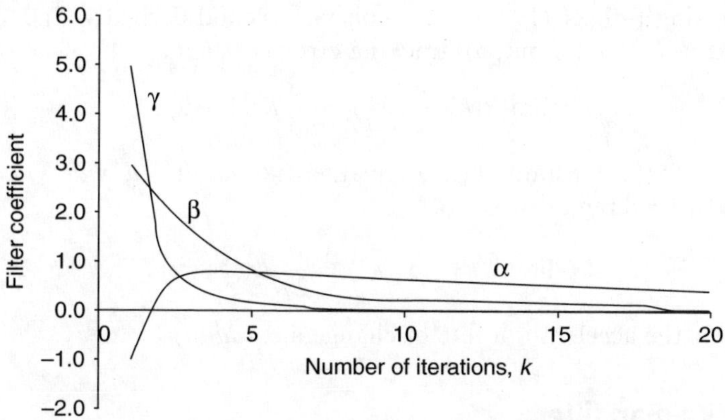

Figure 12.6 αβγ filter coefficient

The converse is also true.

The initial conditions can be written as follows:

$$\hat{x}(1) = x_p(1) = y(1) \tag{12.31a}$$

$$\hat{v}(1) = 0 \tag{12.31b}$$

$$\hat{a}(2) = \hat{a}(1) = 0 \tag{12.31c}$$

$$\hat{v}(2) = \frac{y(2) - y(1)}{T} \tag{12.31d}$$

$$\hat{a}(3) = \frac{y(3) - 2y(2) + y(1)}{T^2} \tag{12.31e}$$

In summary, the αβγ approach is satisfactory if the track is always executing the same manoeuvre. While a constant maintenance of target manoeuvres may be difficult, one should not overlook the effect of simply limiting the maximum value of scans, k, used in the filters' (αβ and/or αβγ) expressions. For instance, if the sensor data rate is high enough – in the light of accuracy, manoeuvre capability, etc.) – the results may be satisfactory. The manoeuvre capability of manned manoeuvrable vehicles such as aircraft, ships and submarines constitute the single feature that makes αβ and αβγ algorithms generally unsuitable for accurate tracking.

12.2.3 Dynamic tracking error

It is obvious from the previous developments that both trackers (αβ and αβγ) will follow an input ramp (plus constant velocity target for the case of αβ and acceleration for the case of αβγ) with no steady-state mean error. However, for a unity detection probability (i.e. $P_d = 1$) when target measurements are made, each dynamic tracker steady-state mean error is determined as follows.

Following Bolgler (1990), for a constant second derivative (i.e. constant accelerator d^2x/dt^2) input, $\alpha\beta$ tracking error is

$$\lim_{\to\infty}(x(k) - x_s(k)) = \frac{a}{\beta}T^2(1 - \alpha) \tag{12.32}$$

However, for a constant third derivative (i.e. $da/dt = d^3x/dt^3$ = constant) input, $\alpha\beta\gamma$ tracking error is

$$\lim_{\to\infty}(x(k) - x_s(k)) = \frac{\dot{a}}{2\gamma}T^3(1 - \alpha) \tag{12.33}$$

where $\dot{a}$ is the acceleration rate of change; i.e. (da/dt).

12.2.4 Kalman filter

Kalman filtering theory has been discussed in Chapter 11, section 11.4.2. The theory assumes that there is a target, which obeys the dynamic model of the filter, and the sensor measurements that update the filter state estimates are from the target. This assumption is loaded: it may not be always satisfied if tracking is performed in a noisy environment. The probability of detection would not be unity: the optimal estimate in (11.53) would require some modification, which will be discussed later in the text.

The missile trajectory example in Chapter 11, Example 11.4, has demonstrated how Kalman filtering theory can be utilized to solve a linear problem. In the example, the noise and measurement variances are assumed to be the same and to have accounted for signal scintillation from the target and environment. In practice, however, the variances tend to vary with sensor and signal processing characteristics. Barton (1988) formulated the relationships between the variances and sensor and signal processing characteristics as:

$$\sigma_r = \frac{c}{4B_w\sqrt{\frac{S}{N}}} \tag{12.34}$$

$$\sigma_\theta = \frac{\lambda}{2D\sqrt{\frac{S}{N}}} \tag{12.35}$$

$$\sigma_{\dot{r}} = \frac{\lambda}{4T_\Delta\sqrt{\frac{S}{N}}} \tag{12.36}$$

where

σ_i = standard deviation of the variable i of interest
c = speed of light (m/s^2)
λ = wavelength of the radar signal (m)
B_w = bandwidth of the radar signal (Hz)
D = radar aperture diameter (m)
T_Δ = duration of the waveform (s)
S/N = radar signal-to-noise ratio.

With experience, the standard deviations of radar-range, azimuth and range-rate measurements may also be expressed as fraction of the sensor's resolution cell in respective domain.

12.2.4.1 Non-manoeuvring target tracking

One of the merits of Kalman filtering algorithms, as noted in Chapter 11, is their applicability to practical prediction problems, in particular for *non-manoeuvring* and *manoeuvring* cases. If a target is following a nominally straight-line constant velocity trajectory, but is subjected to small, random accelerations due to external forces, then the target is said to be *non-manoeuvring*. The application of Kalman filtering theory to these cases requires different initialization conditions. As an illustration, consider the observations of an aircraft range and bearing being made at regular intervals of T seconds. The objective is to track the target and estimate its kinematics: range (r), range rate ($\dot{r}$), bearing (θ) and bearing rate ($\dot{\theta}$). A simple dynamic, discrete model for the target's kinematics can be written as

$$r(k+1) = r(k) + T\dot{r}(k) \tag{12.37}$$

$$\dot{r}(k+1) = \dot{r}(k) + T\ddot{r}(k) \tag{12.38}$$

$$\theta(k+1) = \theta(k) + T\dot{\theta}(k) \tag{12.39}$$

$$\dot{\theta}(k+1) = \dot{\theta}(k) + T\ddot{\theta}(k) \tag{12.40}$$

where $T = 1/\text{scan rate}$.

The measurement equations can be written as

$$y_r(k) = r(k) + \sigma^2_{vr}(k) \tag{12.41}$$

$$y_\theta(k) = \theta(k) + \sigma^2_{v\theta}(k) \tag{12.42}$$

The measurement noises are assumed additive with zero means: only their variances σ^2_{vr} and $\sigma^2_{v\theta}$ are given by (12.41) and (12.42) respectively. Two independent channels, or parallel filters, appear in these equations because target range and bearing are independent variables. If the variances σ^2_{vr} and $\sigma^2_{v\theta}$ are unknown, the use of (12.34) and (12.35) would be appropriate respectively.

Since the range and bearing acceleration components are assumed to be uncorrelated from one radar scan to another, from (12.37) through to (12.40), the target state equations can be arranged in the state space format:

$$\begin{bmatrix} r(k+1) \\ \dot{r}(k+1) \\ \theta(k+1) \\ \dot{\theta}(k+1) \end{bmatrix} = \begin{bmatrix} 1 & T & 0 & 0 \\ 0 & 1 & 0 & 0 \\ 0 & 0 & 1 & T \\ 0 & 0 & 0 & 1 \end{bmatrix} \begin{bmatrix} r(k) \\ \dot{r}(k) \\ \theta(k) \\ \dot{\theta}(k) \end{bmatrix} + \begin{bmatrix} 0 \\ T\ddot{r}(k) \\ 0 \\ T\ddot{\theta}(k) \end{bmatrix} \tag{12.43}$$

Concisely written as

$$\mathbf{x}(k+1) = A\mathbf{x}(k) + B\mathbf{u}(k) \tag{12.44}$$

Given that the radial acceleration is constant, it is useful to assume that random disturbances are uniformly distributed within $\pm\zeta$, since targets (aircraft, missiles, etc.) can accelerate, or decelerate, in either range (r) or bearing (θ) directions to a maximum value $T\zeta$. For example, by letting $\zeta = \Delta g$, and assuming that the disturbance in range and bearing are uncorrelated with zero means, the system noise matrix is:

$$B = \begin{bmatrix} 0 \\ \zeta^2 T^2 \\ 0 \\ \frac{\zeta^2 T^2}{\bar{R}^2} \end{bmatrix} \tag{12.45a}$$

where g is the acceleration due to gravity ($g \approx 9.8\,\mathrm{m/s^2}$), Δ is a value which may be less than or equal to unity and $\bar{R}$ is the average range. Also

$$Q = \begin{bmatrix} 1 & 0 & 0 & 0 \\ 0 & 1 & 0 & 0 \\ 0 & 0 & 1 & 0 \\ 0 & 0 & 0 & 1 \end{bmatrix} = \text{diagonal matrix} \tag{12.45b}$$

And the measurement equations:

$$\begin{bmatrix} y_r(k) \\ y_\theta(k) \end{bmatrix} = \begin{bmatrix} 1 & 0 & 0 & 0 \\ 0 & 0 & 1 & 0 \end{bmatrix} \begin{bmatrix} r(k) \\ \dot{r}(k) \\ \theta(k) \\ \dot{\theta}(k) \end{bmatrix} + \begin{bmatrix} \sigma_{vr}^2 \\ \sigma_{v\theta}^2 \end{bmatrix} \tag{12.46}$$

Employing the first two measurements, the initial conditions of the Kalman filter tracker are

$$\hat{r}(1) = y_r(1) \tag{12.47a}$$

$$\hat{\dot{r}}(1) = \frac{1}{T}(y_r(1) - y_r(0)) \tag{12.47b}$$

$$\hat{\theta}(1) = y_\theta(1) \tag{12.47c}$$

$$\hat{\dot{\theta}}(1) = \frac{1}{T}\{y_\theta(1) - y_\theta(0)\} \tag{12.47d}$$

From these initial conditions, the initial conditions for the error covariance matrix $P(1)$ are

$$P(1) = E\left\{(\hat{x}(1) - x(1))(\hat{x}(1) - x(1))^T\right\} \tag{12.48}$$

which expresses, by definition, to

$$P(1) = \begin{bmatrix} P_{11} & P_{12} & P_{13} & P_{14} \\ P_{21} & P_{22} & P_{23} & P_{24} \\ P_{31} & P_{32} & P_{33} & P_{34} \\ P_{41} & P_{42} & P_{43} & P_{44} \end{bmatrix}_{(1)} \tag{12.49}$$

Note that $E\{.\}$ is the expectance of $\{.\}$. Since there is no coupling between the range and bearing terms, the elements of the error covariance matrix $P(1)$ equate to the following:

$$P_{13} = P_{14} = P_{23} = P_{24} = P_{31} = P_{32} = P_{41} = P_{42} = 0 \tag{12.50}$$

Hence (12.49) reduces to

$$P(1) = \begin{bmatrix} P_{11} & P_{12} & 0 & 0 \\ P_{21} & P_{22} & 0 & 0 \\ 0 & 0 & P_{33} & P_{34} \\ 0 & 0 & P_{43} & P_{44} \end{bmatrix}_{(1)} \tag{12.51}$$

All that is needed is to define the elements in (12.51). Note that the measurement noise in range may be expressed by

$$\hat{x}(1) - x(1) = \hat{r}(1) - r(1) = y_r(1) - r(1) \tag{12.52a}$$

As a result,

$$P_{11} = \sigma_{vr}^2 \tag{12.52b}$$

$$P_{22} = E\left\{[\hat{\dot{r}}(1) - \dot{r}(1)]^2\right\} = \frac{2\sigma_{v\theta}^2}{T^2} + T^2\zeta^2 \tag{12.52c}$$

Following the above procedure other elements can be expressed.

$$P_{12} = \frac{1}{T}\sigma_{vr}^2 = P_{21} \tag{12.52d}$$

$$P_{33} = \sigma_{v\theta}^2 \tag{12.52e}$$

$$P_{34} = \frac{1}{T}\sigma_{v\theta}^2 = P_{43} \tag{12.52f}$$

And finally

$$P_{44} = E\left\{\left[\hat{\dot{\theta}}(1) - \dot{\theta}(1)\right]^2\right\} = \frac{2\sigma_{v\theta}^2}{T^2} + \frac{T^2\zeta^2}{y_\theta^2(1)} \tag{12.52g}$$

Substituting (12.52) in (12.51), the initial condition for the error covariance matrix is written:

$$P(1) = \begin{bmatrix} \sigma_{vr}^2 & \frac{\sigma_{vr}^2}{T} & 0 & 0 \\ \frac{\sigma_{vr}^2}{T} & \left\{\frac{2\sigma_{vr}^2}{T^2} + T^2\zeta^2\right\} & 0 & 0 \\ 0 & 0 & \sigma_{v\theta}^2 & \frac{\sigma_{v\theta}^2}{T^2} \\ 0 & 0 & \frac{\sigma_{v\theta}^2}{T^2} & \left\{\frac{2\sigma_{v\theta}^2}{T^2} + \frac{T^2\zeta^2}{y_\theta^2(1)}\right\} \end{bmatrix} \tag{12.53}$$

Complete information is now available for starting the Kalman tracker, using the algorithms in Table 11.1, Chapter 11.

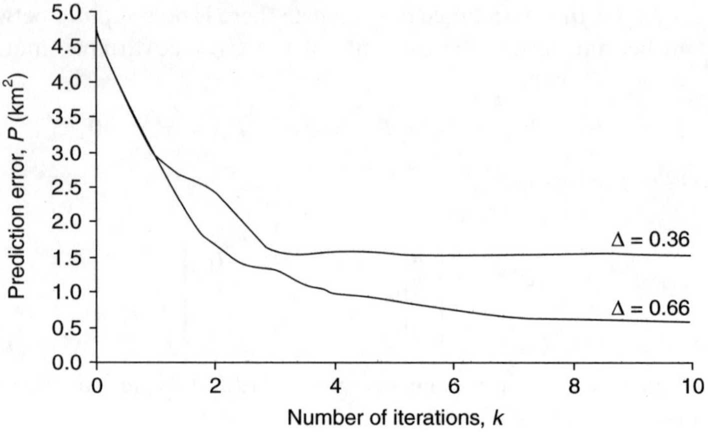

Figure 12.7 Plot of range prediction errors

Example 12.1 A tracking system registers measurement root-mean-squared errors of 1 km and 1° in range and bearing respectively. Assume Δ values of 0.36 and 0.66. Compute the system's range prediction error and Kalman gain when sampled at 5 second intervals.

Figures 12.7 and 12.8 show the computed results of the range prediction error and Kalman gain respectively. It can be seen that increasing the Δ value improves the prediction error by reaching a steady state quickly. Caution must be exercised when attempting to improve overall Kalman filter prediction error by maintaining a balance between practical environment and theory in order to ensure system stability (Kolawole 1994).

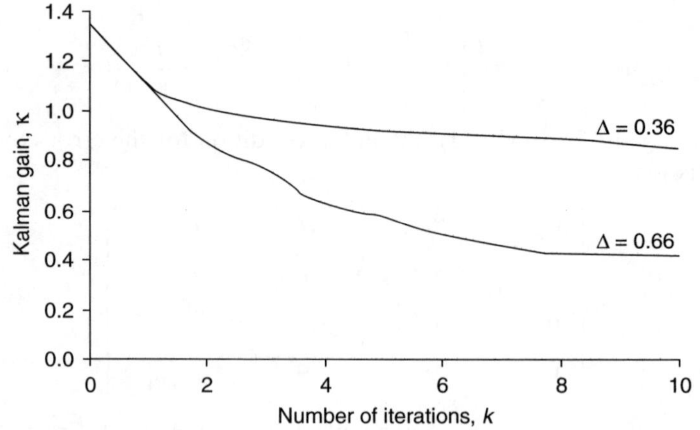

Figure 12.8 Kalman gain

12.2.4.2 Manoeuvring target tracking

Since it is widely assumed that manoeuvres of aircraft tracks are more likely to be due to a heading (bearing) change rather than a speed change, it seems reasonable to consider that there may be some correlation from scan to scan of the acceleration components u_r and u_θ. Instead of a white noise approach to modelling the uncorrelated distances for constant acceleration as in (12.44), a more realistic approach must include the effect of correlation to ensure optimal tracking. The simplest, and arguably the most robust, model for manoeuvres is developed by considering the target acceleration to be exponentially correlated:

$$E\{\mathbf{u}\mathbf{u}^T\}_{k+\Delta s} = aa^{\Delta s} E\{\mathbf{u}\mathbf{u}^T\}_k \tag{12.54a}$$

where '*aa*' relates to the radar and vehicle dynamics and superscript '*T*' denotes transposition. For the purpose of clarity, denote $\Delta s = 1$ and $aa = e^{-\lambda_\Delta T}$, where λ_Δ is the manoeuvre correlation coefficient whose inverse equates to the average manoeuvre duration of the vehicle, and T the sampling interval. So

$$E\{\mathbf{u}\mathbf{u}^T\}_{k+\Delta s} = e^{-\lambda_\Delta T} E\{\mathbf{u}\mathbf{u}^T\}_k \tag{12.54b}$$

Let us explore how the model in (12.54) can be employed in the general framework of the state equation of (12.44). Define

$$\mathbf{x} = \begin{bmatrix} x \\ \dot{x} \\ \ddot{x} \end{bmatrix} \tag{12.55}$$

where $\ddot{x}$ is the acceleration. Like (11.42a), the *continuous-time* state equation is written as

$$\dot{\mathbf{x}}(\mathbf{t}) = \Phi\mathbf{x}(\mathbf{t}) + \mathbf{u}(\mathbf{t}) \tag{12.56}$$

where, in this instance, the system noise matrix B is an identity matrix and the process (plant) noise

$$\mathbf{u} = \begin{bmatrix} 0 \\ \tilde{u} \\ 0 \end{bmatrix} \tag{12.57a}$$

And the state transition vector has the form

$$\Phi = \begin{bmatrix} 0 & 1 & 0 \\ 0 & 0 & 1 \\ 0 & 0 & -\lambda_\Delta \end{bmatrix} \tag{12.57b}$$

The homogeneous solution of equation (12.56) is

$$e^{\Phi t} \tag{12.58a}$$

which is the *fundamental matrix*, A. And when defined by the power series expansion:

$$e^{\Phi t} = \sum_{k=0}^{\infty} \frac{\Phi^k t^k}{k!} \tag{12.58b}$$

Note that $0! = 1$. If the input variables are considered constant within a constant interval of time, that is, $t_k - t_{k-1} = T$, and substituting (12.57b) in (12.58b), the fundamental matrix becomes

$$A = \begin{bmatrix} 1 & T & \frac{\lambda_\Delta T - 1 + e^{-\lambda T}}{\lambda_\Delta^2} \\ 0 & 1 & \frac{1 - e^{-\lambda_\Delta T}}{\lambda_\Delta} \\ 0 & 0 & e^{-\lambda_\Delta T} \end{bmatrix} \tag{12.58c}$$

Following the procedure in section 11.4.1, the differential state equation in (12.56) can be written in the discrete-time format:

$$\mathbf{x}(k+1) = A\mathbf{x}(k) + \mathbf{u}(k) \tag{12.59a}$$

for a constant time, T, between measurement. Define the measurement equation as

$$\mathbf{y}(k+1) = C\mathbf{x}(k+1) + \mathbf{v}(k) \tag{12.59b}$$

Suppressing the time index k, the noise has the covariance defined as

$$E\{\tilde{u}\tilde{u}^T\} = \mathbf{Q} = \sigma_{ac}^2 \begin{bmatrix} q_{11} & q_{12} & q_{13} \\ q_{21} & q_{22} & q_{23} \\ q_{31} & q_{32} & q_{33} \end{bmatrix} \tag{12.60}$$

where σ_{ac}^2 is the variance of target acceleration. Other elements are

$$q_{11} = \frac{1}{\lambda_\Delta^4}\left[1 + 2\lambda_\Delta T - 2\lambda_\Delta^2 T^2 + \frac{2}{3}\lambda_\Delta^3 T^3 - e^{-\lambda_\Delta T}\left(4\lambda_\Delta T + e^{-\lambda_\Delta T}\right)\right] \tag{12.61a}$$

$$q_{12} = q_{21} = \frac{1}{\lambda_\Delta^3}\left[1 - 2\lambda_\Delta T + \lambda_\Delta^2 T^2 + e^{-\lambda_\Delta T}\left(2\lambda_\Delta T - 2 + e^{-\lambda_\Delta T}\right)\right] \tag{12.61b}$$

$$q_{13} = q_{31} = \frac{1}{\lambda_\Delta^2}\left[1 - e^{-\lambda_\Delta T}\left(2\lambda_\Delta T + e^{-\lambda_\Delta T}\right)\right] \tag{12.61c}$$

$$q_{22} = \frac{1}{\lambda_\Delta^2}\left[2\lambda_\Delta T - 3 + e^{-\lambda_\Delta T}\left(4 - e^{-\lambda_\Delta T}\right)\right] \tag{12.61d}$$

$$q_{23} = q_{32} = \frac{1}{\lambda_\Delta}\left[1 - e^{-\lambda_\Delta T}\left(2 - e^{-\lambda_\Delta T}\right)\right] \tag{12.61e}$$

$$q_{33} = 1 - e^{-2\lambda_\Delta T} \tag{12.61f}$$

12.2.4.3 Initiating filter for manoeuvring targets

Like the non-manoeuvring case, the state estimations will need to be initialized first followed by their corresponding error covariance. Thus

$$\hat{x}(1) = y(1) \tag{12.62a}$$

$$\hat{\dot{x}} = \frac{y(1) - y(0)}{T} \tag{12.62b}$$

$$\hat{\ddot{x}}(1) = 0 \tag{12.61c}$$

where $y(0)$ and $y(1)$ correspond to the first and second measurements received. The initial error covariance

$$P(1) = \begin{bmatrix} P_{11} & P_{12} & P_{13} \\ P_{21} & P_{22} & P_{23} \\ P_{31} & P_{32} & P_{33} \end{bmatrix}_{(1)} \tag{12.63a}$$

whose elements are defined thus

$$P_{11}(1) = \sigma_{vr}^2 \tag{12.63b}$$

$$P_{12}(1) = \frac{\sigma_{vr}^2}{T} \tag{12.63c}$$

Since there is no coupling between the position, x, and the acceleration, $\ddot{x}$, components,

$$P_{13}(1) = P_{31}(1) = 0 \tag{12.63d}$$

$$P_{22}(1) = \frac{2}{T^2}\sigma_{vr}^2 - \frac{\sigma_{ac}^2}{\lambda_\Delta^2}\left(1 - \frac{2}{3}T\lambda_\Delta\right) + \frac{2\sigma_{ac}^2}{T^2\lambda_\Delta^4}(1 - e^{-\lambda_\Delta T}[1 + T\lambda_\Delta]) \tag{12.63e}$$

$$P_{23}(1) = \frac{2}{T\lambda}(T\lambda + e^{-T\lambda} - 1)\sigma_m^2 \tag{12.63f}$$

$$P_{33}(1) = \sigma_{ac}^2 \tag{12.63g}$$

By symmetry, $P_{12}(1) = P_{21}(1)$ and $P_{23}(1) = P_{32}(1)$. (12.63h)

As noted earlier in the text, independent channels or parallel filters can process the tracking system because target range and bearing are independent variables. Thus the system can be decoupled, or partitioned, for processing. By extending the estimates to the second derivatives of the range (r, $\dot{r}$, $\ddot{r}$) and bearing (θ, $\dot{\theta}$, $\ddot{\theta}$) terms, the error covariance matrix $\mathbf{P}$ can be written as comprising submatrices of the order corresponding to these terms:

$$\mathbf{P} = \begin{bmatrix} P_r & | & 0 \\ - & | & - \\ 0 & | & P_\theta \end{bmatrix} = \mathrm{diag}(P_r, P_\theta) \tag{12.64}$$

where P_r is a 3 by 3 error covariance matrix of the range terms whose elements are defined by (12.63) and P_θ is the error covariance of the bearing terms. Following a similar procedure to obtaining (12.63), and replacing subscript 'r' with 'θ', the bearing terms can equally be developed. Hence, for the state vector described by

$$\mathbf{x} = \begin{bmatrix} r \\ \dot{r} \\ \ddot{r} \\ \theta \\ \dot{\theta} \\ \ddot{\theta} \end{bmatrix} \tag{12.65}$$

its covariance matrix

$$\mathbf{P} = \begin{bmatrix} P_{11} & P_{12} & P_{13} & 0 & 0 & 0 \\ P_{21} & P_{22} & P_{23} & 0 & 0 & 0 \\ P_{31} & P_{32} & P_{33} & 0 & 0 & 0 \\ 0 & 0 & 0 & P_{44} & P_{45} & P_{46} \\ 0 & 0 & 0 & P_{54} & P_{55} & P_{56} \\ 0 & 0 & 0 & P_{64} & P_{65} & P_{66} \end{bmatrix} \tag{12.66}$$

Complete information has now been assembled to start a Kalman filter tracker for both non-manoeuvring and manoeuvring targets.

12.2.5 Summary of tracking filters

In conclusion, the αβ, αβγ and Kalman filters are of the fading memory type. The choice of tracker to use depends on the complexity, accuracy and requirements of the mission.

The αβγ filters define pre-computed gains for a three-state data filter. This type of filtering is accurate if a tracking is restricted to a constant manoeuvre. Problems associated with transient response and incomplete data could invalidate the tracking performance if gain calculations are not performed adaptively. However, the filters are useful in preliminary systems design and performance prediction stage even if Kalman-based filter tracker is eventually used.

Kalman-based tracking filters are data filters wherein models are postulated for the filter accuracy and whose gains change according to the prescribed models. They handle manoeuvres, provide a better response to initialization transients, measure accuracy fluctuations, and minimize loss of detection. Tracking performance, however, can be degraded if miscorrelation is present in the filter covariance and/or performed in a cluttered environment.

12.3 Tracking with PDA filter in a cluttered environment

In most radar tracking problems, the returns from the target(s) of interest are sought within a time interval determined by the delay corresponding to the anticipated range of the target when it reflects the energy transmitted by the radar. The outline of a PDA filter is shown in Figure 12.9. A procedure that computes the probabilities that detections are from the target of the validation gate measurements and that enables assignments of plots to tracks is called the *probabilistic data association* (PDA). More is said of PDA and gating in the next subsection.

Here the returns from a scanned area are input to a receiver and peak-detection process. The detection process has been discussed in Chapter 10. These returns are assumed to contain no origin identification and are from both clutter and targets. At time k the receiver and detection process forms measurements $y(k) = \{y_1(k), y_2(k), \ldots, y_n(k)\}$. The measurements $y(k)$ from the receiver and detection process are input to a selection process, which also has input from the tracking process acting as feedback from previous track data, $T_t(k-1)$ up to the time $k-1$. For every track $T_t(k)$ in the scanned region, there is a corresponding target. The track is assumed to contain the target's state estimate $\hat{x}(k)$ and covariance $P(k)$. The states of the targets $x(k)$ are assumed independent of one another and of clutter. The selection process takes each track $T_t(k)$ in turn and uses $P(k)$ with the sensor measurement error to define a valid gate $\nu(k)$, centred on $\hat{x}(k)$, upon which the level of

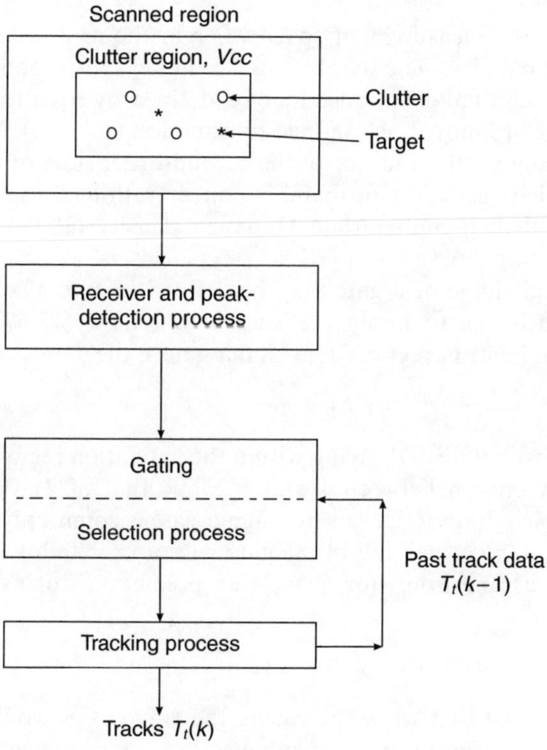

Figure 12.9 Extended sensing and tracking process

confidence placed on the peaks associated with the targets' tracks can be quantified.

Often in practice the radar measurements have multiple components; for instance, range, bearing or azimuth, and elevation. As such, instead of a one-dimensional validation gate, a multi-dimensional gate is set up where multi-dimensional data association with target(s) is made. The selected measurements are used with the previous track data $T(k-1)$ to update the selected target's estimate in the tracking process to give target ith track, $T_i(k)$ with associated states $\hat{x}_i(k)$ and covariance $P_i(k)$. The next task is to formulate the functions that define the selection and track forming processes.

12.3.1 Gating

Gating is a technique of rejecting improbable observation-to-track pairings. The gating, or validation gate, technique needs to be developed that assumes true measurements selected from this gate are made to a high degree of confidence. In the context of Kalman-based tracking filters, a validation gate at time k may be expressed as

$$v(k) = \tilde{\mathbf{y}}(k)S^{-1}(k)\tilde{\mathbf{y}}^T(k) < g_t^2 \tag{12.67}$$

as a region in the measurement space wherein the measurement sought is with high probability. The symbols g_t, S and $\tilde{\mathbf{y}}$ correspond to the gate threshold, the residual covariance matrix (defined by equation (11.64)) and residual vector or innovation (defined by equation (11.56b)). The expression (12.67) is a g-sigma ellipsoid due to the second-order state of the covariance matrix, S, which also is the probability concentration ellipsoid obtained by cutting the tail of a multivariate Gaussian density (already discussed in Chapter 8, section 8.6.3).

The size and shape of a gate may be defined in several ways. A precise method of gating is to apply the statistical test of (12.67). For a two-dimensional validation region ('two-sigma gate'), the gate size is

$$v(k) = \pi g_t^2 \sqrt{|S(k)|} \tag{12.68}$$

with the gate probability, P_g, lying within the validation region. The value of g_t is generally chosen between 1 and 3, while that of P_g depends on the designer's choice. In practice, a three-sigma gate is commonly used ensuring that the measurement will fall in the gate with a probability of 0.998 under the Gaussian assumption. For a one-step prediction, for example, such a gate is within

$$\hat{x}_1(k) \pm 3\sqrt{P_{11}(k)} \tag{12.69}$$

It should be noted that while increasing the value of g_t would increase the probability of associating the correct detections, or plots, with the target, it also increases the probability of associating more false detections with the target. Missed detections may occur if the gate is made too small. Lie (1998) provided a theoretical basis to this practical observation by proposing an enlargement of the filter covariance matrix.

Since the measurements from the validation gate would contain returns from the target(s) and clutter, estimating the target probability with a high degree of confidence poses a major difficulty. This is because targets may be unobservable but produce some returns or may be observable but laden with clutter. Thus the probability is event related, and therefore conditional. It has been coined 'association' or 'event' probability in the literature. Let the ith target event probability be denoted by β_i – an element that is revisited later in this section.

A PDA filter associates all the 'neighbours' to the target of interest. As a result, the information obtained is used subsequently to update the PDA tracks, using detections contained within the validation gate to account for the measurement origin uncertainty. In practice, if more than one measurement is contained within the validation gate or if the gates from two or more tracks overlap, the measurements are assigned to tracks on the basis of 'nearest neighbourhood', as measured by Euclidean distance. The statistical Euclidean distance

$$d^2 = \tilde{y}^T S^{-1} \tilde{y} \tag{12.70}$$

is the weighted norm of the innovation.

Underlying the development of the PDA procedure are two assumptions:

- the event probabilities β_i are known; and
- for every target in a validation region, there are sufficient statistics for its past measurement(s) such that, when sampled at time k, the ith target state estimate $\hat{x}_i(k-1)$ and covariance matrix $P_i(k-1)$ can be determined for all events associated with the target.

For convenience, the events $\Theta_i(k)$ are assumed to have a normal density function, abbreviated as N(mean, variance). Based on the aforelisted assumptions, the PDA procedure is formulated as follows.

All measurements from the validation gate $v(k)$ defined by (12.68) are contained in $y(k)$. At time k, let the following events be conditioned on the measurements from $v(k)$:

$$\Theta_0(k) = \text{none of the measurements originated from the target; in other words, measurements originated from clutter} \tag{12.71}$$

$$\Theta_i(k) = \text{all measurements originated from the target} \tag{12.72}$$

$$\beta_i(k) = \Pr\{\Theta_i(k) \mid y(k)\} = \text{event probability} \tag{12.73}$$

where Pr{.} is the probability of $\{\cdot\}$ and $i = 1, 2, \ldots, m$

The event probability is explored further in section 12.3.2. Note that the symbol 'Pr' is used in this instance to denote probability rather than P, which has been used for error covariance. This expression (12.73) is the *a posteriori* probability that the measurements come from the target.

The above events (12.71) to (12.73) are mutually exclusive and exhaustive. So,

$$\sum_{i=0}^{m} \beta_i(k) = 1 \tag{12.74}$$

At time k, the conditional mean of the target's state can be written as

$$\begin{aligned}\hat{x}(k) &= E\{x(k) \mid y(k)\} \\ &= \sum_{i=0}^{m} E\{x(k) \mid \Theta_i(k), y(k)\} \Pr\{\Theta_i(k) \mid y(k)\}\end{aligned} \tag{12.75}$$

And in view of (12.74)

$$\begin{aligned}\hat{x}(k) &= \sum_{i=0}^{m} \beta_i(k) E\{x(k) \mid \Theta_i(k), y(k)\} \\ &= \sum_{i=0}^{m} \hat{x}_i(k)\beta_i(k)\end{aligned} \tag{12.76}$$

where $\hat{x}_i(k)$ is the updated state estimate conditioned on event $\Theta_i(k)$, on the proviso that the ith validated measurement is truly from the target. The density associated with this instance is $N(\hat{x}_i(k), P(k))$ which reflects that for the Kalman filter. The expression in (12.76) implies that the state estimate is the weighted sum of the estimates of the target's state conditioned on all the target data. The weighting term is the event probability term $\beta_i(k)$. If none of the measurements is from the target (i.e. for $i = 0$), the conditional estimate, by definition, is

$$E\{x(k) \mid \Theta_0(k)\} = \hat{x}_0(k) = \hat{x}(k-1) \tag{12.77}$$

with associated density function $N(\hat{x}_i(k-1), P(k-1))$ because the current target state will be independent of other targets.

For $i = 1, 2, \ldots, m$, the conditional estimate is

$$E\{x(k) \mid \Theta_i(k)\} = E\{x(k) \mid y_i(k)\} = \hat{x}_i(k) \tag{12.78}$$

with its associated density function $N(\hat{x}_i(k), P(k))$ since there are corresponding target measurements available in the validation gate. Following (11.62), the expression in (12.78) can be written as

$$\hat{x}_i(k) = \hat{x}_i(k-1) + \kappa(k)\tilde{y}_i(k) \tag{12.79}$$

where $\tilde{y}_i$ is the ith target innovation.

Substituting (12.77) and (12.78) in (12.76), the conditional state estimate is written as

$$\hat{x}(k) = \hat{x}(k-1) + \kappa(k) \sum_{i=1}^{m} \beta_i(k)\tilde{y}_i \tag{12.80}$$

By comparing (12.80) with (11.62), it is seen that the $\tilde{y}$ in (11.62) is replaced with the *combined innovation* (that is, the sum of the weighted measurements, $\sum_{i=1}^{m} \beta_i(k)\tilde{y}_i$).

The associated covariance $P(k)$ of the conditional mean of the target state can be estimated, which, by definition, is the covariance of the conditional mean of the target state:

$$P(k) = \sum_{i=0}^{m} \beta_i(k) \int [x(k) - \hat{x}(k)][x(k) - \hat{x}(k)]^T \ \Pr(x|\Theta_i(k))dx \tag{12.81}$$

For index $i = 1, 2, \ldots, m$, write $x(k) - \hat{x}(k)$ as $[x(k) - \hat{x}_i(k) + \hat{x}_i(k) - \hat{x}(k)]$ and substitute the expanded terms in (12.81). The integral component of the expanded expression would become

$$\begin{aligned} &\int [x(k) - \hat{x}_i(k)][x(k) - \hat{x}_i(k)]^T N(\hat{x}_i(k), P(k))dx \\ &+ [\hat{x}_i(k) - \hat{x}(k)][\hat{x}_i(k) - \hat{x}(k)]^T \int N(\hat{x}_i(k), P(k))dx \\ &+ [\hat{x}_i(k) - \hat{x}(k)] \int [x(k) - \hat{x}_i(k)]^T N(\hat{x}_i(k), P(k))dx \\ &+ \int [x(k) - \hat{x}_i(k)]N(\hat{x}_i(k), P(k))dx[\hat{x}_i(k) - \hat{x}(k)]^T \end{aligned} \tag{12.82}$$

It may be shown that the first integral evaluates to $P(k)$. The second integral evaluates to unity, while the third and fourth integrals evaluate to zero. For a particular case of $i = 0$, replace the subscript 'i' with '0' and define the normal density function as $N(\hat{x}_0(k), P(k))$. Then follow the same procedure as in (12.82). Subsequently, the first integral term would evaluate to $P(k-1)$ and while the second, third and fourth integrals will evaluate as unity, zero and zero respectively. In view of (12.77) and (12.78), the associated covariance of the conditional mean of the target state can thus be written as

$$\begin{aligned} P(k) = \beta_0(k)P(k-1) + \sum_{i=1}^{m} \beta_i(k)P(k) + \sum_{i=1}^{m} \beta_i(k)\hat{x}_i(k)\hat{x}_i^T(k) \\ - \hat{x}(k)\hat{x}^T(k) \end{aligned} \tag{12.83}$$

The covariance of the conditional mean of the target state may alternatively be written as

$$\begin{aligned} P(k) = \beta_0(k)P(k-1) + [1 - \beta_0(k)]P(k) \\ + \kappa(k)\left[\sum_{i=1}^{m} \beta_i(k)\tilde{y}_i\tilde{y}_i^T - \tilde{y}\tilde{y}^T\right]\kappa^T(k) \end{aligned} \tag{12.84}$$

This expression comprises a sum of three terms, which are modified by the event probabilities:

- the covariance when all measurements are clutter;
- the covariance when all measurements correctly update the filter. Track merit or confidence is derived from the low-pass filtering by the $[1 - \beta_0(k)]$ term; and
- the uncertainty of the weighted innovation $\tilde{y}$.

By comparing (12.84) with the Kalman filter covariance equation (11.61), it can be seen that the conditional covariance in (12.84) is dependent on target measurements while the basic covariance (11.61) is independent of target measurements.

12.3.2 Formulation of the event probability

If one assumes that there are m out of n detections within the validation region and that previous data during previous scans, denoted by $z(k)$, are available and are also in region, then the event probability expression in (12.73) can be restated and expanded as

$$\beta_i(k) = \Pr\{\Theta_i(k)\,|\,m, n, z(k), y(k)\} \tag{12.85}$$

For simplicity, the index k will be omitted in subsequent development. Equation (12.85) can be restructured using the Bayesian rule of Chapter 8, section 8.2.1:

$$\beta_i(k) = \frac{\Pr(y\,|\,\Theta_i, m, n, z)\Pr(\Theta_i\,|\,m, n, z)}{\Pr(y\,|\,m, n, z)} \tag{12.86}$$

Like (8.14), the denominator can be expressed as

$$\Pr(y\,|\,m, n, z) = \sum_{i=0}^{m} \Pr(y\,|\,\Theta_i, m, n, z)\Pr(\Theta_i\,|\,m, n, z) \tag{12.87}$$

It is clear that (12.87) is a normalization constant. It could be said that the numerator of (12.86) contains two probability terms:

probability of the event occurring in the current data, $\Pr(y\,|\,\Theta_i, m, n, z)$; and probability of the event conditioned on number of detections, $\Pr(\Theta_i\,|\,m, n, z)$.

The task now is to define each probability term in the numerator. The developments in the next two subsections draw from Kolawole (1996) and Richards (1992).

12.3.3 Probability of current data

If a set of detections y_j is considered to occur independently under the hypothesis Θ_i, then the probability of the event occurring in the current data set can be expressed as

$$\Pr(y \mid \Theta_i, m, n, z) = \prod_{j=1}^{n} \Pr(y_j \mid \Theta_i, m, n, z) \tag{12.88}$$

In practice, $\Pr(y_j \mid \Theta_i, m, n, z)$ can only be obtained by considering the probability distribution function of the set of detections. If it is possible to distinguish between the target and clutter distributions in the validation volume V_v, so also their volumes can be distinguished. As such, define V_{cc} and V_t as the volume of clutter and target in the validation volume respectively, and their distributions by $f_c(y_j)$ and $f_t(y_j)$ respectively. Thus, the probability of the event occurring in the currently selected data set is

$$\Pr(y_j \mid \Theta_i, m, n, z) = \begin{cases} f_t(y_j) & j = i \\ f_c(y_j) & j \neq i \end{cases} \tag{12.89}$$

Upon an application of the likelihood ratio definition of (9.23), the likelihood ratio of discerning the distributions of the clutter from that of the target in the set is

$$L(y_j) = \frac{f_t(y_j)}{f_c(y_j)} \tag{12.90}$$

By substituting (12.89) and (12.90) in (12.88), two probability states are developed. Specifically,

- when the ith target is observable[1] and not detected, or detected and not observable, which is when $i = 0$; and
- the ith target is observable, detected and selected among the detections, that is, when $i > 0$.

The probability for the two states identified is thus expressed by

$$\Pr(y \mid \Theta_i, m, n, z) = \begin{cases} \prod_{j=1}^{n} f_c(y_j) & i = 0 \\ L(y_j) \prod_{j=1}^{n} f_c(y_j) & i > 0 \end{cases} \tag{12.91}$$

12.3.4 Probability of event conditioned on detection

If the probability of the current data y occurring under the hypothesis Θ_i is assumed to be independent of the previous data, z, then the probability of the event conditioned on the number of detections may be written as

$$\Pr(\Theta_i \mid m, n, z) = \Pr(\Theta_i \mid m, n) \tag{12.92}$$

[1] Another state has been described by Colegrove *et al.* (1986) and Richards (1992) as the event hypothesis when the target is not observable and not detectable, i.e. for Θ_i when $i = -1$. Even with this additional state, equation (12.89) still holds.

It is possible that for m out of n true target detections in the validation region, there may be m_f out of n_f false detections. If so, from the possible values of m_f out of n_f, the probability of the event conditioned on the number of detections may be expressed by

$$\begin{aligned}\Pr(\Theta_i \mid m,n) = &\Pr(\Theta_i \mid m_f = m, n_f = n-m, m, n)\Pr(m_f = m, n_f = n-m \mid m,n) \\ &+ \Pr(\Theta_i \mid m_f = m-1, n_f = n-m, m, n)\Pr(m_f = m-1, n_f = n-m \mid m,n) \\ &+ \Pr(\Theta_i \mid m_f = m, n_f = n-m-1, m, n)\Pr(m_f = m, n_f = n-m-1 \mid m,n)\end{aligned} \tag{12.93}$$

where

(a) $\Pr(\Theta_i \mid m_f = m,\ n_f = n-m,\ m,\ n)$ = the probability that the target is not detected whether it is observable or not
(b) $\Pr(\Theta_i \mid m_f = m-1,\ n_f = n-m,\ m,\ n)$ = the probability that the target is selected from the validation region
(c) $\Pr(\Theta_i \mid m_f = m,\ n_f = n-m-1,\ m,\ n)$ = the probability that the target is not selected.

The task now is to consider each of the probability terms comprising (12.93) to evaluate (12.92). This would require an application of the probability rules discussed in Chapter 8, equation (8.2) through to (8.18).

It is appropriate at this stage to introduce some notations and definitions that will assist in expressing each of the probability terms in terms of tracker settings, namely,

- P_o = the probability that the target can be observed;
- P_d = the probability that the target can be detected;
- P_g = gate probability: the probability that the target is lying within a validation gate.

If an arbitrary divide can be made of the target portion from the clutter portion in the validation gate or region, a model could be made of the target and clutter (false points) distributions with known statistical distribution models. Some samples of known distribution models have already been discussed in Chapter 8, section 8.6. It is useful to assume that the number of detections in the selected target region and clutter region to be independent. In practical situations, the number of false measurements n_f, or detection points n, is large calling into use the Poisson parametric model.

As demonstrated in Chapter 8, section 8.6, a Poisson density with parameter λ can be expressed independently for the clutter and target distributions, such as

$$\text{Target}: \quad \mu_t(m) = \frac{\lambda_t^m}{m!}e^{-\lambda_t} \tag{12.94}$$

$$\text{Clutter}: \quad \mu_c(n) = \frac{\lambda_c^n}{n!}e^{-\lambda_c} \tag{12.95}$$

where λ_c and λ_t correspond to the spatial density of false and target measurements (i.e. the average number per unit volume).

A non-parametric model could also be used as a 'diffuse' prior. In which case

$$\mu_t(m) = \frac{1}{M} \quad m = 0, 1, 2, \ldots, M-1 \tag{12.96a}$$

$$\mu_c(n) = \frac{1}{N} \quad n = 0, 1, 2, \ldots, N-1 \tag{12.96b}$$

The probability terms identified in (12.93) can therefore be defined as follows.

(a) The probability that the target is not detected whether it is observable, or not:

$$\Pr(\Theta_i \mid m_f = m, n_f = n - m, m, n) = \begin{cases} \frac{P_o(1-P_d)}{1-P_oP_d} & i = 0 \\ 0 & \text{otherwise} \end{cases} \tag{12.97}$$

(b) The probability that the target is selected from the validation region:

$$\Pr(\Theta_i \mid m_f = m - 1, n_f = n - m, m, n) = \begin{cases} \frac{1}{m} & 0 < i \leq m \\ 0 & i \leq 0, m < i \leq n \end{cases} \tag{12.98}$$

(c) The probability that the target is not selected:

$$\Pr(\Theta_i \mid m_f = m, n_f = n - m - 1, m, n) = \begin{cases} 0 & i \leq m \\ \frac{1}{n-m} & m < i \leq n \end{cases} \tag{12.99}$$

The other probability terms associated with whether the target is observable, selected, or not selected in the validation region are expressed using the Bayesian rule as follows.

$$\Pr(m_f = m, n_f = n - m \mid m, n) = \frac{\Pr(m, n \mid m_f = m, n_f = n - m)\Pr(m_f = m, n_f = n - m)}{\Pr(m, n)} \tag{12.100}$$

$$\Pr(m_f = m - 1, n_f = n - m \mid m, n) = \frac{\Pr(m, n \mid m_f = m - 1, n_f = n - m)\Pr(m_f = m - 1, n_f = n - m)}{\Pr(m, n)} \tag{12.101}$$

$$\Pr(m_f = m, n_f = n - m - 1 \mid m, n) = \frac{\Pr(m, n \mid m_f = m, n_f = n - m - 1)\Pr(m_f = m, n_f = n - m - 1)}{\Pr(m, n)} \tag{12.102}$$

where the denominator $\Pr(m, n)$ is expressed by

$$\begin{aligned}\Pr(m, n) = {} & \Pr(m_f = m, n_f = n - m \mid m, n) + \Pr(m_f = m - 1, n_f = n - m \mid m, n) \\ & + \Pr(m_f = m, n_f = n - m - 1 \mid m, n)\end{aligned} \tag{12.103}$$

By taking the number of detections and false measurements as independent, their combined probability becomes the product of their density functions. Specifically,

$$\Pr(m_f = m, n_f = n) = \mu_t(m)\mu_c(n) \tag{12.104a}$$
$$\Pr(m_f = m, n_f = n - m) = \mu_t(m)\mu_c(n - m) \tag{12.104b}$$
$$\Pr(m_f = m, n_f = n - m - 1) = \mu_t(m)\mu_c(n - m - 1) \tag{12.104c}$$

The other terms, comprising the components in (12.100) through to (12.102), are defined as follows:

$$\Pr(m, n \mid m_f = m, n_f = n - m) = 1 - P_o P_d \tag{12.105a}$$
$$\Pr(m, n \mid m_f = m - 1, n_f = n - m) = P_g P_o P_d \tag{12.105b}$$
$$\Pr(m, n \mid m_f = m, n_f = n - m - 1) = (1 - P_g) P_o P_d \tag{12.105c}$$

And finally, by substituting (12.104) and (12.105) in (12.103) yields

$$\begin{aligned}\Pr(m, n) = {} & \mu_t(m)\mu_c(n - m)(1 - P_o P_d) + \mu_t(m - 1)\mu_c(n - m) P_o P_d P_g \\ & + \mu_t(m)\mu_c(n - m - 1) P_o P_d (1 - P_g)\end{aligned} \tag{12.106}$$

By careful rearrangement and substitution of equations (12.88) through to (12.106) in (12.86), and taking the spatial density of false and target measurements to be the same (i.e. $\lambda_c = \lambda_t = \lambda_v \cong n/V_v$), a complete expression of the event probability β_i (i.e. (12.73)) is readily obtained for all events:

$$\beta_0 = \frac{\lambda_v}{\Lambda_\Delta}\left[P_o(1 - P_d) + \frac{P_o P_d (1 - P_g)}{V_{cc}\lambda_v} \sum_{j=m+1}^{n} L(y_j)\right] \quad i = 0 \tag{12.107}$$

$$\beta_i = \frac{P_o P_d P_g}{\Lambda_\Delta V_t} \sum_{j=1}^{m} L(y_j) \quad i > 0 \tag{12.108}$$

where

$$\Lambda_\Delta = \frac{n(1 - P_o P_d)}{V_v} + \frac{P_o P_d P_g}{V_t} \sum_{j=1}^{m} L(y_j) + \frac{P_o P_d (1 - P_g)}{V_{cc}} \sum_{j=1}^{m+1} L(y_j) \tag{12.109}$$

The second term of (12.107) elevates the probability term β_0 in those cases when the distance to the mth measurement is small and when V_{cc} is also small. In the PDA filter based on selection with a gate, a large gate has to be used to provide a sample of the clutter conditions, which may complicate the covariance calculation had it not been thresholded.

The development so far has provided the necessary ingredients for the implementation of a real radar tracking system. They attempt to overcome discrepancies between theory and practice. The development of probability procedures conditioned on available data raises the sophistication of the basic PDA method to a new paradigm, called improved PDA or joint PDA (JPDA) in the literature. A good tracker should have a good track initiator: a filter that caters for non-uniform clutter density.

12.3.5 General initiation techniques

Track initiation is an important function of a tracking system, particularly where multiple target tracking is done in cluttered environments. Essentially, a track initiator must be capable of starting, or initiating, a track whenever a new target appears in the scanned (or surveillance) region while minimizing the number of false tracks due to clutter. Three groupings of track initiation techniques have been reported in the literature: rule-based, logic-based and (modified) Hough transform. These methods are summarized below.

12.3.5.1 Logic-based track initiation technique

Let the target position of the kth measurement at the ith radar scan be denoted by $r_k^{(i)}$. The logic-based track initiation technique is carried out as follows (Bar-Shalom and Fortmann 1988):

(i) Initialize on the first two scans of the measurements and estimate the apparent velocity, with every pair of the measurements. Let the velocity be denoted by $v^{(2)}$. Hence

$$v^{(2)} = \frac{1}{T}\left(r_j^{(2)} - r_k^{(1)}\right) \tag{12.110}$$

If this expression satisfies the speed gating criterion, i.e.,

$$\vartheta_{\min} \leq \|v^{(2)}\| < \vartheta_{\max} \tag{12.111}$$

then a track is initiated. Then predict the position of the track for the third scan as

$$r^{(3)} = r_j^{(2)} + Tv^{(2)} \tag{12.112}$$

Set an acceptable gate around the predicted position using (12.68) to (12.70).

(ii) On the third scan, any measurement $r_k^{(3)}$ that falls within the gate, i.e.,

$$\left|r_k^{(3)} - r^{(3)}\right| < \nu(k) \tag{12.113}$$

will extend the initiated track. If more than one track satisfies the gating criterion, the track will be split. If none falls into the gate the initiated track will be terminated.

Next compute the velocity and acceleration

$$\begin{aligned} v^{(3)} &= \frac{1}{T}\left(r_j^{(3)} - r_k^{(2)}\right) \\ a^{(3)} &= \frac{1}{T}\left(v^{(3)} - v^{(2)}\right) \end{aligned} \tag{12.114}$$

Then predict the next scan's position as

$$r^{(4)} = r_j^{(3)} + Tv^{(3)} + \frac{1}{2}T^2 a^{(3)} \tag{12.115}$$

Measurements that are not associated with any track at any scan, together with those not used at the previous scan, are used to initiate new tracks.

(iii) The procedure in (ii) is repeated for a predetermined number of scans. Every initiated track that remains at the end of the process will start a new track.

It should be noted that in a clutter environment, the track splitting technique in step (ii) is highly likely to produce a large number of false tracks. Given that *m* out of *n* detections are selected for the PDA tracking technique, and upon application of the nearest neighbour approach, the track splitting will be limited to *m* measurements. Further on in the tracking process, the split tracks would be grouped or clustered together to ascertain their origins. Those from the same source would be merged as one and those from other sources would be independently tracked and labelled.

12.3.5.2 Rule-based track initiation technique

The rule-based technique is similar to methods used for initiating the αβ tracking filter (12.31) and the Kalman-based tracking filter (12.47). It is sequential and limits the extent of tracking to the assumed target minimum ($\vartheta_{\min}$) and maximum ($\vartheta_{\max}$) speed, i.e.,

$$\vartheta_{\min} < \frac{\|r_{i+1} - r_i\|}{T} < \vartheta_{\max} \quad i = 1, 2, \ldots, N-1 \tag{12.116}$$

where r_i is the target position at the *i*th scan of *N*-scan initiator and *T* is the measurement time interval.

12.3.5.3 Hough transform track initiation technique

The Hough transform maps points in the Cartesian coordinate to the θ–ρ plane by (Carlson *et al.* 1994)

$$\rho = x\cos\theta + y\sin\theta \tag{12.117}$$

where ρ is the distance from the line through (*x*, *y*) to the origin, and θ is the angle to the normal with the *x*-axis restricted to values between 0° and 180°. The value of ρ can be positive or negative. Each point in the *x*–*y* plane defines a curve in the θ–ρ plane, and the family of curves generated by a set of collinear points intersect at a point (θ_0, ρ_0). To initiate a straight-line target track in the *x*–*y* plane is equivalent to searching the intersect points in the (θ–ρ) plane (Hu *et al.* 1997). The procedure is as follows:

(i) Divide the parameter θ into N_θ equal segments, each with $\Delta\theta = \pi/N_\theta$ in length, and the centres of these intervals are given as

$$\theta_n = \left(n - \frac{1}{2}\right)\Delta_\theta \quad n = 1, 2, \ldots, N_\theta \tag{12.118}$$

The values of ρ for each observation (x_i, y_i) are then calculated at all θ_n points.

(ii) Compute for N measurements from N consecutive scans, resulting in a set of ρ values. Denote these values by

$$\rho_i(\theta_n) = x_i \cos\theta_n + y_i \sin\theta_n \tag{12.119}$$

where $i = 1, 2, \ldots, N$ and $n = 1, 2, \ldots, N_\theta$.

(iii) Calculate the average of ρ over all i at each of θ_n points. Denote this average by $<\rho(\theta_n)>$. Calculate the maximum deviation of ρ from their average by

$$\Delta\rho(\theta_n) = \max\{\rho_i(\theta_n) - \langle\rho_i(\theta_n)\rangle\} \tag{12.120}$$

(iv) Search over all θ_ns and obtain the minimum of the deviation

$$\Delta\rho = \min\{\Delta\rho(\theta_n)\} \tag{12.121}$$

If Δρ is less than a predetermined threshold, say g_0, the detection of a straight-line trajectory is claimed, and a new track is initiated, otherwise discarded.

12.3.6 Conclusion

Complete information has now been assembled to start a PDA filter tracker for both non-manoeuvring and manoeuvring targets in cluttered environments. It has been demonstrated in the preceding developments that the PDA tracking technique is a suboptimal Bayesian algorithm whose formulation includes *a priori* probability of obtaining a target measurement. The *a priori* probability is also the detection probability: the probability that the target is detected. The absence of certainty in sensor measurement data ensures the inclusion of the event probability as a means of introducing qualified weightings on the measurement data. This assures a reasonable interpretation of resulting tracks.

12.4 Summary

This chapter has examined the basic tracking principles including the commonly utilized tracking filters: αβ, αβγ and Kalman. These filters are of the fading memory type, which can be implemented recursively.

For operational purposes, higher accuracy is demanded of the tracking system. The use of 'improved probability data association' filters attempts to produce a workable and efficient tracker in a cluttered environment. An important developmental area that requires further attention is the development of a robust initiation algorithm that considerably reduces clutter-initiated tracks within a reasonable specified time. This chapter is intended to provide a significant basis for an enhanced development of a practical tracker, which is clearly an active area of investigation.

Problems

1. A rigid communication satellite is to be tracked in all weather. The satellite has moment inertia of I_m with an applied torque T_r acting along the direction of rotation. The equation of motion is written as

$$I_m\ddot{\theta}(t) = T_r(t) + w(t)$$

where

$\ddot{\theta}$ = rotational angular acceleration of the satellite
θ = angle of the satellite
w = process noise with zero mean and unity variance; that is $\mu_w = 0$ and $\sigma_w^2 = E\left[w(t)w^T(t)\right] = 1$

By first-order state-space form

$$\underbrace{\begin{bmatrix}\dot{\theta}(t)\\ \ddot{\theta}(t)\end{bmatrix}}_{\dot{x}(t)} = \underbrace{\begin{bmatrix}0 & 1\\ 0 & 0\end{bmatrix}}_{A_*}\underbrace{\begin{bmatrix}\theta(t)\\ \dot{\theta}(t)\end{bmatrix}}_{x(t)} + \underbrace{\begin{bmatrix}0\\ 1\end{bmatrix}}_{B_*}\underbrace{\frac{T_r(t)}{I_m}}_{u(t)} + \underbrace{\begin{bmatrix}0\\ 1\end{bmatrix}\frac{w(t)}{I_m}}_{w_*(t)}$$

Or

$$\dot{x}(t) = A_*x(t) + B_*u(t) + w_*(t)$$

If the angle θ is sampled on every time interval Δt and the applied torque remains constant over the sampled period, then the continuous-time mode can be written in discrete form as

$$x(k+1) = Ax(k) + Bu(k) + Q$$

And the measurement equation:

$$y(k) = Cx(k) + R$$

where the measurement noise $v(k)$ is assumed random with zero mean and variance

$$R = E\left[v(k)v^T(k)\right] = \sigma_m^2$$

$$Q = E\left[w_* w_*^T\right] = \frac{\sigma_w^2}{I_m}\begin{bmatrix} \frac{\Delta t^3}{3} & \frac{\Delta t^2}{2} \\ \frac{\Delta t^2}{2} & \Delta t \end{bmatrix}$$

$$A = e^{\Delta t A_*} = \begin{bmatrix} 1 & \Delta t \\ 0 & 1 \end{bmatrix}$$

$$B = \int_0^{\Delta t} B_* e^{\Delta \tau A_*} d\tau = \begin{bmatrix} \Delta t & \frac{\Delta t^2}{2} \\ 0 & \Delta t \end{bmatrix}\begin{bmatrix} 0 \\ 1 \end{bmatrix}$$

$$= \begin{bmatrix} \frac{\Delta t^2}{2} \\ \Delta t \end{bmatrix}$$

Suppose $\Delta t = 1$ s, $\sigma_m^2 = 10^4$ and $I_m = 1$, plot the satellite trajectory for the next 40 s.
If σ_m^2 is varied between 10^2 and 10^6, what significant changes will you observe in the target's trajectory?
If the measurement noise variance is made constant, i.e. $\sigma_m^2 = 10^4$, and the process noise variance is not unity, i.e. $\sigma_w^2 \neq 1$, but $\sigma_w^2 = 5, 10^2, 10^4$, what effects will the process noise changes have on the target trajectory?

2. If in the process of tracking, you noticed some discontinuity in the tracks (missing detections), what improvement do you need to make to the model in Question 1 to reduce, or eliminate, misses?
3. If it is possible for you to collect real-life radar data from any responsible agencies (e.g. defence department, airport, etc.) write a computer program based on the improved PDA to track the data and determine any identified target(s) speed and coordinates in the region where the data are collected.

References

Aboaba, A. (1975). *Lecture note in mathematics*. Yaba College of Technology, Lagos, Nigeria.

Abramowitz, M. and Stegun, I.A. (1968). *Handbook of mathematical functions*. Dover.

Adomian, G. (1983). *Stochastic systems*. Academic.

Ahmed, N. (1987). Adaptive filtering, in *Handbook of digital signal processing: engineering application* (D.F. Elliott, ed.). Academic.

Anderson, S.J. (1986). Remote sensing with Jindalee Skywave radar. *IEEE Trans. Oceanic Eng.*, **2**, 158–163.

Barghausen, A.F., Finney, J.W., Proctor, L.L. and Schultz, L.D. (1969). *Predicting long-term operational parameters of high-frequency sky-wave telecommunications systems*. US Govt Printing Office: ESSA Tech. Rep., ERL 110-ITS 78.

Barnum, J. (1969). *The effect of polarisation rotation on the amplitude of ionospherically propagated sea backscatter*. Stanford Electronics Laboratories USA: Tech. Rep. No. 157.

Bar-Shalom, Y. and Fortmann, T.E. (1988). *Tracking and data association*. Academic.

Barton, D.K. (1988). *Modern radar systems analysis*. Artech House.

Barton, P. (1980). Digital beam forming for radar. *IEE Proc.*, **127**, F, 266–277.

Beauchamp, K.G. (1973). *Signal processing: using analog and digital techniques*. Allen & Unwin.

Beck, J.V. and Arnold, J.K. (1977). *Parameter estimation in engineering and science*. John Wiley.

Beckmann, P. and Spizzichino, R. (1987). *The scattering of electromagnetic waves from rough surfaces*. Artech House.

Bellanger, M.C. (1987). *Adaptive digital filters and signal analysis*. Marcel Dekker.

Benedict, T. and Bordner, G. (1962). Synthesis of an optimal set of radar track-while-scan smoothing equations. *IRE Trans. Automatic Control*, **7**, 27–32.

Bent, R.B., Lepofsky, J.R., Llewellyn, S.K. and Schmid, P.G. (1978). Ionospheric range-rate effects in satellite-to-satellite tracking. *AGARD Conference Proc.*, I:238, 9-1.

Bergland, G.D. (1969). A guided tour of the fast Fourier Transform. *IEEE Spectrum*, July, 41–52.

Berkowitz, R.S. (1965). *Modern radar: analysis, evaluation, and system design*. John Wiley.

Bilitza, D. (ed.) (1990). *International reference ionosphere*. Greenbelt, Maryland: NSSDC 90-22.

Bilitza, D., Sheikh, N.M. and Eyfrig, R. (1979). A global model for the height of the F2-peak using M(3000)F2 values from the CCIR numerical map. *Telecommunications J.*, **46**, 549.

Blackman, S.S. (1986). *Multiple-target tracking with radar applications*. Artech House.

Blackman, R.B. and Tukey, J.W. (1958). *The measurement of power spectrum from the point view of communication engineering*. Dover.

Blickmore, R.W. (1958). A note of the effective aperture of electrically scanned arrays. *IRE Trans.*, **AP-6**, 194–196.

Boashash, P., Peters, E.J. and Zoubir, A.M. (1995). *Higher-order statistical signal processing*. Longman.

Bolgler, P.L. (1990). *Radar principles with applications to tracking systems*. John Wiley.

Brace, L.H. and Theis, R.F. (1978). An empirical model of the interrelationship of electron temperature and density in the daytime temperature of solar minimum. *Geophys. Res. Letters*, **5**, 275.

Bradley, P.A. and Dudeney, J.R. (1973). A simple model for vertical distribution of electron concentration in the ionosphere. *J. Atmos. Terres. Physics*, **35**, 2131.

Broida, T.J., Chandrashekhar, S. and Chellappa, R. (1990). Recursive 3-D motion estimation from a monocular image sequence. *IEEE Trans. Aeros. Electr. Systems*, **26**, 639–656.

Carlson, B.D., Evans, E.D. and Wilson, S.L. (1994). Search detection and track with Hough Transform, part I: System concept. *IEEE Trans. Aeros. Electr. Systems*, **30**, 102–108.

CCIR atlas of ionospheric characteristics propagation in ionized media. Geneva: CCIR Recommendations and Reports, Rep. 340–4, VI.

Chapman, S. (1931). Some phenomena of the upper atmosphere. *Proc. Royal Society of London*, **A132**, 353.

Chapman, S. and Bartels, J. (1940). *Geomagnetism*. Volume 1. Oxford University.

Childers, D. and Durling, A. (1981). *Digital filtering and signal processing*. West.

Chrzanowski, E.J. (1990). *Active radar electronic counter measures*. Artech House.

Claerbout, J.F. (1976). *Fundamentals of geoscience data processing*. Blackwell.

Colegrove, S.B., Davis, A.W. and Ayliffe, J.K. (1986). Track initiation and nearest neighbours incorporated into probabilistic data association. *Journal Elect. Electronics Eng.* (Australia), **6**, 191–198.

Cook, C.E. and Bernfield, M. (1967). *Radar signals – an introduction to theory and application*. Academic.

Cooley, J.W. and Tukey, J.W. (1965). An algorithm for the machine calculation of complex Fourier series. *Mathematics of Computation*, **19**, 297–301.

Cover, T.M. and Thomas, J.A. (1991). *Elements of information theory*. John Wiley.

Crispin, J.W. and Siegel, K.M. (1968). *Methods of radar cross-section analysis*. Academic.

Croft, T. (1972). Skywave backscatter: a means for observing our environment at great distances. *Rev. Geophys. Space Phys*, **10**, 73–155.

Davies, K. (1990). *Ionospheric radio*. Peter Peregrinus.

Drabowitch, S. and Ancona, C. (1988). *Antennas applications*. Academic.
Drukarev, G. (1946). *J. Physics* (USSR), **10**, 81.
Ducharme, E.D., Petrie, L.E. and Eyfrig, R. (1971). A method of predicting the F1 layer critical frequency. *Radio Sci.*, **6**, 369.
Earl, G.F. and Ward, B.D. (1986). Frequency management support for remote sea-state sensing using the Jindalee skywave radar. *IEEE Trans. Oceanic Eng.*, **11**, 164–173.
Fano, R.M. (1963). A heuristic discussion of probabilistic decoding. *IEEE Trans. Infor. Theory*, **9**, 64–74.
Fraser, D. (1979). *Optimized mass storage FFT program, programs in digital signal processing*. IEEE Press.
Gabor, D. (1946). Theory of communication. *J. IEE*, **Part III, 93**, 429–457.
Galati, G. (1993). *Advanced radar techniques and systems*. Peter Peregrinus.
Gallager, R.G. (1978). Variations on a theme by Huffman. *IEEE Trans. Infor. Theory*, **24**, 668–674.
Gangolli, R.A. and Ylvisaker, D. (1967). *Discrete probability*. Harcourt Brace Jovanovich.
Gauss, K.G. (1963). *Theory of motion of heavenly bodies*. Dover.
George, S.F. and Zamanakos, A. (1954). Comb filters for pulse radar use. *Proc. IRE*, **42**, 1159–1165.
Georges, T.M., Harlan, J.A., Leben, R.R. and Lematta, R.A. (1998). A test of ocean surface-current mapping with over-the-horizon radar. *IEEE Trans. Geos. Remote Sens.*, **36**, 101–110.
Georges, T.M., Harlan, J.A., Meyer, L.R. and Peer, R.G. (1993). Tracking Hurricane Claudette with the U.S. Air Force over-the-horizon radar. *J. Atmos. Oceanic Tech.*, **10**, 441–451.
Giraud, A. and Petit, M. (1978). *Ionospheric techniques and phenomena*. Reidel.
Guinard, N.W. and Daley, J.C. (1970). An experimental study of a sea clutter model. *Proc. IEEE*, **58**, 543–550.
Hansen, R.C. (1990). Evaluation of the large array method. *Proc. IEE*, **137**, **H**, 94–98.
Headrick, J.M. (1990). Looking over the horizon. *IEEE Spectrum*, 8–11.
Heering, P. (1977). Modelling and detection, in *Aspects of signal process* (A. Tacconi, ed.). Reidel.
Helms, H.D. and Rabiner, L.R. (1972). *Literature in digital signal processing*. IEEE.
Hovanessian, S.A. (1988). *Introduction to sensor systems*. Artech House.
Hu, Z., Leung, H. and Blanchette, M. (1997). Statistical performance analysis of track initiation techniques. *IEEE Trans. Signal Proc.*, **45**, 445–456.
Huffman, D.A. (1951). A method for the construction of minimum redundancy codes. *Proc. IRE*, **40**, 1098–1101.
IEEE Standard 145–1983. IEEE standard definitions of terms for antennas. *IEEE Trans. Antenna Propag.*, **31**, Part II, 5–29.
IEEE Standard Dictionary of Electrical and Electronic Terms ANSI/IEEE Std. 100-1984, 1984.
IPS (Australia). http://www.ips.gov.au/asfc/current/
Jarrott, R.K. and Soame, T.A. (1994). The processing of HF skywave radar signals. *Proc. IEEE Acoust., Speech Signal Proc., Conf.*, Adelaide, Australia, 165–168.
Jelalian, A.V. (1992). *Laser radar systems*. Artech House.
Jenkins, G.M. and Watts, D.G. (1968). *Spectral analysis and its applications*. Holden-Day.

Johnson, J.B. (1928). Thermal agitation of electricity in conductors. *Physics Rev.*, **32**, 97–109.

Johnson, R.C. and Jasik, H. (1984). *Antenna Engineering Handbook*. McGraw-Hill.

Jones, N.B. (1962). *Digital signal processing*. IEE Control Engineering Series.

Kalman, R.E. (1960). A new approach to linear filtering and prediction problems. *J. Basic Eng.*, **82**, 33–45.

Kassam, S.A. and Poor, H.V. (1985). Robust techniques for signal processing: a survey. *Proc. IEEE*, **73**, 433–461.

Kato, S. (1980). *Dynamics of the upper atmosphere*. Academic.

Katzin, M. (1957). On mechanisms of radar sea clutter. *Proc. IRE*, **45:1**, 44–54.

Kerr, D.E. (1951). *Propagation of short waves*. McGraw-Hill.

Keydel, W. (1976). An experimental model for average scattering cross section computation for land and sea surfaces. *AGARD Conf., Proc.*, **208**, 6.1–6.15.

Kingsley, S.P. and Quegan, S. (1992). *Understanding radar systems*. McGraw-Hill.

Kolawole, M.O. (1994). Stabilization of tracks with multiple model filters. *Proc. IEEE Acoust., Speech Signal Proc., Conf.*, Adelaide, **3**, 45–411.

Kolawole, M.O. (1996). *Event probability: an application to target state's estimation*. Telstra (Jindalee) TSG Seminar Series, 15 Feb., 1–11.

Kolawole, M.O. (2002). *Satellite communication engineering*. Marcel Dekker.

Kramer, E. (1986). Polarization loss probability. *IEEE AP Newsletter*, 10–11.

Krolik, J.L. and Anderson, R.H. (1997). Maximum likelihood coordinate registration for over-the-horizon radar. *IEEE Trans. Signal Proc.*, **45**, 945–959.

Laing, A.C. (1967). *ASW target motion analysis*. TRW Systems Report, TR 3838-6013-R0–000.

Lees, M.L. (1987). An overview of signal processing for an Over-the-horizon radar. *Proc. Inter. Symp. Signal Proc. Applications*, 491–494.

Leftin, M. (1976). *Numerical representation of monthly median critical frequencies of the regular E region (foE)*. US Printing Office: OT Rep., 76–88.

Lehman, E.L. (1959). *Testing statistics hypotheses*. John Wiley.

Levanon, N. (1988). *Radar principles*. John Wiley.

Lewis, R. and Newell, A. (1985). *An efficient and accurate method for calculating and representing power density in the near zone of microwave antenna*. National Bureau of Standards: NBSIR 85S-3036.

Lie, X.R. (1998). Tracking in clutter with strongest neighbour measurements: I–theoretical analysis. *IEEE Trans. Autom. Control*, **43**, 1560–1579.

Litva, J. and Lo, T.K. (1996). *Digital beamforming in wireless communications*. Artech House.

Long, M.W. (1975). *Radar reflectivity of land and sea*. Heath.

Lynn, P.A. (1982). *An introduction to the analysis and processing of signals*. Macmillan.

Mao, Y.H. (1993). MTI, MTD and adaptive clutter cancellation, in *Advanced radar techniques and systems* (G. Galati, ed.). Peter Peregrinus.

Matsushita, S. (1967). Solar quite and lunar daily variation fields, in *Physics of geomagnetic phenomena: I* (S. Matsushita and W.D. Campbell, eds): Academic.

McEliece, R.J. (1977). *The theory of information and coding in volume 3 of Encyclopedia of Mathematics and its applications*. Addison-Wesley.

McNamara, L.F. (1991). *The ionosphere: communications, surveillance and direction finding*. Krieger.

Melsa, J.L. and Cohn, D.L. (1978). *Decision and estimation theory*. McGraw-Hill.

Mitra, S.K. (1952). The upper atmosphere. *The Asiatic Society of Calcutta*.

Millan, G.H. (1965). Atmospheric effects on radio wave propagation, in *Modern radar: analysis, evaluation, and system design* (R.S. Berkowitz, ed.). John Wiley.

Morris, G.V. (1988). *Airborne pulsed Doppler radar*. Artech House.

Morchin, W. (1993). *Radar engineer's source*. Artech House.

Morton, A.H. (1966). *Advanced electrical engineering*. Pitman.

Murtagh, T.B. (1965). A study of navigation measurement schedules for lunar excursion module rendezvous. *Proc. AIAA/ION Guidance and Control Conf.*

Nakagami, M. (1960). Statistical methods, in *Radio wave propagation* (Hoffman, ed.). Pergamon.

NASA. http://umbra.nascom.nasa.gov/images/latest.html

Nathanson, F.E. (1969). *Radar design and principles*. McGraw-Hill.

Netherway, D.J., Ewing, G.E. and Anderson, S.J. (1989). Reduction of some environmental effects that degrade the performance of HF skywave radar. *Proc. Symp. Signal Proc. and Applications*, 288–292.

Neuvy, J. (1970). An aspect of determining the range of radar detection. *IEEE Trans. Aeros. Electronic Syst.*, **6**, 514–521.

Nickisch, L.J. and Hausman, M. (1996). CREDO: Coordinate registration enhancement by dynamic optimisation. *Proc. Ionosph. Effects Symp.*, IB 2–1–2–9.

Oppenheim, A.V. and Schafer, R.W. (1975). *Digital signal processing*. Prentice-Hall.

Orfanidis, S.J. (1996). *Introduction to signal processing*. Prentice-Hall.

Parsen, E. (1962). *Stochastic Processes*. Holden-Day.

Rabiner, L.R. and Gold, B. (1975). *Theory and application of digital signal processing*. Prentice-Hall.

Rabiner, L.R., McCellan, J.H. and Parks, T.W. (1974). FIR digital filter design techniques using weighted Chebyshev approximation. *Proc. IEEE*, **63**, 595–610.

Rice, S.O. (1944). Mathematical analysis of random noise. *Bell System Tech. J.*, **23**, 282–332.

Richards, G.A.R. (1992). *Extended event probabilities*. Telstra (Jindalee) Tech. Rep., J0976.

Rihaczek, A.W. (1969). *Principles of high-resolution radar*. McGraw-Hill.

Rishbeth, H. (1988). Basic physics of the ionosphere: a tutorial review. *J. IERE*, **58**, 207–223.

Rosenblatt, M. (1985). *Stationary sequences and random fields*. Birkauser.

Rosich, R.K. and Jones, W.B. (1973). *The numerical representation of the critical frequency of the F1 region of the ionosphere*. US Printing Office: OT Report 73-22.

Rush, C.M. (1986). Ionospheric radio propagation models and predictions – a mini review. *IEEE Trans. Antennas Propag.*, **34**, 1163.

Ruck, G.T., Barrick, D.E., Stuart, W.D. and Krichbaum, C.K. (1970). *Radar cross section handbook. Vol. 1*. Plenum.

Rush, C.M., Pokemoner, M., Anderson, D.N., Perry, J., Stewart, F.G. and Reasoner, R. (1984). Maps of foF2 derived from observations and theoretical data. *Radio Sci.*, **19**, 1083.

Schekunoff, S.A. (1943). *Electromagnetic waves*. Van Nostrand.

Schoenberger, J.G., Forrest, J.R. and Pell, C. (1982). Active array receiver studies for bistatic/multistatic radar. *IEEE Inter. Radar Conf.*, 174–178.

Schutte, K. (1940). Die transformation beliebiger spharischer koordinatensysteme nit einer einzigen immerwahrenden hilfstafel. *Astr. Nachr.*, **270**, 76.

Shannon, C.E. (1959). Coding theorems for a discrete source with a fidelity criterion. *IRE National Convention Record*, Part 4, 142–163.

Siebert, W.M. (1956). A radar detection philosophy. *IRE Trans. Infor. Theory*, **2**, 204–221.

Silver, S. (1949). *Microwave antenna theory and design*. McGraw-Hill.

Simpson, H.R. (1963). Performance measures and optimization conditions for a third-order sampled data tracker. *IEEE Trans. Auto. Control*, **8**, 182–183.

Sinnott, D.H. (1987). Jindalee – DSTO's Over-the-horizon radar project. *Proc. IREECON*, 661–664.

Sinnott, D.H. (1989). Antenna requirements for Over-the-horizon radar. *Proc. IREECON*, 677–680.

Skolnik, M.I. (1980). *Introduction to radar systems*. McGraw-Hill.

Southcott, M.L., Kolawole, M.O. and Jarrott, R.K. (1998). Distortion correction for skywave radar signals via deconvolution. *Proc. IASTED Conf. Signal Proc. and Comm.*, Las Palmas de Gran Canaria, Spain, 48–52.

Stanley, W.D. (1975). *Digital signal processing*. Reston.

Stehel, R.H. and Hagn, G.M. (1991). HF channel occupancy and band congestion: the other user interference problem. *Radio Sci.*, **26**, 959–970.

Steyskai, H. (1978). Digital beam-forming antennas. *Microwave J.*, 107.

Storer, J.A. (1976). *Data compression, methods and theory*. Computer Science Press.

Stratton, J.A. (1941). *Electromagnetic theory*. McGraw-Hill.

Strejc, V. (1981). *State space theory of discrete linear control*. John Wiley.

Stutzman, W.L. and Thiele, G.A. (1981). *Antenna theory and design*. John Wiley.

Swerling, P. (1960). Probability of detection for fluctuating targets. *IRE Trans. Infor. Theory*, **6**, 269–304.

Terman, F.E. (1949). *Radio engineers' handbook*. McGraw-Hill.

Tiehan, M. and Peinan, J. (1996). *Ionospheric studies and sounding at the CRIRP*. IEEE Press.

Treichler, J.R., Johnson, C.R. and Larimore, M.G. (1987). *Theory and design of adaptive filters*. John Wiley.

Trunk, G.V. (1972). Radar properties of non-Rayleigh sea clutter. *IEEE Trans.* AES-8, 196–204.

Ulaby, F.T., Moore, R.K. and Fung, A.K. (1986). *Microwave remote sensing active and passive, Vol. 3*. Artech House.

University of Massachusetts. http://digisonde.haystack.edu

Vakman, D.E. (1968). *Sophisticated signals and the uncertainty principle in radar*. Pergamon.

Van Trees, H.L. (1971). *Detection, estimation, and modulation theory. Part III*. John Wiley.

Van Veen, B.D. and Buckley, K.M. (1988). Beamforming: a versatile approach to spatial filtering. *IEEE ASSP Magazine*, 4–24.

Vizmuller, P. (1995). *RF design guide: systems, circuits, and equations*. Artech House.

Ward, K.D. (1982). A radar sea clutter model and its application to performance assessment. *Inter. Conf., Radar-82*.

Weaver, H.J. (1983). *Applications of discrete and continuous Fourier analysis*. John Wiley.

Weiner, M.M. (1991). Noise factor and antenna gains in the signal/noise equations for over-the-horizon radar. *IEEE Trans. Aeros. Electronic Syst.*, **27**, 886–890.

Whner, D.R. (1995). *High-resolution radar*. Artech House.

Widrow, B. and Stearns, S. (1985). *Adaptive signal processing*. Prentice-Hall.

Wilkes, M.V. (1954). A table of Chapman grazing incidence integral ch(x, ξ). *Proc. Physics Society*, **67B**, 304.

Woodman, R.F. and Chau, J.L. (2001). Antenna compression method using binary phase coding. *Radio Sci.*, **36**, 45–51.

Woodward, P.M. (1953). *Probability and information theory, with applications to radar*. Pergamon.

Xie, X. and Evans, R.J. (1991). Multiple target tracking and multi-frequency line tracking using Hidden Markov models. *IEEE Trans. Signal Proc.*, **39**, 2659–2676.

Zadek, L.A. and Ragazzini, J.R. (1952). Optimum filters for the detection of signals in noise. *Proc. IRE*, **40**, 1123–1131.

Zollo, A.O. and Anderson, S.J. (1992). Accurate skywave radar coordinate registration based on morphological processing of ground clutter maps. *Proc. Inter. Symp. Signal Proc. Applications*, **2**, 459–462.

Glossary

Algorithm A systematic technique of performing a series of computations in sequence.

Antenna The interface between a free-space electromagnetic wave and a guided wave.

Aperture The surface area of an antenna, which is exposed to radio frequency (RF) signals.

Array A collection of radiators or antennas.

Bandwidth The frequency range of a data transmitting or receiving device, which dictates how much data can flow per unit time.

Beamwidth The width of the sent beam measured in degrees after discounting sidelobes.

Budget (time, power) A set of bounds, or allocations, inherent to radar design and operation.

Clutter The returns from the Earth's surface, electromagnetic interference, meteor, lightning and even other objects in the vicinity of the target(s) of interest that could mask the true identification and quantification of the target(s) signatures.

Compression The process of converting an input data stream (the source stream, or the original raw data) into a smaller data stream (the output, or the compressed stream).

Correlation A process of determining the mutual relationships that exist between several functions or signals. If the measurement is the average self that exists within a signal, then the correlation is called the autocorrelation function. But measurement that exists between different signals is called the cross-correlation function.

Critical frequency The limiting frequency at which the reflections of radio waves begin to disappear for a specific ionospheric layer.

Cumulant A random variable whose moment properties cannot be described about the origin.

Data association The procedure used to imply the origin of measurement uncertainties.

Data conditioning A means of bringing the spectrum of the signal close to that of white noise by rejecting any unwanted data from the signal before analysis starts.

Data processing The transformation of a set of coordinated physical measurements into decision statistics for some hypotheses.

Detection The technique by which the signature of a target can be discerned among various background features.

Entropy The quantity of data transmitted per second, or the average self-information per transmitted symbol.

Error The difference between the estimate and the actual.

Estimate An arithmetic mean of a set of observations: the parallel of decision.

Estimator A formula, or a procedure, for deriving from a sample or set of observations to generate an estimate.

EUV Stands for the extreme ultraviolet. The spectral band that is responsible for ionization in the E and F regions of the ionosphere.

Event A combination of possible outcomes.

Extraordinary wave One of the two magneto-ionic components associated with a characteristic wave that propagates through the ionosphere having a polarization property – the second component is called the ordinary wave.

FFT Stands for the 'fast Fourier transform'. It is an efficient algorithm for the numerical computation of discrete Fourier transform (DFT) with a minimum computation time.

Filtering A process of understanding the status of a system at a particular instant.

FOM Stands for 'figure of merit'. It is a measure of radar capability.

Gating A technique of rejecting unlikely observation-to-track pairings.

Geographic latitude Latitude measured from 0° at the Earth's equator up to 90° at its pole, positive to the north, negative to the south.

Gyromagnetic frequency The electron frequency above the Earth's magnetic field.

Hour angle Sun angle measured westward from apparent noon.

Hypothesis A supposition from which to draw conclusions.

Innovation A sequence that provides an easy check for the optimality of a filtering system, or the difference between measured and predicted quantities.

Interpolation A process of calculating approximately a system's attributes from past parameters and current values of parameters.

Ionogram Recorded tracings of reflected HF radio pulses generated by a sounder (also called ionosonde).

Ionosphere A region of the outer atmosphere, starting at a height of 50 km, which contains many ions and free electrons and is capable of reflecting radio waves.

Kalman gain The ratio between the uncertainty in the state estimates and the uncertainty in the measurements.

Measurement equations Recursive equations that are linearly related to measurements variables.

MUF The maximum-usable-frequency for a specific magnetic index.

MUF(3000) The highest frequency that, refracted in the ionosphere, can be received at a distance of 3000 km.

Ordinary wave One of the two magneto-ionic components associated with a characteristic wave that propagates through the ionosphere having a polarization property – the second component is called the extraordinary wave.

PDA A procedure that computes the probabilities that detections are from the target of the validation gate measurements and that enables assignments of plots to tracks.

Pixel A dot on a raster output device that represents one picture element. A pixel may be round, square, oval, or rectangular – whatever shape most appropriate and convenient for the specific output device manufacturer.

Plasma A phenomenon that occurs when an atom has been stripped of its electron resulting in a net positive electrically charged gas.

Prediction A stated expectation about a given attribute that may be verified by subsequent observation.

Preprocessing A method of conditioning a signal, or a number of signals, into a form suitable for analysis.

Prewhitening Same as Data conditioning.

Probability A notion of chance.

Radar An active electromagnetic surveillance device that transmits a burst of electromagnetic energy necessary to allow detection of targets intercepting the energy by its receiver.

Redundancy The difference between the entropy and the smallest entropy.

Refraction The bending associated with a signal beamed from a transmitter sufficient for the signal to return to the Earth's surface. Reflection and refraction are sometimes difficult to separate.

Residual Same as Innovation.

Residue Same as Error.

Signal processing A technique used for performing certain functions, namely, signal enhancement, clutter suppression, interference suppression, target detection or extraction, target classification estimation and imaging.

Skywave radar A type of radar that sees beyond the horizon because it makes use of the ionosphere to refract the radar wave propagated back to Earth.

Solar zenith angle Angle measured at the Earth's surface between the Sun and the zenith.

Splitting (tracks) A process of separating tracks formed on closely spaced targets.

Sporadic E A transient or irregular layer of the ionosphere, which can occur in patches about 100 km wide and can reflect radio waves up to frequencies of about 100 MHz.

State transition A process by which a form of the state is transformed into another as time passes.

Stationarity A situation when the mean, expected, or ensemble average value of a signal is constant at different times.

Sunspots Dark spots that appear and disappear with time which occur, on the average, with an 11-year cycle.

Target A physical object that can produce sensor measurements.

Track The symbolic representation of a target, formed from successive detected positions.

Tracking A process of determining the speed and direction of a target and which enables monitoring of the target throughout the radar cover area.

Unbiased estimate An estimate is said to be unbiased if the expectance of the error vector is zero or the expectance of the estimate is equal to the actual.

Virtual height The point of reflection of radio pulses generated by an ionospheric sounder or ionosonde.

Zenith angle An angle measured at the Earth's surface between the Sun and the zenith.

Index